# THE STRUCTURE
## OF
# SCIENTIFIC INFERENCE

*by the same author*

Science and the Human Imagination
Forces and Fields
Models and Analogies in Science

# THE STRUCTURE
# OF
# SCIENTIFIC
# INFERENCE

## MARY HESSE

*Woolfson College*
*and University of Cambridge*

UNIVERSITY OF CALIFORNIA PRESS
Berkeley and Los Angeles 1974

University of California Press
Berkeley and Los Angeles, California

ISBN: 0-520-02582-2

Library of Congress Catalog Card Number: 73-85373

© M. B. Hesse 1974

pages 210-222 © University of Minnesota Press 1970

*Printed in Great Britain*

# Contents

# Acknowledgments

Some previously published material has been incorporated in this book: chapter 1 is reprinted from *The Nature and Function of Scientific Theories*, ed. R. G. Colodny (1970), 35–77, with permission of the editor and Pittsburgh University Press; chapter 3 is adapted from 'Ramifications of "grue"', *Brit. J. Phil. Sci.*, **20** (1969), 13–25, with permission of the Cambridge University Press; parts of chapters 5 and 6 will appear in volume 6 of *Minnesota Studies in the Philosophy of Science*, ed. G. Maxwell (University of Minnesota Press); part of chapter 8 is reprinted from 'Confirmation of laws', in *Philosophy, Science, and Method*, ed. S. Morgenbesser, P. Suppes and M. White (1969), 74–91, with permission of St Martin's Press; part of chapter 9 is reprinted from 'An inductive logic of theories', *Minnesota Studies in the Philosophy of Science*, vol. 4, ed. M. Radner and S. Winokur (1970), 164–180, with permission of the University of Minnesota Press; and chapter 11 is reprinted from *Foundations of Scientific Method*, ed. R. Giere and R. S. Westfall (1973), 86–114, with permission of Indiana University Press. All these permissions to reprint are gratefully acknowledged.

Formal and informal discussions with many friends and colleagues have contributed decisively to anything that is of value in this book. Among many others I should like specially to acknowledge the help and continuing collaboration of Gerd Buchdahl, Hugh Mellor and Robert Young. David Bloor and Allan Birnbaum saved me from many pitfalls in chapters 1 and 5–7 respectively, and I am indebted to Elliot Sober for long discussions on the subject matter of chapter 10. I am deeply grateful to the University of Notre Dame Philosophy Department for enabling me to spend a peaceful and intellectually active semester on their beautiful campus in the spring of 1972, and especially to Ernan McMullin, Gary Gutting and Vaughan McKim, and to the members of the graduate seminar who were exposed to much of the book in embryo. My gratitude is also due to Mrs Verna Cole and Mrs B. Reeson for doing the typing. Responsibility for the finished product is, of course, all my own.

*Cambridge*
*April 1973*

# Introduction

## THE TASK OF A LOGIC OF SCIENCE

The modern analytic style in philosophy of science is the product of an era in which confidence in natural science was at its height. In both nineteenth-century positivism and twentieth-century logical positivism the only recognized mode of knowledge was the scientific, together with those extensions of the scientific which could plausibly be claimed to be modelled on its methods. Classic problems of the theory of knowledge were brushed aside by appeal to the scientific consensus of 'observation'; metaphysics was dismissed as unintelligible because untestable by experience; experience itself was reduced to the narrowest and most superficial domain of agreed sense-experience; and introspection, insight and creativity had their place only in subjective psychology, not in philosophy or logic.

Doubts about the adequacy of this positivist epistemology arose within the analysis of science itself, first with regard to its theoretical components, and then increasingly with regard to its observational basis. It was soon found to be impossible to devise logical criteria of 'verifiability' for scientific *theories* which do not admit either too little or too much. Either all concepts not derived directly from observation were excluded by such criteria—manifestly distorting the nature of theoretical science—or else the attempt to liberalize the notion of verification in order to include theoretical concepts was found to readmit as meaningful the metaphysics and the nonsense which verification had been designed to exclude.

Popper concluded from this development that the problem should be seen as one of *demarcation* of scientific from other forms of knowledge,[1] thus significantly breaking with the positivist tradition that *only* science constitutes knowledge. His introduction in this context of the criterion of 'falsifiability' appeared to solve the immediate problem of demarcation, and led to the replacement of verificationism by the so-called *hypothetico-deductive model* of science (or the *deductive model* for short). Here theories are held to be deductive in structure: their postulates and concepts are not necessarily

---

[1] See especially K. R. Popper, *The Logic of Scientific Discovery* (London, 1959; first published 1934), chap. 1, and *Conjectures and Refutations* (London, 1963), chap. 1.

directly derived from observation, but they must be connected with observables by deductive (or, more generally, by deductive and statistical) chains of inference, in such a way that a theory can be falsified by *modus tollens* argument if any of its consequences turn out observationally to be false.[1] Problems remained, however, about the character and function of theoretical postulates and concepts. Are postulates derived merely by imaginative conjecture, as Popper maintained, or by some form of inductive reasoning? If theoretical concepts are not part of ordinary descriptive language, or immediately definable in terms of such language, how are they to be understood? What is it to be a theoretical explanation?

These problems were, however, quickly overtaken by a more fundamental issue. All positivist analyses had presupposed a common and easily intelligible 'observation language' which was at once the basis of, and the court of appeal for, all scientific theorizing. As early as 1906 Pierre Duhem[2] had implicitly questioned this basis by pointing out that experimental laws derived from observation form part of a network of mutually supporting laws, the whole of which may be used to correct and reinterpret even parts of its own observational basis. The same point was made by N. R. Campbell,[3] who went on to propose a solution to the problem of the derivation and meaning of theoretical postulates and terms by regarding them as analogies or models drawn from the observationally familiar. Quine[4] has pursued Duhem's network analogy in increasingly far-reaching analyses of logical and scientific systems, culminating in influential studies of the problem of translation from the language of one total theory, including its observational component, to another, and from one total culture to another.

Other philosophers have drawn even more radical conclusions from the 'holistic' character of theory and interpretations of observation in the light of theory. In some of the most influential of these discussions, P. K.

---

[1] For the deductive model of scientific explanation (also called the covering-law model), see especially C. G. Hempel and P. Oppenheim, 'Studies in the logic of explanation', *Phil. Sci.*, **15** (1948), 135, reprinted in Hempel, *Aspects of Scientific Explanation* (New York, 1965), 245; R. B. Braithwaite, *Scientific Explanation* (Cambridge, 1953), chaps. 1, 2; and E. Nagel, *The Structure of Science* (New York and London, 1961), chap. 2.

The special problems for this model of statistical theories which, if true, do not *exclude* any observation but only make some observations very improbable, and therefore cannot be conclusively falsified by observation, are considered by Popper, *The Logic of Scientific Discovery*, chap. 8; Braithwaite, *Scientific Explanation*, chaps. 5–7; and Hempel, 'Inductive inconsistencies', *Aspects of Scientific Explanation*, 53.

[2] P. Duhem, *The Aim and Structure of Physical Theory* (Eng. trans. of 2nd edition, Princeton, 1954; first published as *La Théorie physique*, Paris, 1906), part II, chaps. 4–7.

[3] N. R. Campbell, *Foundations of Science* (New York, 1957; first published as *Physics: the Elements*, Cambridge, 1920), part I.

[4] W. v. O. Quine, 'Two dogmas of empiricism', *From a Logical Point of View* (Cambridge, Mass., 1953), 20; *Word and Object* (New York, 1960), chaps. 1–3; *Ontological Relativity and other Essays* (New York, 1969).

Feyerabend[1] has attacked what he calls the 'pragmatic' and the 'phenomeno-logical' theories of meaning which are presupposed by those who wish to rest science respectively upon easily agreed observational reports, or upon 'sense data'. Feyerabend goes on to reject the alleged stability and theory-independence of the observation language, as had been universally pre-supposed in empiricist philosophy of science, and also to reject the claim of the deductive model that observation statements are strictly deducible from theories (even when these are non-statistical) and hence that if the observation statements are false they necessarily act as falsifiers of the theory from which they are deduced. On the contrary, Feyerabend argues, all interpretation of observation is 'theory-laden', and this includes even the common descriptive terms with which we form observation statements in natural language. For example, 'light' and 'heavy' mean something different for the Aristotelian and for the Newtonian, and 'mass' means something different in classical and in relativistic physics.

Feyerabend's analysis is supported by historical examples intended to illustrate the 'meaning variance' of observational terms when infected by the concepts of radically different theories and cosmologies. The history of science has been used in a similar way by other writers to arrive at similar conclusions. For example, T. S. Kuhn[2] interprets the history of science as a succession of 'paradigms' imposed on nature and interspersed by 'revolutions' which are explainable not by a logic of scientific change but by the increasing complexity and anomaly of a previous paradigm and the impatience with it of a new generation, helped by cultural and social discontinuities. N. R. Hanson has given analyses of historical cases in science in terms of Wittgen-steinian 'duck–rabbit', or *gestalt*, switches, and S. E. Toulmin in terms of an evolutionary model of mutations and natural selection of scientific theories.[3]

Such studies are generally characterized not only by rejection of most of the presuppositions of positivist philosophy of science, but also by explicit rejection of the logical and analytic *style* in philosophy of science in favour of persuasive argument from historical examples, on the grounds that in the past logical formalism has grossly distorted the natural of 'real science'. Some of these studies are also increasingly characterized by appeal to pragmatic, intuitive, subjective and ultimately polemical explanations and justifications of the development of science, and abandonment of the search for logical criteria of the empirical truth or falsity of science. Thus discussions of truth

[1] P. K. Feyerabend, 'An attempt at a realistic interpretation of experience', *Proc. Aris. Soc.*, 58 (1957–8), 143; 'Explanation, reduction and empiricism', *Minnesota Studies in the Philosophy of Science*, vol. III, ed. H. Feigl and G. Maxwell (Minneapolis, 1962), 28; 'How to be a good empiricist', *Philosophy of Science: the Delaware Seminar*, vol. 2, ed. B. Baumrin (New York, 1963), 3; and many subsequent papers.

[2] T. S. Kuhn, *The Structure of Scientific Revolutions* (Chicago, 1962; second edition 1970).

[3] N. R. Hanson, *Patterns of Discovery* (Cambridge, 1958); S. E. Toulmin, *Foresight and Understanding* (London, 1961), and *Human Understanding* (Oxford, 1972).

criteria are often replaced by descriptions of science wholly within its own context, without 'external' judgments of validity, or by judgments of science relative to the consensus of a scientific elite, or even by avowedly aesthetic or 'hedonistic' criteria.[1]

Romantic excesses in reaction to excessive logical pedantry and analytic subtlety are not unknown in the history of philosophy, and their lifetime is usually brief. The essays which follow in this book are not in a rigorously formal mode, but they are in a looser sense analytic in style and are intended to indicate a *via media* between the extremes of both formalism and historical relativism. In a series of papers over the last six or seven years I have attempted to develop a model of scientific theorizing which takes account of recent radical criticism but also retains the notions of empirical truth-value and of logical inference, particularly inductive inference. Some of the chapters which follow are reprinted from earlier papers, some are substantially new, and others provide linking and systematizing arguments.

The model of science developed here is essentially *inductive*, and it owes much to the *network model* first outlined by Duhem and adopted by Quine. Briefly, the model interprets scientific theory in terms of a network of concepts related by laws, in which only pragmatic and relative distinctions can be made between the 'observable' and the 'theoretical'. Some lawlike statements of the system can be tested relatively directly by observation, but which these are may depend on the current state of the whole theory, and whether an observation statement is accepted as 'true' or 'false' in any given case may also depend on the coherence of the observation statement with the rest of the currently accepted theory. Both coherence and correspondence aspects of truth are involved here. The correspondence aspect requires that, at any given time in any given descriptive language, most but not necessarily all statements made on the basis of observation must be taken to be true, but at that time we shall not usually be able to identify *which* are the true statements. Which statements are *taken* to be true depends on coherence with a whole theoretical network. In this account theoretical concepts are introduced by analogy with the observational concepts constituting the natural descriptive language. Scientific language is therefore seen as a dynamic system which constantly grows by metaphorical extension of natural language, and which

---

[1] An interpretation of science in terms of 'intuition' and 'tacit knowledge' is given by M. Polanyi, *Personal Knowledge* (London, 1958). For the 'consensus' interpretation, see for example J. M. Ziman, *Public Knowledge* (Cambridge, 1968), and for 'hedonism' and all kinds of other polemic, see P. K. Feyerabend, 'Consolations for the specialist', *Criticism and the Growth of Knowledge*, ed. I. Lakatos and A. Musgrave (Cambridge, 1970), 197, and 'Against method', *Minnesota Studies*, vol. IV, ed. M. Radner and S. Winokur (Minneapolis, 1970), 17.

Almost any recent contribution of quality to the 'history of scientific ideas' illustrates the tendency to judge and interpret past science in the categories of its own time, without raising the question of 'truth'. For some references, and some misgivings about this tendency, see my 'Hermiticism and historiography', *Minnesota Studies*, vol. V, ed. R. Stuewer (Minneapolis, 1970), 134.

also changes with changing theory and with reinterpretation of some of the concepts of the natural language itself. In this way an empirical basis of science is maintained, without the inflexibility of earlier views, where the observation language was assumed to be given and to be stable.

The main problems that arise from the network model concern its analysis of the 'meaning' of observational and theoretical concepts, and of the 'truth' of statements containing them. In chapter 1 I attack some epistemological problems of the theses of theory-ladenness and meaning variance by explicitly directing attention to the problematic character of the observation language itself. On the basis of a reinterpretation of the notion of 'observable' in terms of the theoretical network, I try to reconstruct and provide solutions in general terms for some problems bequeathed by positivism and deductivism, namely the meaning and justification of theoretical terms, the relation between theory and observation, the role of models in theories, and the nature of explanation and reduction. In chapter 2 I relate the reinterpretation of observation terms to more traditional discussions of universals, sense and reference, intensions and extensions, and correspondence and coherence accounts of truth. The theory of universals assumed is essentially a *resemblance* theory, and depends on non-extensional recognition of similarities; therefore some defence is given of the notion of 'similarity' against recent critics, especially Goodman. Goodman's 'grue' paradox stands in the way of any attempt to construe the observation language as an empirical basis of science, and so in chapter 3 I suggest a solution to this paradox which fits well with the network model, in that what counts as an 'observation term' is shown to be partially dependent on the whole context of theory. Anyone who wishes to claim that 'gruified' predicates form a basis for descriptive language or for inductive inference that is as valid as our usual predicates, is challenged to justify his choice in relation to a total physics which is non-trivially distinct from our physics.

The logical system that immediately suggests itself for the explication of inference in a network of theory and observation is the theory of probability. This is an obvious generalization of the deductive model of science, which, though inadequate as it stands, may certainly be seen as a first approximation to adequacy as a logic of science. In the deductive model, theory and observation are seen in terms of a hierarchy ordered by deductive inferences from theory to observation, but the deductive model as such gives no account of logical inference from observation to theory. In a probabilistic inductive model, on the other hand, the hierarchy is replaced by a system in which all statements are reciprocally related by conditional probability, of which deductive entailment is the limiting case. Probabilistic inference can therefore be seen as a generalization of deduction, permitting inductive and analogical as well as deductive forms of reasoning in the theoretical network.

Chapters 4, 5 and 6 lay the foundations of a probabilistic confirmation theory which can be used to analyse the network model in more detail. The intention here is to show that commonly accepted scientific methods, including parts of the traditional logic of induction, can be explicated in a probabilistic theory, where probability is interpreted in personalist and Bayesian terms as 'degree of rational belief'. The explicatory relation between probability and induction is neither a justification of induction in terms of probability—for this would lead to a further demand for non-inductive justification of the probability postulates of induction themselves—nor does it require that all intuitively adopted inductive methods are necessarily acceptable in the light of systematic reconstruction in a probabilistic theory. Explication consists rather of analysing and systematizing intuitive methods and assumptions about the logic of science, and hence opens up possibilities of clarifying and modifying some of these in the light of others. In particular, and this is one of the more controversial outcomes of this study, the explication suggests that the interpretation of scientific theory as consisting essentially of strictly universal laws in potentially infinite domains is mistaken, and should be replaced by a view of science as *strictly finite*. That is to say, I shall argue that in so far as scientific theories and laws can be reasonably believed to be true, this reasonable belief is, and needs to be, non-zero only for statements whose domain of application is finite. Such a reinterpretation of the nature of scientific laws is argued in chapters 7 and 8. In chapters 9 and 10 finite probability methods are extended to account for theoretical inference which depends on analogical argument from models and on simplicity criteria. Chapter 11 is a detailed case history to illustrate the method of analogy, taken from the electrodynamics of J. C. Maxwell. In chapter 12 the consequences of the network model for a realistic interpretation of science are discussed.

Since I am here attempting, against all current odds and fashions, to develop a *logic* of science, a little more must be said about the nature of such an enterprise, and it must be distinguished from other types of study with which it may be confused.

Firstly, a logic of science differs from a descriptive study of methodology, whether historical or contemporary, since it should supplement mere description with normative considerations. This is because it presupposes that there are norms or criteria of 'good science' in terms of which groups of scientists judge scientific theories, and that these have some elements, perhaps tacit, of internal logical coherence and rationality. Obviously such criteria are not timeless, and they may not even be the same for different groups of scientists at the same time. In almost all periods, for example, there have been opposing tendencies towards *a priori* or speculative science on the one hand, and instrumentalist or positivist science on the other. But it does not follow that all sets of criteria are logically arbitrary. For each such set it is possible to explore the rational or normative connections and consequences of principles explicitly or implicitly adopted. Some of these may have been known to the group, and some may not have been known; this approach therefore makes possible a *critical history* of a group's methodology. For example, it will be my contention that William Whewell missed an important element of his own analysis of 'consilience of inductions' when he used it to interpret the history of science, and that Maxwell misunderstood the significance of his own Newtonian claim that he had 'deduced' his electromagnetic theory from experiments, and hence underestimated the analogical element in his thinking. It follows from this normative character of the logic of science, and from the possibility of critical history, that there will be no simple process of testing of a proposed logic against historical examples—the relation of logic and cases will rather be one of mutual comparison and correction.

Secondly, a normative critique of method may disclose the implicit aims of a methodology and judge the appropriateness of its means to its ends. A science whose aim is application and prediction may have different normative requirements from one which desires truth, beauty or morality. Sometimes comprehensive theories of maximum empirical content are appropriate, sometimes instrumentalist predictions, sometimes inductive inferences. It is a naïve reading of the history of science to suppose that different methodologies are necessarily in conflict, given their different aims. The logic of science should provide a comparative study of such methodologies, rather than a partisan polemic on behalf of some against others.

Thirdly, it is often claimed that philosophical studies of science are irrelevant to the practice of science and unrecognizable as accounts of its methods. This objection, however, rests on a misunderstanding of the primary purpose of a logic of science. A study of the logical structure of science is not intended in the first place as an aid to scientific research, much less as a descriptive manual of experimental method. For example, although probabilistic methods have been suggested here as a handy means of explicating the inductive criteria of a scientific theory, it is unlikely that professional statisticians will find much that is relevant to their own technical problems, except perhaps incidentally. But the logic of science is in the first place a branch of *philosophy*, specifically of epistemology or the theory of knowledge, and also of ontology or the theory of what kinds of things there are. It is not surprising if the practice of science largely passes these questions by. They have, after all, been controversial for many centuries, and modern science in part developed intentionally as an enterprise that was neutral with respect to them and could afford to ignore them. Science has been remarkably successful in pursuing its own aims independently of philosophical disputes. But that is not to say that the philosophical critique of the foundations of science itself can ultimately be ignored, for that critique is concerned both with the understanding and justification of the aims of science itself, and with the existence and character of modes of knowledge other than the scientific.

In particular, the logic of science is very relevant to two matters of current interest and controversy. Firstly, there are rising doubts about the value of natural science itself as this has been traditionally understood. Far from being the paradigm of all knowledge, its aims to make true discoveries about the natural world and to exploit these discoveries are both being increasingly questioned.[1] And secondly, the question of the character of other modes of knowledge, particularly in history, and the human and social sciences generally, has attained a new importance.[2] These *Geisteswissenschaften* no longer look automatically to natural science for methodological guidance, and are badly in need of analysis and systematization of their own aims and methodologies. In relation to such questions a search for better understanding of the logic of the natural sciences themselves is clearly very relevant. A logic of science may also, of course, have some fallout in terms of direct usefulness to research in natural science, for self-understanding is often a desirable supplement to any enterprise, but the logic of science does not have to justify itself primarily on these grounds.

[1] See for example H. Marcuse, *One Dimensional Man* (London, 1964), chap. 6; and J. R. Ravetz, *Scientific Knowledge and its Social Problems* (Oxford, 1971).

[2] For careful comparative analyses of method, see especially J. Habermas, *Knowledge and Human Interests* (Eng. trans., London, 1972; first published 1968); and C. Taylor, 'Interpretation and the sciences of man', *Rev. Met.*, **25** (1971), 3. I have discussed some of Habermas's conclusions in 'In defence of objectivity', *Proc. Brit. Academy*, **57** (1972).

CHAPTER ONE

# Theory and Observation

## I. Is there an independent observation language?

Rapidity of progress, or at least change, in the analysis of scientific theory structure is indicated by the fact that only a few years ago the natural question to ask would have been 'Is there an independent theoretical language?' The assumption would have been that theoretical language in science is parasitic upon observation language, and probably ought to be eliminated from scientific discourse by disinterpretation and formalization, or by explicit definition in or reduction to observation language. Now, however, several radical and fashionable views place the onus on believers in an observation language to show that such a concept has any sense in the absence of a theory. It is time to pause and ask what motivated the distinction between a so-called theoretical language and an observation language in the first place, and whether its retention is not now more confusing than enlightening.

In the light of the importance of the distinction in the literature, it is surprisingly difficult to find any clear statement of what the two languages are supposed to consist of. In the classic works of twentieth-century philosophy of science, most accounts of the observation language were dependent on circular definitions of observability and its cognates, and the theoretical language was generally defined negatively as consisting of those scientific terms which are not observational. We find quasi-definitions of the following kind: "Observation-statement" designates a statement 'which records an actual or possible observation'; 'Experience, observation, and cognate terms will be used in the widest sense to cover observed facts about material objects or events in them as well as directly known facts about the contents or objects of immediate experience'; 'The observation language uses terms designating observable properties and relations for the description of observable things or events'; 'observables, i.e., . . . things and events which are ascertainable by direct observation'.[1] Even Nagel, who gives the most thorough account of the alleged distinction between theoretical and observation terms, seems to

---

[1] A. J. Ayer, *Language, Truth, and Logic*, 2nd edition (London, 1946), 11; R. B. Braithwaite, *Scientific Explanation* (New York, 1953), 8; R. Carnap, 'The methodological character of theoretical concepts', in *Minnesota Studies in the Philosophy of Science*, vol. I, ed. H. Feigl and M. Scriven (Minneapolis, 1956), 38; C. G. Hempel, 'The theoretician's dilemma', in *Minnesota Studies in the Philosophy of Science*, vol. II, ed. H. Feigl, M. Scriven and G. Maxwell (Minneapolis, 1958), 41.

presuppose that there is nothing problematic about the 'direct experimental evidence' for observation statements, or the 'experimentally identifiable instances' of observation terms.[1]

In contrast with the allegedly clear and distinct character of the observation terms, the meanings of theoretical terms, such as 'electron', 'electromagnetic wave' and 'wave function',[2] were held to be obscure. Philosophers have dealt with theoretical terms by various methods, based on the assumption that they have to be explained by means of the observation terms as given. None of the suggested methods has, however, been shown to leave theoretical discourse uncrippled in some area of its use in science. What suggests itself, therefore, is that the presuppositions of all these methods themselves are false, namely

(a) that the meanings of the observation terms are unproblematic;
(b) that the theoretical terms have to be understood by means of the observation terms; and
(c) that there is, in any important sense, a distinction between two *languages* here, rather than different kinds of uses within the same language.

In other words, the fact that we somehow understand, learn and use observation terms does not in the least imply that the way in which we understand, learn and use them is either different from or irrelevant to the way we understand, learn and use theoretical terms. Let us then subject the observation language to the same scrutiny which the theoretical language has received.

Rather than attacking directly the dual language view and its underlying empiricist assumptions, my strategy will be first to attempt to construct a different account of meaning and confirmation in the observation language. This project is not the ambitious one of a general theory of meaning, nor of the learning of language, but rather the modest one of finding conditions for understanding and use of terms in science—some specification, that is to say, in a limited area of discourse, of the 'rules of usage' which distinguish meaningful discourse from mere vocal reflexes. In developing this alternative account I shall rely on ideas which have become familiar particularly in connection with Quine's discussions of language and meaning and the replies of his critics, whose significance for the logic of science seems not yet to have been exploited nor even fully understood.

I shall consider in particular the predicate terms of the so-called observation language. But first something must be said to justify considering the problem as one of 'words' and not of 'sentences'. It has often been argued that it is

---

[1] E. Nagel, *The Structure of Science*, chap. 5.

[2] It would be possible to give examples from sciences other than physics: 'adaptation', 'function', 'intention', 'behaviour', 'unconscious mind'; but the question whether these are theoretical terms in the sense here distinguished from observation terms is controversial, and so is the question whether, if they are, they are eliminable from their respective sciences. These questions would take us too far afield.

sentences that we learn, produce, understand and respond to, rather than words; that is, that in theoretical discussion of language, sentences should be taken as units. There are, however, several reasons why this thesis, whether true or false, is irrelevant to the present problem, at least in its preliminary stages. The observation language of science is only a segment of the natural language in which it is expressed, and we may for the moment assume that rules of sentence formation and grammatical connectives are already given when we come to consider the use of observation predicates. Furthermore, since we are interested in alleged distinctions between the observation and theoretical languages, we are likely to find these distinctions in the characteristics of their respective predicates, not in the connectives which we may assume that they share. Finally, and most importantly, the present enterprise does not have the general positive aim of describing the entire structure of a language. It has rather the negative aim of showing that there are no terms in the observation language which are sufficiently accounted for by 'direct observation', 'experimentally identifiable instances' and the like. This can best be done by examining the hardest cases, that is, predicates which do appear to have direct empirical reference. No one would seriously put forward the direct-observation account of grammatical connectives; and if predicates are shown not to satisfy the account, it is likely that the same arguments will suffice to show that sentences do not satisfy it either.

So much for preliminaries. The thesis I am going to put forward can be briefly stated in two parts.

(*i*) All descriptive predicates, including observation and theoretical predicates, must be introduced, learned, understood and used, either by means of direct empirical associations in some physical situations, or by means of sentences containing other descriptive predicates which have already been so introduced, learned, understood and used, or by means of both together. (Introduction, learning, understanding and use of a word in a language will sometimes be summarized in what follows as the *function* of that word in the language.)

(*ii*) No predicates, not even those of the observation language, can function by means of direct empirical associations alone.

The process of functioning in the language can be spelled out in more detail.

*A.* Some predicates are initially learned in empirical situations in which an association is established between some aspects of the situation and a certain word. Given that any word with extralinguistic reference is ever learned, this is a necessary statement and does not presuppose any particular theory about what an association is or how it is established. This question is one for psychology or linguistics rather than philosophy. Two necessary remarks can, however, be made about such learning.

(1) Since every physical situation is indefinitely complex, the fact that the

particular aspect to be associated with the word is identified out of a multiplicity of other aspects implies that degrees of physical similarity and difference can be recognized between different situations.

(2) Since every situation is in detail different from every other, the fact that the word can be correctly reused in a situation in which it was not learned has the same implication.

These remarks would seem to be necessarily implied in the premise that some words with reference are learned by empirical associations. They have not gone unchallenged, however, and it is possible to distinguish two sorts of objections to them. First, some writers, following Wittgenstein, have appeared to deny that physical similarity is necessary to the functioning of *any* word with extralinguistic reference. That similarity is not *sufficient*, I am about to argue, and I also agree that not all referring words need to be introduced in this way, but if *none* were, I am unable to conceive how an intersubjective descriptive language could ever get under way. The onus appears to rest upon those who reject similarity to show in what other way descriptive language is possible. For example, Donald Davidson claims that there is no need for a descriptive predicate to be learned in the presence of the object to which it is properly applied, since, for example, it might be learned in 'a skilfully faked environment'.[1] This possibility does not, however, constitute an objection to the thesis that it must be learned in *some* empirical situation, and that this situation must have some similarity with those situations in which the predicate is properly used. Chomsky, on the other hand, attacks what he regards as Quine's 'Humean theory' of language acquisition by resemblance of stimuli and conditioned response.[2] But the necessity of the *similarity* condition for language learning does not depend on the particular empirical mechanism of learning. Learning by patterning the environment in terms of a set of 'innate ideas' would depend equally upon subsequent application of the same pattern to similar features of the environment. Moreover, 'similar' cannot just be *defined as* 'properly ascribed the same descriptive predicate in the same language community', since for one thing similarity is a matter of degree and is a non-transitive relation, whereas 'properly ascribed the same descriptive predicate' is not, or not obviously. The two terms can therefore hardly be synonymous. I therefore take it as a necessary *a priori* condition of the applicability of a language containing universal terms that *some* of these terms presuppose primitive causal recognitions of physical similarities.

A different sort of objection to the appeal to similarity is made by Popper, who argues that the notion of repetition of instances which is implied by

[1] D. Davidson, 'Theories of meaning and learnable languages', *Logic, Methodology and Philosophy of Science*, ed. Y. Bar-Hillel (Amsterdam, 1965), 386.

[2] N. Chomsky, 'Quine's empirical assumptions', *Synthese*, **19** (1968), 53. See also Quine's reply to Chomsky, *ibid.*, 274.

1 and 2 is essentially vacuous, because similarity is always similarity *in certain respects*, and 'with a little ingenuity' we could always find similarities in *some* same respects between all members of any finite set of situations. That is to say, 'anything can be said to be a repetition of anything else, if only we adopt the appropriate point of view'.[1] But if this were true, it would make the learning process in empirical situations impossible. It would mean that however finitely large the number of presentations of a given situation-aspect, that aspect could never be identified as the desired one out of the indefinite number of other respects in which the presented situations are all similar. It would, of course, be possible to eliminate some other similarities by presenting further situations similar in the desired respect but not in others, but it would then be possible to find other respects in which all the situations, new and old, are similar—and so on without end.

However, Popper's admission that 'a little ingenuity' may be required allows a less extreme interpretation of his argument, namely that the physics and physiology of situations already give us some 'point of view' with respect to which some pairs of situations are similar in more obvious respects than others, and one situation is more similar in some respect to another than it is in the same respect to a third. This is all that is required by the assertions 1 and 2. Popper has needlessly obscured the importance of these implications of the learning process by speaking as though, before any repetition can be recognized, we have to take thought and *explicitly* adopt a point of view. If this were so, a regressive problem would arise about how we ever learn to apply the predicates in which we explicitly express that point of view. An immediate consequence of this is that there must be a stock of predicates in any descriptive language for which it is impossible to *specify* necessary and sufficient conditions of correct application. For if any such specification could be given for a particular predicate, it would introduce further predicates requiring to be learned in empirical situations for which there was no specification. Indeed, such unspecified predicates would be expected to be in the majority, for those for which necessary and sufficient conditions can be given are dispensable except as a shorthand and hence essentially uninteresting. We must therefore conclude that the primary process of recognition of similarities and differences is necessarily *unverbalizable*. The emphasis here is of course on *primary*, because it may be perfectly possible to give empirical descriptions of the conditions, both psychological and physical, under which similarities are recognized, but such descriptions will themselves depend on further undescribable primary recognitions.

*B.* It may be thought that the primary process of classifying objects according to recognizable similarities and differences will provide us with

---

[1] *The Logic of Scientific Discovery*, appendix *x, 422.

exactly the independent observation predicates required by the traditional view. This, however, is to overlook a logical feature of relations of similarity and difference, namely that they are not *transitive*. Two objects *a* and *b* may be judged to be similar to some degree in respect to predicate *P*, and may be placed in the class of objects to which *P* is applicable. But object *c* which is judged similar to *b* to the same degree may not be similar to *a* to the same degree or indeed to any degree. Think of judgments of similarity of three shades of colour. This leads to the conception of some objects as being more 'central' to the *P*-class than others, and also implies that the process of classifying objects by recognition of similarities and differences is necessarily accompanied by some loss of (unverbalizable) information. For if *P* is a predicate whose conditions of applicability are dependent on the process just described, it is impossible to *specify* the degree to which an object satisfies *P* without introducing more predicates about which the same story would have to be told. Somewhere this potential regress must be stopped by some predicates whose application involves loss of information which is present to recognition but not verbalizable. However, as we shall see shortly, the primary recognition process, though necessary, is not sufficient for classification of objects as *P*, and the loss of information involved in classifying leaves room for changes in classification to take place under some circumstances. Hence primary recognitions do not provide a stable and independent list of primitive observation predicates.

*C.* It is likely that the examples that sprang to mind during the reading of the last section were such predicates as 'red', 'ball' and 'teddy bear'. But notice that nothing that has been said rules out the possibility of giving the same account of apparently much more complex words. 'Chair', 'dinner' and 'mama' are early learned by this method, and it is not inconceivable that it could also be employed in first introducing 'situation', 'rule', 'game', 'stomach ache' and even 'heartache'. This is not to say, of course, that complete fluency in using these words could be obtained by this method alone; indeed, I am now going to argue that complete fluency cannot be obtained in the use of *any* descriptive predicate by this method alone. It should only be noticed here that it is possible for any word in natural language having some extralinguistic reference to be introduced in suitable circumstances in some such way as described in section *A*.

*D.* As learning of the language proceeds, it is found that some of these predicates enter into general statements which are accepted as true and which we will call *laws*: 'Balls are round'; 'In summer leaves are green'; 'Eating unripe apples leads to stomach ache'. It matters little whether some of these are what we would later come to call analytic statements; some, perhaps most, are synthetic. It is not necessary, either, that every such law should be *in fact* true, only that it is for the time being accepted as true by the language community. As we shall see later, any one of these laws may be *false* (although not all could be false at once). Making explicit these general laws is only a continuation and extension of the process already described as identifying and reidentifying proper occasions for the use of a predicate by means of physical similarity. For knowledge of the laws will now enable the language user to apply descriptions correctly in situations other than those in which he learned them, and even in situations where nobody could have learned them in the absence of the laws—for example, 'stomach ache' of an absent individual known to have consumed a basketful of unripe apples, or even 'composed of diatomic molecules' of the oxygen in the atmosphere. In other words, the laws enable generally correct inferences and predictions to be made about distant ('unobservable') states of affairs.

*E.* At this point the system of predicates and their relations in laws has become sufficiently complex to allow for the possibility of internal misfits and even contradictions. This possibility arises in various ways. It may happen that some of the applications of a word in situations turn out not to satisfy the laws which are true of other applications of the word. In such a case, since degrees of physical similarity are not transitive, a reclassification may take place in which a particular law is preserved in a subclass more closely related by similarity, at the expense of the full range of situations of application which are relatively less similar. An example of this would be the application of the word 'element' to water, which becomes incorrect in order to preserve the truth of a system of laws regarding 'element', namely that elements cannot be chemically dissociated into parts which are themselves elements, that elements always enter into compounds, that every substance is constituted by one or more elements, and so on. On the other hand, the range of applications may be widened in conformity with a law, so that a previously incorrect application becomes correct. For example, 'mammal' is correctly applied to whales, whereas it was previously thought that 'Mammals live only on land' was a well-entrenched law providing criteria for correct use of 'mammal'. In such a case it is not adequate to counter with the suggestion that the correct use of 'mammal' is *defined* in terms of animals which suckle their young, for it is conceivable that if other empirical facts had been different, the classification in terms of habitat would have been more useful and comprehensive than that in terms of milk production. And in regard to the first example, it cannot be maintained that it is the *defining* characteristics of 'element' that are preserved at the expense of its application to water because, of the conditions mentioned, it is not clear that any particular one is, or ever has been, taken as *the* defining characteristic; and since the various characteristics are logically independent, it is empirically possible that some might be satisfied and not others. *Which* is preserved will always depend on what system of laws is most convenient, most coherent and most comprehensive. But the most telling objection to the suggestion that correct application is decided by definition is of course the general point made at the end of section *A* that there is always a large number of predicates for which *no* definition in terms of necessary and sufficient conditions of application can be given. For these predicates it is possible that the primary recognition of, for example, a whale as being sufficiently similar to some fish to justify its inclusion in the class of fish may be explicitly overridden in the interests of preserving a particular set of laws.

Properly understood, the point developed in the last paragraph should lead to a far-reaching reappraisal of orthodoxy regarding the theory–observation distinction. To summarize, it entails that no feature in the total landscape of functioning of a descriptive predicate is exempt from modification under pressure from its surroundings. That any empirical law may be abandoned in the face of counterexamples is trite, but it becomes less trite when the functioning of every predicate is found to depend essentially on some laws or other and when it is also the case that any 'correct' situation of application— *even that in terms of which the term was originally introduced*—may become incorrect in order to preserve a system of laws and other applications. It is in this sense that I shall understand the 'theory dependence' or 'theory-ladenness' of all descriptive predicates.

One possible objection to this account is easily anticipated. It is not a *conventionalist* account, if by that we mean that any law can be assured of truth by sufficiently meddling with the meanings of its predicates. Such a view does not take seriously the systematic character of laws, for it contemplates preservation of the truth of a given law irrespective of its coherence with the rest of the system, that is, the preservation of simplicity and other desirable internal characteristics of the system. Nor does it take account of the fact that not all primary recognitions of empirical similarity can be overridden in the interest of preserving a given law, for it is upon the existence of some such recognitions that the whole possibility of language with empirical reference rests. The present account on the other hand demands both that laws shall remain connected in an economical and convenient system and that at least most of its predicates shall remain applicable, that is, that they shall continue to depend for applicability upon the primary recognitions of similarity and difference in terms of which they were learned. That it is possible to have such a system with a given set of laws and predicates is not a convention but a fact of the empirical world. And although this account allows that *any* of the situations of correct application may change, it cannot allow that *all* should change, at least not all at once. Perhaps it would even be true to say that only a small proportion of them can change at any one time, although it is conceivable that over long periods of time most or all of them might come to change piecemeal. It is likely that almost all the terms used by the alchemists that are still in use have now changed their situations of correct use quite radically, even though at any one time chemists were preserving most of them while modifying others.

## II. Entrenchment

It is now necessary to attack explicitly the most important and controversial question in this area, namely the question whether the account of predicates that has been given really applies to all descriptive predicates whatsoever, or whether there are after all some that are immune to modification in the light of further knowledge and that might provide candidates for a basic and independent observation language. The example mentioned at the end of the last paragraph immediately prompts the suggestion that it would be possible at any time for both alchemists and chemists to 'withdraw' to a more basic observation language than that used in classifying substances and that this language would be truly primitive and theory-independent. The suspicion that this may be so is not even incompatible with most of the foregoing account, for it may be accepted that we often do make words function without reflecting upon more basic predicates to which we could withdraw if challenged. Thus, it may not be disputed that we learn, understand and use words like 'earth', 'water', 'air' and 'fire' in empirical situations and that their subsequent functioning depends essentially upon acceptance of some laws; and yet it may still be maintained that there are some more basic predicates for which cash value is payable in terms of empirical situations alone. Let us therefore consider this argument at its strongest point and take the case of the putative observation predicate 'red'. Is this predicate subject to changes of correct application in the light of laws in the way that has been described? The defence of our account at this point comes in two stages. First, it must be shown that *no* predicate of an observation language can function by mere empirical situations alone, independently of any laws. Second, it must be shown that there is no set of observation predicates whose interrelating laws are absolutely invariant to changes in the rest of the network of laws.

When a predicate such as 'red' is claimed to be 'directly' descriptive, this claim is usually made in virtue of its use as a predicate of immediate experience —a sensation of a red postage stamp, a red spectral line, a red after-image. It is unnecessary here to enter into the much discussed questions of whether there are any such 'things' as sensations for 'red' to be a predicate of, whether such predicates of sensations could be ingredients of either a public or a private language, and whether there is indeed any sense in the notion of a private language. The scientific observation language at least is not private but must be intersubjective; and whether some of its predicates are predicates of sensations or not, it is still possible to raise the further question: in *any* intersubjective language can the functioning of the predicates be independent of accepted laws? That the answer is negative can be seen by considering the original account of the empirical situations given in section I.*A* and by adopting one generally acceptable assumption. The assumption is that in using a public language, the correctness of any application of a predicate in a given situation must in principle be capable of intersubjective test.[1] Now if my careful response of 'red' to each of a set of situations were all that were involved in my correct use of 'red', this response would not be sufficient to ensure intersubjectivity. It is possible, in spite of my care, that I have responded mistakenly, in which case the laws relating 'red' to other predicates can be appealed to in order to correct me (I can even correct myself by this method): 'It can't have been red, because it was a sodium flame, and sodium flames are not red'. If my response 'red' is intended to be an ingredient of a public observation language, it carries at least the implication that disagreements can be publicly resolved, and this presupposes laws conditioning the function of 'red'. If this implication is absent, responses are mere verbal reflexes having no intersubjective significance (unless of course they are part of a physiological–psychological experiment, but then I am subject, not observer). This argument does not, I repeat, purport to show that there could not be a sense-datum language functioning as the observation language of science—only that if this were so, its predicates would share the double aspect of empirical situation and dependence on laws which belongs to all putative observation predicates.

Now consider the second stage of defence of our account. The suggestion to be countered here is that even if there are peripheral uses of 'red' which might be subject to change in the event of further information about laws, there is nevertheless a central core of function of 'red', with at least some laws which ensure its intersubjectivity, which remains stable throughout all extensions and modifications of the rest of the network of accepted laws. To illustrate the contrast between 'periphery' and 'core' take the following examples: we might come to realize that when 'red' is applied to a portion of the rainbow, it is not a predicate of an object, as in the paradigm cases of 'red', or that the ruddy hue of a distant star is not the colour of the star but an effect of its recession. But, it will be said, in regard to cherries, red lips and the colour of a strontium compound in a Bunsen flame, 'red' is used entirely independently of the truth of or knowledge of the great majority of laws in our network. We might of course be mistaken in application of 'red' to situations of this central kind, for we may mistake colour in a bad light or from defects of vision; but there are enough laws whose truth cannot be in doubt to enable us to correct mistakes of this kind, and by appealing to them we are always able to come to agreement about correct applications. There is no sense, it will be argued, in supposing that in cases like this we could all be mistaken all the time or that we might, in any but the trivial sense of deciding to use another word equivalent to 'red', come to change our usage in these central situations.

---

[1] See, for example, L. Wittgenstein, *Philosophical Investigations* (London, 1953), sec. 258ff: A. J. Ayer, *The Concept of a Person* (London, 1963), 39ff; K. R. Popper, *The Logic of Scientific Discovery*, 44–5.

One possible reply[1] is to point out that the admission that there are *some* situations in which we might change our use even of a predicate like 'red' is already a significant one, especially in the examples given above. For the admission that the 'red' of a rainbow or a receding star is not the colour of an object is the admission that in these cases at least it is a *relational* predicate, where the relata, which may be quite complex, are spelled out by the laws of physics. Now no doubt it does not *follow* that 'red' ascribed to the book cover now before me is also a relational predicate, unless we take physics to provide the real truth about everyday objects as well as those that are more remote. The schizophrenia induced by not taking physics seriously in this way raises problems of its own which we cannot pursue here. But suppose our critic accepts the realist implication that 'red' is, on all occasions of its use as a predicate of objects, in fact a relational predicate, and then goes on to discount this admission by holding that such a relatively subtle logical point is irrelevant to the ordinary function of 'red' in the public language. Here we come near the heart of what is true in the critic's view. The truth might be put like this: Tom, Dick and Mary do indeed use the word 'red' with general indifference to logical distinctions between properties and relations. Even logicians and physicists continue to use it in such a way that in ordinary conversation it need never become apparent to others, or even to themselves, that they 'really believe' that colour predicates are relational. And more significantly for the ultimate purpose of this essay, the conversation of a Newtonian optician about sticks and stones and rolls of bread need never reveal a difference of function of 'red' from the conversation of a post-relativity physicist.

Such a concession to the critic with regard to invariance of function in limited domains of discourse is an important one, but it should be noticed that its force depends not upon fixed stipulations regarding the use of 'red' in particular empirical situations, but rather upon empirical facts about the way the world is. Physically possible situations can easily be envisaged in which even this central core of applicability of 'red' would be broken. Suppose an isolated tribe all suffered a congenital colour blindness which resulted in light green being indistinguishable from red and dark green from black. Communication with the outside world, or even the learning of physics without any such communication, might very well lead them to revise the function of 'red' and 'black' even in paradigm cases of their own language.

A more realistic and telling example is provided by the abandonment of Newtonian time simultaneity. This is an especially striking case, because time concepts are among the most stable in most languages and particularly in a physics which has persistently regarded spatial and temporal qualities as primary and as providing the indispensable framework of a mechanistic

---

[1] *Cf.* P. K. Feyerabend, 'An attempt at a realistic interpretation of experience', *Proceedings of the Aristotelian Society*, **58** (1957–8), 143, 160.

science. As late as 1920 N. R. Campbell, usually a perceptive analyst of physical concepts, wrote:

> Is it possible to find any judgement of sensation concerning which all sentient beings whose opinion can be ascertained are always and absolutely in agreement? . . . I believe that it is possible to obtain absolutely universal agreement for judgements such as, 'The event A happened at the same time as B, or A happened between B and C.[1]

Special relativity had already in 1905 shown this assumption to be false. This means that at any time before 1905 the assumption was one from which it was certainly possible to withdraw; it was in fact 'theory-laden', although it had not occurred to anybody that this was the case. Now let us cast Einstein in the role of the 'operationist' physicist who, wiser than his contemporaries, has detected the theory-ladenness and wishes to withdraw from it to a 'level of direct observation', where there are no theoretical implications, or at least where these are at a minimum.[2] What can he do? He can try to set up an operational definition of time simultaneity. When observers are at a distance from each other (they are always at *some* distance), and when they are perhaps also moving relatively to each other, he cannot assume that they will agree on judgments of simultaneity. He will assume only that a given observer can judge events that are simultaneous in his own field of vision, provided they occur close together in that field. The rest of Einstein's operational definition in terms of light signals between observers at different points is well known. But notice that this definition does not carry out the programme just proposed for an operationist physicist. For far from withdrawal to a level of direct observation where theoretical implications are absent or at a minimum, the definition requires us to assume, indeed to postulate, that the velocity of light *in vacuo* is the same in all directions and invariant to the motions of source and receiver. This is a postulate which is logically prior in special relativity to any experimental measurement of the velocity of light, because it is used in the very definition of the time scale at distant points. But from the point of view of the operationist physicist before 1905, the suggestion of withdrawing from the assumption of distant absolute time simultaneity to this assumption about the velocity of light could not have appeared to be a withdrawal to more direct observation having fewer theoretical implications, but rather the reverse. This example illustrates well the impossibility of even talking sensibly about 'levels of more direct observation' and 'degrees of theory-ladenness' *except in the context of some framework of accepted laws.* That such talk is dependent on this context is enough to refute the thesis that the contrast between 'direct observation' and 'theory-ladenness' is itself theory-independent. The example also illustrates the fact that at any given

---

[1] *Foundations of Science*, 29.

[2] That this way of putting it is a gross distortion of Einstein's actual thought processes is irrelevant here.

stage of science it is never possible to know *which* of the currently entrenched predicates and laws may have to give way in the future.

The operationist has a possible comeback to this example. He may suggest that the process of withdrawal to the directly observed is not a process of constructing another theory, as Einstein did, but properly stops short at the point where we admitted that at least one assumption of Newtonian physics is true and must be retained, namely, that 'a given observer can judge events that are simultaneous in his own field of vision, provided they occur close together in that field'—call this assumption *S*. This, it may be said, is a genuine withdrawal to a less theory-laden position, and all that the rest of the example shows is that there is in fact no possibility of advance again to a more general conception of time simultaneity without multiplying insecure theoretical assumptions. Now, of course, the game of isolating some features of an example as paradigms of 'direct observation', and issuing a challenge to show how *these* could ever be overthrown, is one that can go on regressively without obvious profit to either side. But such a regress ought to stop if either of the following two conditions is met:

(*a*) that it is logically possible for the alleged paradigm to be overthrown and that its overthrow involves a *widening* circle of theoretical implications; or

(*b*) that the paradigm becomes less and less suitable as an observation statement, because it ceases to have the required intersubjective character.

The time simultaneity example made its point by illustrating condition *a*. The assumption *S* to which it is now suggested we withdraw can be impaled on a dilemma between *a* and *b*. Suppose it were shown that an observer's judgment of simultaneity in his field of sensation were quite strongly dependent on the strength of the gravitational field in his neighbourhood, although this dependence had not yet been shown up in the fairly uniform conditions of observation on the surface of the earth. Such a discovery, which is certainly conceivable, would satisfy condition *a*. As long as the notion of simultaneity is so interpreted as to allow for intersubjective checking and agreement, there is always an indefinite number of such possible empirical situations whose variation might render the assumption *S* untenable. The only way to escape this horn of the dilemma is to interpret *S* as referring to the direct experience of simultaneity of a single observer, and this is intersubjectively and hence scientifically useless, and impales us on the horn of condition *b*.

The comparative stability of function of the so-called observation predicates is logically speaking an accident of the way the world is. But it may now be suggested that since the way the world is is not likely to alter radically during the lifetime of any extant language, we might define an observation language to be just that part of language which the facts allow to remain stable. This, however, is to take less than seriously the effects of scientific knowledge on our ways of talking about the world and also to underestimate the tasks that ordinary language might be called upon to perform as the corpus of scientific knowledge changes. One might as well hold that the ordinary language of Homer, which identifies life with the breath in the body and fortuitous events with interventions of divine personages, and was no doubt adequate to discourse before the walls of Troy, should have remained stable in spite of all subsequent changes in physics, physiology, psychology and theology. Our ordinary language rules for the use of 'same time', which presuppose that this concept is independent of the distance and relative motion of the spatial points at which time is measured, are not only contradicted by relativity theory, but would possibly need fundamental modification if we all were to take habitually to space travel. Another point to notice here is that the comparatively stable area within which it is proposed to define an observation language itself is partly known to us because its stability is explained by the theories we now accept. It is certainly not sufficiently defined by investigating what observation statements have in fact remained stable during long periods of time, for this stability might be due to accident, prejudice or false beliefs. Thus any attempted definition itself would rely upon current theories and hence not be a definition of an observation language which is theory-independent. Indeed, it might justly be concluded that we shall know what the most adequate observation language is only when, if possible, we have true and complete theories, including theories of physiology and physics which tell us what it is that is most 'directly observed'. Only then shall we be in a position to make the empirical distinctions that seem to be presupposed by attempts to discriminate theoretical and observation predicates.

The upshot of all this may be summarized by saying that although there is a nucleus of truth in the thesis of invariance of the observation language and, hence, of the theory–observation distinction among predicates, this truth has often been located in the wrong place and used to justify the wrong inferences. The invariance of observation predicates has been expressed in various ways, not all equivalent to one another and not all equally valid.[1] Let us summarize the discussion so far by examining some of these expressions.

(i) 'There are some predicates that are *better entrenched* than others, for instance, "red" than "ultra-violet", "lead" than "π-meson".'

If by 'better entrenched' is meant less subject to change of function in ordinary discourse and therefore less revelatory of the speaker's commitments to a system of laws or of his relative ignorance of such systems, then (i) is true. But this is a *factual* truth about the relative invariance of some empirical laws to increasing empirical information, not about the *a priori* features of a peculiar set of predicates, and it does not entail that any predicate is *absolutely*

---

[1] The many-dimensional character of the theory–observation distinction has been discussed by P. Achinstein in 'The problem of theoretical terms', *Am. Phil. Quart.*, **2** (1965), 193, and *Concepts of Science* (Baltimore, 1968), chaps. 5, 6.

entrenched, nor that any subsystems of predicates and the laws relating them are immune to modification under pressure from the rest of the system.

(ii) 'There are some predicates that refer to aspects of situations more *directly observable* than others.'

If this means that their function is more obviously related to empirical situations than to laws, (ii) is true, but its truth does not imply that a line can be drawn between theoretical and observation predicates in the place it is usually desired to draw it. For it is not at all clear that highly complex and even theoretical predicates may not sometimes be directly applicable in appropriate situations. Some examples were given in section I.*C*; other examples are thinkable where highly theoretical descriptions would be given directly: 'particle-pair annihilation' in a cloud chamber, 'glaciation' of a certain landscape formation, 'heart condition' of a man seen walking along the street. To the immediate rejoinder that these examples leave open the possibility of withdrawal to less 'theory-laden' descriptions, a reply will be given in (v) below. Meanwhile it should be noticed that this sense of 'observable' is certainly not coextensive with that of (i).

(iii) 'There are some predicates that are learnable and applicable in a *pragmatically* simpler and quicker manner than others.'

This is true, but does not necessarily single out the same set of predicates in all language communities. Moreover, it does not necessarily single out all or only the predicates that are 'observable' in senses (i) and (ii).

(iv) 'There are some predicates in terms of which others are *anchored to the empirical facts*.'

This may be true in particular formulations of a theory, where the set of anchor predicates is understood as in (i), (ii) or (iii), but little more needs to be said to justify the conclusion that such a formulation and its set of anchoring predicates would not be unique. In principle it is conceivable that any predicate could be used as a member of the set. Thus, the commonly held stronger version of this assumption is certainly false, namely that the anchor predicates have unique properties which allow them to endow theoretical predicates with empirical meaning which these latter would not otherwise possess.

(v) The most important assumption about the theory–observation distinction, and the one which is apparently most damaging to the present account, can be put in a weaker and a stronger form:

(*a*) 'There are some predicates to which we could always *withdraw* if challenged in our application of others.'

(*b*) 'These form a unique subset in terms of which "pure descriptions" free from "theory-loading" can be given.'

Assumption *a* must be accepted to just the extent that we have accepted the assumption that there are degrees of entrenchment of predicates, and for the same reasons. It is indeed sometimes possible to withdraw from the implications of some ascriptions of predicates by using others better entrenched in the network of laws. To use some of the examples already mentioned, we may withdraw from 'particle-pair annihilation' to 'two white streaks meeting and terminating at an angle'; from 'heart condition' to a carefully detailed report of complexion, facial structure, walking habits and the like; and from 'epileptic fit' to a description of teeth-clenching, falling, writhing on the floor and so on. So far, these examples show only that some of the lawlike implications that are in mind when the first members of each of these pairs of descriptions are used can be withdrawn from and replaced by descriptions which do not have *these* implications. They do not show that the second members of each pair are free from lawlike implications of their own, nor even that it is possible to execute a series of withdrawals in such a way that each successive description contains fewer implications than the description preceding it. Far less do they show that there is a unique set of descriptions which have *no* implications; indeed the arguments already put forward should be enough to show that this assumption, assumption *b*, must be rejected. As in the case of entrenchment, it is in principle possible for any particular lawlike implication to be withdrawn from, although not all can be withdrawn from at once. Furthermore, although in any given state of the language some descriptive predicates are more entrenched than others, it is not clear that withdrawal to those that are better entrenched is withdrawal to predicates which have *fewer* lawlike implications. Indeed, it is likely that better entrenched predicates have in fact far more implications. The reason why these implications do not usually seem doubtful or objectionable to the observational purist is that they have for so long proved to be true, or been believed to be true, in their relevant domains that their essentially inductive character has been forgotten. It follows that when well-entrenched predicates and their implications are from time to time abandoned under pressure from the rest of the network, the effects of such abandonment will be more far-reaching, disturbing and shocking than when less well-entrenched predicates are modified.

## III. The network model

The foregoing account of theories, which has been presented as more adequate than the deductive two-language model, may be dubbed the *network model* of theories. It is an account that was first explicit in Duhem and more recently reinforced by Quine. Neither in Duhem nor in Quine, however, is it quite clear that the netlike interrelations between more directly observable predicates and their laws are in principle just as subject to modifications from the rest of the network as are those that are relatively theoretical. Duhem seems sometimes to imply that although there is a network of relatively phenomenological representations of facts, once established this network remains stable with respect to the changing explanations. This is indeed one reason why he rejects the view that science aims at explanation in terms of unobservable entities and restricts theorizing to the articulation of mathematical representations which merely systematize but do not explain the facts. At the same time, however, his analysis of the facts is far subtler than that presupposed by later deductivists and instrumentalists. He sees that what is primarily significant for science is not the precise nature of what we directly observe, which in the end is a *causal* process, itself susceptible of scientific analysis. What is significant is the interpretative expression we give to what is observed, what he calls the *theoretical facts*, as opposed to the 'raw data' represented by *practical facts*. This distinction may best be explained by means of his own example. Consider the theoretical fact 'The temperature is distributed in a certain manner over a certain body'.[1] This, says Duhem, is susceptible of precise mathematical formulation with regard to the geometry of the body and the numerical specification of the temperature distribution. Contrast the practical fact. Here geometrical description is at best an idealization of a more or less rigid body with a more or less indefinite surface. The temperature at a given point cannot be exactly fixed, but is only given as an average value over vaguely defined small volumes. The theoretical fact is an imperfect translation, or interpretation, of the practical fact. Moreover, the relation between them is not one-to-one but rather many-to-many, for an infinity of idealizations may be made to more or less fit the practical fact, and an infinity of practical facts may be expressed by means of one theoretical fact.

Duhem is not careful in his exposition to distinguish *facts* from *linguistic expressions of facts*. Sometimes both practical and theoretical facts seem to be intended as linguistic statements (for instance, where the metaphor of 'translation' is said to be appropriate). But even if this is his intention, it is clear that he does not wish to follow traditional empiricism into a search for forms of expression of practical facts which will constitute the basis of science. Practical facts are not the appropriate place to look for such a basis— they are imprecise, ambiguous, corrigible, and on their own ultimately meaningless. Moreover, there is a sense in which they are literally inexpressible. The absence of distinction between fact and linguistic expression here is not accidental. As soon as we begin to try to capture a practical fact in language, we are committed to some theoretical interpretation. Even to say of the solid body that 'its points are more or less worn down and blunt' is to commit ourselves to the categories of an ideal geometry.

What, then, is the 'basis' of scientific knowledge for Duhem? If we are to use this conception at all, we must say that the basis of science is the set of theoretical facts in terms of which experience is interpreted. But we have just seen that theoretical facts have only a more or less loose and ambiguous relation with experience. How can we be sure that they provide a firm empirical

---

[1] P. Duhem, *The Aim and Structure of Physical Theory*, 133.

B

foundation? The answer must be that we cannot be sure. There is no such foundation. Duhem himself is not consistent on this point, for he sometimes speaks of the persistence of the network of theoretical facts as if this, once established, takes on the privileged character ascribed to observation statements in classical positivism. But this is not the view that emerges from his more careful discussion of examples. For he is quite clear, as in the case of the correction of the 'observational' laws of Kepler by Newton's theory, that more comprehensive mathematical representations may show particular theoretical facts to be false.

However, we certainly seem to have a problem here, because if it is admitted that subsets of the theoretical facts may be removed from the corpus of science, and if we yet want to retain some form of empiricism, the decision to remove them can be made only by reference to *other* theoretical facts, whose status is in principle equally insecure. In the traditional language of epistemology some element of correspondence with experience, though loose and corrigible, must be retained but also be supplemented by a theory of the coherence of a network. Duhem's account of this coherence has been much discussed but not always in the context of his complete account of theoretical and practical facts, with the result that it has often been trivialized. Theoretical facts do not stand on their own but are bound together in a network of laws which constitutes the total mathematical representation of experience. The putative theoretical fact that was Kepler's third law of planetary motion, for example, does not fit the network of laws established by Newton's theory. It is therefore modified, and this modification is possible without violating experience because of the many-to-one relation between the theoretical fact and that practical fact understood as the ultimately inexpressible situation which obtains in regard to the orbits of planets.

It would seem to follow from this (although Duhem never explicitly draws the conclusion) that there is no theoretical fact or lawlike relation whose truth or falsity can be determined in isolation from the rest of the network. Moreover, many conflicting networks may more or less fit the same facts, and which one is adopted must depend on criteria other than the facts: criteria involving simplicity, coherence with other parts of science, and so on. Quine, as is well known, has drawn this conclusion explicitly in the strong sense of claiming that any statement can be maintained true in the face of any evidence.

> Any statement can be held true come what may, if we make drastic enough adjustments elsewhere in the system. . . . Conversely, by the same token, no statement is immune to revision.[1]

In a later work, however, he does refer to 'the philosophical doctrine of infallibility of observation sentences' as being sustained in his theory.

---

[1] W. v. O. Quine, *From a Logical Point of View*, 43.

Defining the stimulus meaning of a sentence as the class of sensory stimulations that would prompt assent to the sentence, he regards observation sentences as those sentences whose stimulus meanings remain invariant to changes in the rest of the network and for which 'their stimulus meanings may without fear of contradiction be said to do full justice to their meanings'.[1] This seems far too conservative a conclusion to draw from the rest of the analysis, for in the light of the arguments and examples I have presented it appears very dubious whether there are such invariant sentences if a long enough historical perspective is taken.

There are other occasions on which Quine seems to obscure unnecessarily the radical character of his own position by conceding too much to more traditional accounts. He compares his own description of theories to those of Braithwaite, Carnap and Hempel in respect of the 'contextual definition' of theoretical terms. But his own account of these terms as deriving their meaning from an essentially *linguistic* network has little in common with the formalist notion of 'implicit definition' which these deductivists borrow from mathematical postulate systems in which the terms need not be interpreted empirically. In this sense the implicit definition of 'point' in a system of Riemannian geometry is entirely specified by the formal postulates of the geometry and does not depend at all on what would count empirically as a realization of such a geometry.[2] Again, Quine refers particularly to a net analogy which Hempel adopts in describing theoretical predicates as the knots in the net, related by definitions and theorems represented by threads. But Hempel goes on to assert that the whole 'floats . . . above the plane of observation' to which it is anchored by *threads of a different kind*, called 'rules of interpretation', *which are not part of the network itself*.[3] The contrast between this orthodox deductivism and Quine's account could hardly be more clear. For Quine, and in the account I have given here, there is indeed a network of predicates and their lawlike relations, but it is not floating above the domain of observation; it is attached to it at some of its knots. *Which* knots will depend on the historical state of the theory and its language and also on the way in which it is formulated, and the knots are not immune to change as science develops. It follows, of course, that 'rules of interpretation' disappear from this picture: *all* relations become laws in the sense defined above, which, it must be remembered, includes near analytic definitions and conventions as well as empirical laws.

---

[1] *Word and Object*, 42, 44.

[2] *Ibid.*, 11. For an early and devastating investigation of the notion of 'implicit definition' in a formal system, see G. Frege, 'On the foundations of geometry', trans. M. E. Szabo, *Philosophical Review*, 69 (1960), 3 (first published 1903), and in specific relation to the deductive account of theories, see C. G. Hempel, 'Fundamentals of concept formation in empirical science', *International Encyclopedia of Unified Science*, vol. II, no. 7 (Chicago, 1952), 81.

[3] C. G. Hempel, 'Fundamentals of concept formation', 36.

## IV. Theoretical predicates

So far it has been argued that it is a mistake to regard the distinction between theoretical and observational predicates either as providing a unique partition of descriptive predicates into two sets or as providing a simple ordering such that it is always possible to say of two predicates that one is under all circumstances more observational than or equally observational with the other. Various relative and noncoincident distinctions between theoretical and observational have been made, none of which is consistent with the belief that there is a unique and privileged set of observation predicates in terms of which theories are related to the empirical world. So far in the network model it has been assumed that any predicate may be more or less directly ascribed to the world in some circumstances or other, and that none is able to function in the language by means of such direct ascription alone. The second of these assumptions has been sufficiently argued; it is now necessary to say more about the first. Are there any descriptive predicates in science which could not under any circumstances be directly ascribed to objects? If there are, they will not fit the network model as so far described, for there will be nothing corresponding to the process of classification by empirical associations, even when this process is admitted to be fallible and subject to correction by laws, and they will not be connected to other predicates by laws, since a law presupposes that the predicates it connects have all been observed to co-occur in some situation or other.

First, it is necessary to make a distinction between theoretical *predicates* and theoretical *entities*, a distinction which has not been sufficiently considered in the deductivist literature. Theoretical entities have sometimes been taken to be equivalent to unobservable entities. What does this mean? If an entity is unobservable in the sense that it never appears as the subject of observation reports, and is not in any other way related to the entities which do appear in such reports, then it has no place in science. This cannot be what is meant by 'theoretical' when it is applied to such entities as electrons, mesons, genes and the like. Such applications of the terms 'theoretical' and 'unobservable' seem rather to imply that the entities do not have predicates ascribed to them in observation statements, but only in theoretical statements. Suppose the planet Neptune had turned out to be wholly transparent to all electromagnetic radiation and, therefore, invisible. It might still have entered planetary theory as a theoretical entity in virtue of the postulated force relations between it and other planets. Furthermore, the monadic predicate 'mass' could have been inferred of it, although mass was never ascribed to it in an observation statement. Similarly, protons, photons and mesons have monadic and relational predicates ascribed to them in theoretical but not in observation statements, at least not in prescientific language. But this distinction, like others between the theoretical and the observational domains, is relative; for once a theory is accepted and further experimental evidence obtained for it, predicates may well be ascribed directly to previously unobservable entities, as when genes are identified with DNA molecules visible in micrographs or when the ratio of mass to charge of an elementary particle is 'read off' the geometry of its tracks in a magnetic field.

In contrasting theoretical with observable entities, I shall consider that theoretical entities are sufficiently specified as being those to which monadic predicates are not ascribed in relatively observational statements. It follows from this specification that relational predicates cannot be ascribed to them in observation statements either, for in order to recognize that a relation holds between two or more objects, it is necessary to recognize the objects by means of at least some monadic properties. ('The tree is to the left of $x$' is not an observation statement; 'the tree is to the left of $x$ and $x$ is nine stories high' may be.) A theoretical entity must, however, have some postulated relation with an observable entity in order to enter scientific theory at all, and both monadic and relational predicates may be postulated of it in the context of a theoretical network. It must be emphasized that this specification is not intended as a close analysis of what deductivists have meant by 'theoretical entity' (which is in any case far from clear), but rather as an explication of this notion in terms of the network account of theories. At least it can be said that the typical problems that have seemed to arise about the existence of and reference to theoretical entities have arisen only in so far as these entities are not subjects of monadic predicates in observation statements. If a monadic predicate were ascribed to some entity in an observation statement it would be difficult to understand what would be meant by calling such an entity 'unobservable' or by questioning its 'existence'. The suggested explication of 'theoretical entity' is, therefore, not far from the apparent intentions of those who have used this term, and it does discriminate electrons, mesons and genes on the one hand from sticks and stones on the other.

When considering the relatively direct or indirect ascription of predicates to objects, it has already been argued that the circumstances of use must be attended to before the term 'unobservable' is applied. In particular it is now clear that a predicate may be observable of some kinds of entities and not of others. 'Spherical' is observable of baseballs (entrenched and directly and pragmatically observable), but not of protons; 'charged' is observable in at least some of these senses of pith balls but not of ions; and so on. No monadic predicate is observable of a theoretical entity; some predicates may be observable of some observable entities but not of others: for example, 'spherical' is not directly or pragmatically observable of the earth. The question whether there are absolutely theoretical *predicates* can now be seen to be independent of the question of theoretical entities; if there are none, this does not imply that there are no theoretical entities, nor that predicates ascribed to them may not also be ascribed to observable entities.

How is a predicate ascribed to theoretical entities or to observable entities of which it is not itself observable? If it is a predicate that has already been ascribed directly to some observable entity, it may be inferred of another entity by analogical argument. For example, stones released near the surface of Jupiter will fall towards it because Jupiter is in other relevant respects like the earth. In the case of a theoretical entity, the analogical argument will have to involve relational predicates: high energy radiation arrives from a certain direction; it is inferred from other instances of observed radiation transmission between pairs of objects that there is a body at a certain point of space having a certain structure, temperature, gravitational field and so on.

But it is certain that some predicates have been introduced into science which do not appear in the relatively entrenched observation language. How are they predicated of objects? Consistently with the network model, there seem to be just two ways of introducing such newly minted predicates. First, they may be introduced as new observation predicates by assigning them to recognizable empirical situations where descriptions have not been required in prescientific language. Fairly clear examples are 'bacteria' when first observed in microscopes and 'sonic booms' first observed when aircraft 'broke the sound barrier'. Such introductions of novel terms will of course share the characteristic of all observation predicates of being dependent for their functions on observed associations or laws as well as direct empirical recognitions. In some cases it may be difficult to distinguish them from predicates introduced by *definition* in terms of previously familiar observation predicates. Fairly clear examples of this are 'molecule', defined as a small particle with certain physical and chemical properties such as mass, size, geometrical structure, and combinations and dissociations with other molecules, which are expressible in available predicates (most *names* of theoretical entities seem to be introduced this way); or 'entropy', defined quantitatively and operationally in terms of change of heat content divided by absolute temperature. In intermediate cases, such as 'virus', 'quasar' and 'Oedipus complex', it may be difficult to decide whether the function of these predicates is exhausted by logical equivalence with certain complex observation predicates or whether they can be said to have an independent function in some empirical situations where they are relatively directly observed. Such ambiguities are to be expected because, in the network model, laws which are strongly entrenched may sometimes be taken to be definitional, and laws introduced as definitions may later be regarded as being falsifiable empirical associations.

Notice that in this account the view of the function of predicates in theories that has been presupposed is explicitly nonformalist. The account is in fact closely akin to the view that all theories require to be interpreted in some relatively observable model, for in such a model their predicates are ascribed in observation statements. It has been assumed that when familiar predicates such as 'charge', 'mass' and 'position' are used of theoretical entities, these predicates are the 'same' as the typographically similar predicates used in observation statements. But it may be objected that when, say, elementary particles are described in terms of such predicates, the predicates are not used in their usual sense, for if they were, irrelevant models and analogies would be imported into the theoretical descriptions. It is important to be clear what this objection amounts to. If it is the assertion that a predicate such as 'charge' used of a theoretical entity has a sense related to that of 'charge' used of an observable entity only through the apparatus of formal deductive system plus correspondence rules, then the assertion is equivalent to a formal construal of theories, and it is not clear why the word 'charge' should be used at all. It would be less conducive to ambiguity to replace it with an uninterpreted sign related merely by the theoretical postulates and correspondence rules to observation predicates. If, however, the claim that it is used of theoretical entities in a different sense implies only that charged elementary particles are different kinds of entities from charged pith balls, this claim can easily be admitted and can be expressed by saying that the predicate co-occurs and is co-absent with different predicates in the two cases. The fact that use of the predicate has different lawlike implications in relatively theoretical contexts from those in observation contexts is better represented in the network model than in most other accounts of theories, for it has already been noticed that in this model the conditions of correct application of a predicate depend partly on the other predicates with which it is observed to occur. This seems sufficiently to capture what is in mind when it is asserted that 'charge' 'means' something different when applied to elementary particles and pith balls, or 'mass' when used in Newtonian and relativistic mechanics.

Since formalism has been rejected, we shall regard predicates such as those just described as retaining their identity (and hence their logical substitutivity) whether used of observable or theoretical entities, though they do not generally retain the same empirical situations of direct application. But the formalist account, even if rejected as it stands, does suggest another possibility for the introduction of new theoretical predicates, related to observation neither by assignment in recognizable empirical situations nor by explicit definition in terms of old predicates. Can the network model not incorporate new predicates whose relations with each other and with observation predicates are 'implicit', not in the sense intended by formalists, but rather as a new predicate might be coined in myth or in poetry, and understood in terms of its context, that is to say, of its asserted relations with both new and familiar predicates? This suggestion is perhaps nearer the intentions of some deductivists than is pure formalism from which it is insufficiently discriminated.[1]

[1] It certainly represents what Quine seems to have *understood* some deductive accounts to be (*cf.* p. 27 above).

It is not difficult to see how such a suggestion could be incorporated into the network model. Suppose instead of relating predicates by known laws, we *invent a myth* in which we describe entities in terms of some predicates already in the language, but in which we introduce other predicates in terms of some mythical situations and mythical laws. In other words we build up the network of predicates and laws partly imaginatively, but not in such a way as to contradict known laws, as in a good piece of science fiction.[1] It is, moreover, perfectly possible that such a system might turn out to have true and useful implications in the empirical domain of the original predicates, and in this way the mythical predicates and laws may come to have empirical reference and truth. This is not merely to repeat the formalist account of theoretical predicates as having meaning only in virtue of their place within a postulate system, because it is not necessary for such a formal system to have any interpretation, whereas here there is an interpretation, albeit an imaginary one. Neither are the predicates introduced here by any mysterious 'implicit definition' by a postulate system; they are introduced by the same two routes as are all other predicates, except that the laws and the empirical situations involved are imaginary.

Whether any such introduction of new predicates by mythmaking has ever occurred in science may be regarded as an open question. The opinion may be hazarded that no convincing examples have yet been identified. All theory construction, of course, involves an element of mythmaking, because it makes use of *familiar* predicates related in new ways by postulated laws not yet accepted as true. Bohr's atom, for example, was postulated to behave as no physical system had ever been known to behave; however, the entities involved were all described in terms of predicates already available in the language. There is, moreover, a reason why the mythical method of introducing new predicates is not likely to be very widespread in science. The reason is that use of known predicates which already contain some accepted lawlike implications allow inductive and analogical inference to further as yet unknown laws, which mythical predicates do not allow. There could be no prior inductive confidence in the implications of predicates and laws which were wholly mythical, as there can be in the implications of predicates at least some of whose laws are accepted. How important such inductive confidence is, however, is a controversial question which will be pursued later. But it is sufficient to notice that the network model does not demand that theories should be restricted to use of predicates already current in the language or observable in some domain of entities.

---

[1] We build up the network in somewhat the same way that M. Black (*Models and Metaphors* (Ithaca, 1962), 43) suggests a poet builds up a web of imagined associations within the poem itself in order to make new metaphors intelligible. He might, indeed, in this way actually coin and give currency to wholly new words.

# V. Theories

Under the guise of an examination of observational and theoretical predicates, I have in fact described a fully-fledged account of theories, observation and the relation of the one to the other. This is of course to be expected, because the present account amounts to a denial that there is a fundamental distinction between theoretical and observation predicates and statements, and implies that the distinction commonly made is both obscure and misleading. It should not therefore be necessary to say much more about the place of theories in this account. I have so far tried to avoid the term 'theory', except when describing alternative views, and have talked instead about laws and lawlike implications. But a theory *is* just such a complex of laws and implications, some of which are well entrenched, others less so, and others again hardly more than suggestions with as yet little empirical backing. A given theory may in principle be formulated in various ways, and some such formulations will identify various of the laws with postulates; others with explicit definitions; others with theorems, correspondence rules or experimental laws. But the upshot of the whole argument is that these distinctions of function in a theory are relative not only to the particular formulation of the 'same' theory (as with various axiomatizations of mechanics or quantum theory) but also to the theory itself, so that what appears in one theory as an experimental law relating 'observables' may in another be a high-level theoretical postulate (think of the chameleon-like character of the law of inertia, or the conservation of energy). It is one of the more misleading results of the deductive account that the notion of 'levels', which has proper application to proofs in a formal postulate system in terms of the order of deducibility of theorems, has become transferred to an ordering in terms of more and less 'theory-laden' constituents of the theory. It should be clear from what has already been said that these two notions of 'level' are by no means co-extensive.

So much is merely the immediate application to theories of the general thesis here presented about descriptive predicates. But to drive the argument home, it will be as well to consider explicitly some of the problems which the theory–observation relation has traditionally been felt to raise and how they fare in the present account.

## A. *The circularity objection*

An objection is sometimes expressed as follows: if the use of all observation predicates carries theoretical implications, how can they be used in descriptions which are claimed to be evidence for these same theories? At least it must be possible to find terms in which to express the evidence which are not laden with the theory for which they express the evidence.

This is at best a half-truth. If by 'theory-laden' is meant that the terms used in the observation report presuppose the *truth* of the very theory under test, then indeed this observation report cannot contribute evidence for this theory. If, for example, 'motion in a straight line with uniform speed' is *defined* (perhaps in a complex and disguised fashion) to be equivalent to 'motion under no forces', this definition implies the truth of the law of inertia, and an observation report to the effect that a body moving under no forces moves in a straight line with uniform speed does not constitute evidence for this law. The logic of this can be expressed as follows.

$$\text{Definition:} \quad P(x) \equiv {}_{df} Q(x)$$
$$\text{Theory:} \quad (x)[P(x) \supset Q(x)]$$
$$\text{Observation:} \quad P(a)\&Q(a)$$

Clearly neither theory nor observation report states anything empirical about the relation of $P$ and $Q$.

Contrast this with the situation where the 'theory-loading' of $P(a)$ is interpreted to mean 'Application of $P$ to an object $a$ implies acceptance of the truth of some laws into which $P$ enters, and these laws are part of the theory under test', or, colloquially, 'The meaning of $P$ presupposes the truth of some laws in the theory under test'. In the inertia example the judgment that $a$ is a body moving in a straight line with uniform speed depends on the truth of laws relating measuring rods and clocks, the concept of 'rigid body', and ultimately on the physical truth of the postulates of Euclidean geometry, and possibly of classical optics. All these are part of the theory of Newtonian dynamics and are confirmed by the very same kinds of observation as those which partially justify the assertion $P(a)\&Q(a)$. The notion of an observation report in this account is by no means simple. It may include a great deal of other evidence besides the report that $P(a)\&Q(a)$, namely the truth of other implications of correct application of $P$ to $a$ and even the truth of universal laws of a high degree of abstractness. It is, of course, a standard objection to accounts such as the present, which have an element of 'coherence' in their criteria of truth, that nothing can be known to be true until everything is known. But although an adequate confirmation theory for our account would not be straightforward, it would be perfectly possible to develop one in which the correct applicability of predicates, even in observation reports, is strongly influenced by the truth of some laws into which they enter, and only vanishingly influenced by others. The notion of degrees of entrenchment relative to given theories would be essential to expressing the total evidence in such a confirmation theory.[1]

---

[1] That no simple formal examples can be given of the present account is perhaps one reason why it has not long ago superseded the deductive account. I have made some preliminary suggestions towards a confirmation theory for the network account in my chapter on 'Positivism and the logic of scientific theories', in *The Legacy of Logical Positivism for the Philosophy of Science*, ed. P. Achinstein and S. Barker (Baltimore, 1969), 85, and in 'A self-correcting observation language'. in *Logic, Methodology and Philosophy of Science*, ed. B. van Rootselaar and J. F. Stahl (Amsterdam, 1968), 297. These suggestions are developed below (chaps. 5–9).

The reply to the circularity objection as it has been stated is, then, that although the 'meaning' of observation reports is 'theory-laden', the truth of particular theoretical statements depends on the coherence of the network of theory and its empirical input. The objection can be put in another way, however: if the meaning of the terms in a given observation report is even partially determined by a theory for which this report is evidence, how can the same report be used to decide between two theories, as in the classic situation of a crucial experiment? For if this account is correct the same report cannot have the same meaning as evidence for two different theories.

This objection can be countered by remembering what it is for the 'meaning' of the observation report to be 'determined by the theory'. This entails that ascription of predicates in the observation report implies acceptance of various other laws relating predicates of the theory, and we have already agreed that there may be a hard core of such laws which are more significant for determining correct use than others. Now it is quite possible that two theories which differ very radically in most of their implications still contain some hard-core predicates and laws which they both share. Thus, Newtonian and Einsteinian dynamics differ radically in the laws into which the predicate 'inertial motion' enters, but they share such hard-core predicates as 'acceleration of falling bodies near the earth's surface', 'velocity of light transmitted from the sun to the earth', and so on, and they share some of the laws into which these predicates enter. It is this area of *intersection* of laws that must determine the application of predicates in the report of a crucial experiment. The situation of crucial test between theories is not correctly described in terms of 'withdrawal to a neutral observation language', because, as has already been argued, there is no such thing as an absolutely neutral or non-theory-laden language. It should rather be described as exploitation of the area of intersection of predicates and laws between the theories; this is, of course, entirely relative to the theories in question.

An example due originally to Feyerabend[1] may be developed to illustrate this last point. Anaximenes and Aristotle are devising a crucial experiment to decide between their respective theories of free fall. Anaximenes holds that the earth is disc-shaped and suspended in a non-isotropic universe in which there is a preferred direction of fall, namely the parallel lines perpendicular to and directed towards the surface of the disc on which Greece is situated. Aristotle, on the other hand, holds that the earth is a large sphere, much larger than the surface area of Greece, and that it is situated at the centre of a universe organized in a series of concentric shells, whose radii directed towards the centre determine the direction of fall at each point. Now clearly the word 'fall' as used by each of them is, in a sense, loaded with his own theory. For Anaximenes it refers to a preferred direction uniform throughout space; for

---

[1] P. K. Feyerabend, 'Explanation, reduction and empiricism', 85.

Aristotle it refers to radii meeting at the centre of the earth. But equally clearly, while they both remain in Greece and converse on non-philosophical topics, they will use the word 'fall' without danger of mutual misunderstanding. For each of them the word will be correlated with the direction from his head to his feet when standing up, and with the direction from a calculable point of the heavens toward the Acropolis at Athens. Also, both of them will share most of these lawlike implications of 'fall' with ordinary Greek speakers, although the latter probably do not have any expectations about a preferred direction throughout universal space. This is not to say, of course, that the ordinary Greek speaker uses the word with *fewer* implications, for he may associate it with the passage from truth to falsehood, good to evil, heaven to hell—implications which the philosophers have abandoned.

Now suppose Anaximenes and Aristotle agree on a crucial experiment. They are blindfolded and carried on a Persian carpet to the other side of the earth. That it is the other side might be agreed upon by them, for example, in terms of star positions—this would be part of the intersection of their two theories. They now prepare to let go of a stone they have brought with them. Anaximenes accepts that this will be a test of his theory of fall and predicts, 'The stone will fall'. Aristotle accepts that this will be a test of his theory of fall and predicts, 'The stone will fall'. Their Persian pilot performs the experiment. Aristotle is delighted and cries, 'It falls! My theory is confirmed'. Anaximenes is crestfallen and mutters, 'It rises; my theory must be wrong'. Aristotle now notices that there is something strange about the way in which they have expressed their respective predictions and observation reports, and they embark upon an absorbing analysis of the nature of observation predicates and theory-ladenness.

The moral of this tale is simply that confirmation and refutation of competing theories does not depend on all observers using their language with the same 'meaning', nor upon the existence of any neutral language. In this case, of course, they could, if they thought of it, agree to make their predictions in terms of 'moves from head to foot', instead of 'falls', but *this* would have presupposed that men naturally stand with their feet on the ground at the Antipodes, and this is as much an uncertain empirical prediction as the original one. Even 'moves perpendicularly to the earth' presupposes that the Antipodes is not a series of steeply sloping enclosed caves and tunnels in which it is impossible to know whether the stars occasionally glimpsed are reflected in lakes or seen through gaps in thick clouds. In Anaximenes' universe, almost anything might happen. But we are not trying to show that in any particular example there are *no* intersections of theories, only that the intersection does not constitute an independent observation language, and that some predicates of the observation reports need not even lie in the intersection in order for testing and mutual understanding to be possible. The final analysis undertaken by Anaximenes and Aristotle will doubtless include the learning of each other's theories and corresponding predicates or the devising of a set of observation reports in the intersection of the two theories, or, more probably, the carrying out of both together.

As a corollary of this account of the intersections of theories, it should be noted that there is no *a priori* guarantee that two persons brought up in the same language community will use their words with the same meanings in all situations, even when each of them is conforming to standard logic within his own theory. If they discourse only about events which lie in the intersection of their theories, that they may have different theories will never be behaviourally detected. But such behavioural criteria for 'same meaning' may break down if their theories are in fact different and if they are faced with new situations falling outside the intersection. Misunderstanding and logical incoherence cannot be logically guarded against in the empirical use of language. The novelty of the present approach, however, lies not in that comparatively trivial remark, but in demonstrating that rational communication can take place in intersections, even when words are being used with 'different meanings', that is, with different implications in areas remote from the intersection.

## B. *The two-language account and correspondence rules*

Many writers have seen the status of the so-called rules of interpretation, or correspondence rules, as the key to the proper understanding of the problem of theory and observation. The concept of correspondence rules presupposes the theory–observation distinction, which is bridged by the rules and therefore seems to have been bypassed in the present account. But there are cases where it seems so obvious that correspondence rules are both required and easily identifiable that it is necessary to give some attention to them, in case some features of the theory–observation relation have been overlooked.

These cases arise most persuasively where it seems to be possible to give two descriptions of a given situation, one in theoretical and one in observation terminology, and where the relation between these two descriptions is provided by the set of correspondence rules. Take, for example, the ordinary-language description of the table as hard, solid and blue, and the physicist's description of the same table in terms of atoms, forces, light waves and so on— or the familiar translation from talk of the pressure, volume and temperature of a gas to talk of the energy and momentum of random motions of molecules. It seems clear that in such examples there is a distinction between theoretical and observational descriptions and also that there are correspondence rules which determine the relations between them. How does the situation look in the network account?

It must be accepted at once that there is something more 'direct' about describing a table as hard, solid and blue than as a configuration of atoms exerting forces. 'Direct' is to be understood on our account in terms of the better entrenchment of the predicates 'hard', 'solid' and 'blue' and the laws which relate them, and in terms of the practical ease of learning and applying these predicates in the domain of tables, compared with the predicates of the physical description. This does not imply, however, that 'atom', 'force' and 'light wave' function in a distinct theoretical language nor that they require to be connected with observation predicates by extraneous and problematic correspondence rules. Consider as a specific example, usually regarded as a correspondence rule: ' "This exerts strong repulsive forces" implies "This is hard".' Abbreviate this as 'Repulsion implies hardness', and call it $C$. What is the status of $C$? Various suggestions have been made, which I shall now examine.[1]

(*a*) It is an analytic definition. This is an uninteresting possibility, and we shall assume it to be false, because 'repulsion' and 'hardness' are not synonymous in ordinary language. They are introduced in terms of different kinds of situation and generally enter into different sets of laws. Furthermore, in this domain of entities 'hardness' has the pragmatic characteristics of an observation predicate, and 'repulsion' of a theoretical predicate, and hence they cannot be synonymous here. Therefore, $C$ is a synthetic statement.

How, then, do 'hard' and 'repulsion' function in the language? Consistently with our general account we should have to say something like this: meaning is given to 'hard' by a complex process of learning to associate the sound with certain experiences and also by accepting certain empirical correlations between occurrences reported as 'This is hard', 'This exerts pressure' (as of a spring or balloon), 'This is an area of strong repulsive force' (as of iron in the neighbourhood of a magnet), 'This is solid, impenetrable, undeformable . . .', 'This bounces, is elastic . . .'. Similarly, 'repulsion' is introduced in a set of instances including some of those just mentioned and also by means of Newton's second law and all its empirical instances. Granted that this is how we *understand* the terms of $C$, what kind of synthetic statement is it?

(*b*) It may be suggested that it is a theorem of the deductive system representing the physical theory. This possibility has to be rejected by two-language philosophers because for them 'hard' does not occur in the language of the theory and, hence, cannot appear in any theorem of the theory. But for us the possibility is open, because both terms of $C$ occur in the same language, and it is perfectly possible that having never touched tables, but knowing all that physics can tell us about the forces exerted by atoms and knowing also analogous situations in which repulsive forces are in fact correlated with the property of hardness (springs, balloons, etc.), we may be able to deduce $C$ as a theorem in this complex of laws.

(*c*) More simply, $C$ may be not so much a deductive inference from a system

---

[1] I owe several of these suggestions to a private communication from Paul E. Meehl. See also E. Nagel, *The Structure of Science*, 354ff.

of laws as an inductive or analogical inference from other accepted empirical correlations of repulsive force and hardness. This is a possibility two-language philosophers are prone to overlook, because they are wedded to the notion that 'repulsion' is a theoretical term in the context of tables and therefore not a candidate for directly observed empirical correlations. But it does not follow that it is not comparatively observable in other domains—springs, magnets, and the like. Observation predicates are, as we have remarked, relative to a domain of entities.

(*d*) Unable to accept *a*, *b* or *c*, the two-language philosopher is almost forced to adopt yet another alternative in his account of correspondence rules, namely, that *C* is an independent empirical postulate,[1] which is added to the postulates of the theory to make possible the deduction of observable consequences from that theory. There is no need to deny that this possibility may sometimes be exemplified. It should only be remarked that if all correspondence rules are logically bound to have this status, as a two-language philosopher seems forced to hold, some very strange and undesirable consequences follow. If there are no deductive, inductive or analogical reasons other than the physicist's fiat why particular theoretical terms should be correlated with particular observation terms, how is it possible for the 'floating' theory ever to be refuted? It would seem that we could always deal with an apparent refutation at the observation level by arbitrarily modifying the correspondence rules, for since on this view these rules are logically and empirically quite independent of the theory proper, they can always be modified without any disturbance to the theory itself. It may be replied that considerations of simplicity would prevent such arbitrary salvaging of a theory. But this objection can be put in a stronger form: it has very often been the case that a well-confirmed theory has enabled predictions to be made in the domain of observation, where the deduction involved one or more *new* correspondence rules, relating theoretical to observation terms in a new way. If these correspondence rules were postulates introduced for no reason intrinsic to the theory, it is impossible to understand how such predictions could be made with confidence.

On the present account, then, it need not be denied that there is sometimes a useful distinction to be made between comparatively theoretical and comparatively observational descriptions, nor that there are some expressions with the special function of relating these descriptions. But this does not mean that the distinction is more than pragmatically convenient, nor that the correspondence rules form a logically distinct class of statements with unique status. Statements commonly regarded as correspondence rules may in different circumstances function as independent theoretical postulates, as theorems, as inductive inferences, as empirical laws, or even in uninteresting cases as

[1] *Cf.* Carnap's 'meaning postulates', *Philosophical Foundations of Physics* (New York and London, 1966), chap. 27.

analytic definitions. There is no one method of bridging a logical gap between theory and observation. There is no such logical gap.

## C. Replaceability

Granted that there is a relative distinction between a set of less entrenched (relatively theoretical) predicates and better entrenched observation predicates, and that correspondence rules do not form a special class of statements relating these two kinds of predicates, there still remains the question: What is the relation between two descriptions of the same subject matter, one referring to theoretical entities and the other observation entities?

First of all, it follows from the present account that the two descriptions are not equivalent or freely interchangeable. To describe a table as a configuration of atoms exerting forces is to use predicates which enter into a system of laws having implications far beyond the domain of tables. The description of the table in ordinary language as hard and solid also has implications, which may not be fewer in number but are certainly different. One contrast between the two descriptions which should be noted is that the lawlike implications of the theoretical descriptions are much more explicit and unambiguous[1] than those of ordinary-language predicates like 'hard' and 'solid'. Because of this comparative imprecision it is possible to hold various views about the status of an observational description. It is sometimes argued that an observational description is straightforwardly *false*, because it carries implications contradicted by the theoretical description, which are probably derived from out-of-date science. Thus it is held that to say a table is hard and solid implies that it is a continuum of material substance with no 'holes' and that to touch it is to come into immediate contact with its substance. According to current physics these implications are false. Therefore, it is claimed, in all honesty we must in principle replace all our talk in observation predicates by talk in theoretical predicates in which we can tell the truth.

This view has a superficial attraction, but as it stands it has the very odd consequence that most of the descriptions we ever give of the world are not only false but known to be false in known respects. While retaining the spirit of the replaceability thesis, this consequence can be avoided in two ways. First, we can make use of the notion of intersection of theories to remark that there will be a domain of discourse in which there is practical equivalence between some implications of observational description and some implications of theoretical description. In this domain the observation language user is

---

[1] This is not to say, as formalists are prone to do, that theoretical terms are completely unambiguous, precise or exact, like the terms of a formal system. If they were, this whole account of the functioning of predicates would be mistaken. The question of precision deserves more extended discussion. It has been investigated by Stephan Körner, *Experience and Theory* (London, 1966); D. H. Mellor, 'Experimental error and deducibility', *Philosophy of Science*, **32** (1965), 105, and his 'Inexactness and explanation', *Philosophy of Science*, **33** (1966), 345.

telling the truth so long as he is not tempted to make inferences outside the domain. This is also the domain in which pragmatic observation reports provide the original evidence for the theory. Within this domain ordinary conversation can go on for a long time without its becoming apparent that an observation language user, an ordinary linguist and a theoretician are 'speaking different languages' in the sense of being committed to different implications outside the domain. It may even be the case that an ordinary linguist is not committed to *any* implications outside the domain which conflict with those of the theoretician. For example, and in spite of much argument to the contrary, it is not at all clear that the user of the ordinary English word 'solid' *is*, or ever was, committed to holding that a table is, in the mathematically infinitesimal, a continuum of substance. The question probably never occurred to him, either in the seventeenth century or the twentieth, unless he had been exposed to some physics. Secondly, the network account of predicates makes room for change in the function of predicates with changing knowledge of laws. In this case it may very well be that use of 'hard' and 'solid' in the observation language comes to have whatever implications are correct in the light of the laws of physics or else to have built-in limitations on their applicability, for example, to the domain of the very small.

These suggestions help to put in the proper light the thesis that it is in principle possible to replace the observation language by the theoretical, and even to teach the theoretical language initially to children without going through the medium of the observation language.[1] Such teaching may indeed be *in principle* possible, but consider what would happen if we assume that the children are being brought up in normal surroundings without special experiences devised by physicists. They will then learn the language in the intersection of physics and ordinary language; and though they may be taught to mouth such predicates as 'area of strong repulsive force' where other children are taught 'hard', they will give essentially the same descriptions in this intersection as the ordinary linguist, except that every observation predicate will be replaced by a string of theoretical predicates. Doubtless they will be better off when they come to learn physics, much of which they will have learned implicitly already; and if they were brought up from the start in a highly unnatural environment, say in a spaceship, even ordinary discourse might well be more conveniently handled in theoretical language. But all these possibilities do not seem to raise any special problems or paradoxes.

## D. *Explanation and reduction*
It has been presupposed in the previous section that when an observational description and a theoretical description of the same situation are given, both have reference to the same entities, and that they can be said to contradict or

---

[1] *Cf.* W. Sellars, 'The language of theories', *Current Issues in the Philosophy of Science*, ed. H. Feigl and G. Maxwell (New York, 1961), 57.

to agree with one another. Furthermore, it has been suggested that there are circumstances in which the two descriptions may be equivalent, namely when both descriptions are restricted to a certain intersection of theoretical implications and when the implications of the observation predicates have been modified in the light of laws constituting the theory. Sometimes the objection is made to this account of the relation of theoretical and observational descriptions that, far from their being potentially equivalent descriptions of the same entities, the theory is intended to *explain* the observations; and explanation, it is held, must be given in terms which are different from what is to be explained. And, it is sometimes added, explanation must be a description of *causes* which are distinct from their observable effects.

It should first be noticed that this argument cannot be used in defence of the two-language view. That explanations are supposed to refer to entities different from those referred to in the explananda does not imply that these sets of entities have to be described in different languages. Explanation of an accident, a good crop or an economic crisis will generally be given in the same language as that of the explananda.

It does seem, however, that when we give the theoretical description of a table as a configuration of atoms exerting repulsive forces, we are saying something which *explains* the fact that the table is hard and states the *causes* of that hardness. How then can this description be in any sense *equivalent* to the observational description of the table as hard? It does of course follow from the present account that they are not equivalent in the sense of being *synonymous*. That much is implied by the different function of the theoretical and observation predicates. Rather, the descriptions are equivalent in the sense of having the same reference, as the morning star is an alternative description of the evening star and also of the planet Venus. It is possible for a redescription in this sense to be explanatory, for the redescription of the table in theoretical terms serves to place the table in the context of all the laws and implications of the theoretical system. It is not its reference to the *table* that makes it explanatory of the observation statements, which also have reference to the table. It is rather explanatory because it says of the table that in being 'hard' ('exerting repulsive force') it is *like* other objects which are known to exert repulsive force and to feel hard as the table feels hard, and that the table is, therefore, an instance of general laws relating dynamical properties with sensations. And in regard to the *causal* aspects of explanation, notice that the repulsive forces are not properly said to be the causes of the table having the property 'hard', for they *are* the property 'hard'; but rather repulsive forces are causes of the table *feeling* hard, where 'hard' is not a description of the table but of a sensation. Thus the cause is not the same as the effect, because the referents of the two descriptions are different; and the explanans is not the same as the explanandum, because although their referents are the same, the theoretical description explains by relating the explanandum to other like entities and to a system of laws, just in virtue of its use of relatively theoretical predicates. There is some truth in the orthodox deductive account of explanation as deducibility in a theoretical system, but there is also truth in the contention that explanation involves stating as well what the explanandum *really* is and, hence, relating it to other systems which are then seen to be essentially similar to it. Initial misdescription of the function of descriptive predicates precludes the deductive account from doing justice to these latter aspects of explanation, whereas in the present account they are already implied in the fact that redescription in theoretical predicates carries with it lawlike relations between the explanandum and other essentially similar systems.

## VI. Conclusion

In this chapter I have outlined a network model of theoretical science and argued that it represents the structure of science better than the traditional deductivist account, with its accompanying distinction between the theoretical and the observational.

First, I investigated some consequences of treating the theoretical and the observational aspects of science as equally problematic from the point of view of truth conditions and meaning. I described the application of observation in empirical situations as a classificatory process, in which unverbalized empirical information is lost. Consequently, reclassification may in principle take place in any part of the observational domain, depending on what internal constraints are imposed by the theoretical network relating the observations. At any given stage of science there are *relatively* entrenched observation statements, but any of these may later be rejected to maintain the economy and coherence of the total system.

This view has some similarity with other non-deductivist accounts in which observations are held to be 'theory-laden', but two familiar objections to views of this kind can be answered more directly in the network account. First, it is not a conventionalist account in the sense that any theory can be imposed upon any facts regardless of coherence conditions. Secondly, there is no vicious circularity of truth and meaning, for at any given time *some* observation statements result from correctly applying observation terms to empirical situations according to learned precedents and independently of theories, although the relation of observation and theory is a self-correcting process in which it is not possible to know at the time which of the set of observation statements are to be retained as correct in this sense, because subsequent observations may result in rejection of some of them.

Turning to the relatively theoretical aspects of science, I have argued that a distinction should be made between theoretical *entities* and theoretical *predicates*. I have suggested that if by theoretical predicates is meant those which are never applied in observational situations to any objects, and if the open-ended character of even observation predicates is kept in mind, there are no occasions on which theoretical predicates are used in science, although of course there are many theoretical entities to which predicates observable in other situations are applied. It follows that there is no distinction in kind between a theoretical and an observation language. Finally, I have returned to those aspects of scientific theories which are analysed in the deductive view in terms of the alleged theory–observation distinction and shown how they can be reinterpreted in the network model. Correspondence rules become empirical relations between relatively theoretical and relatively observational parts of the network; replaceability of observational descriptions by theoretical descriptions becomes redescription in more general terms in which the 'deep' theoretical similarities between observationally diverse systems are revealed; and theoretical explanation is understood similarly as redescription and not as causal relationship between distinct theoretical and observable domains of entities mysteriously inhabiting the same space–time region. Eddington's two tables are one table.

CHAPTER TWO

# A Network Model of Universals

## I. The problem of universals

In the previous chapter I have given an account of scientific theory structure which depends crucially on the concept of recognizable similarities and differences, and hence on what is essentially a resemblance theory of universal terms. (In this chapter, in order to accommodate classic terminology, I shall use the terms 'resemblance' and 'similarity' interchangeably.) The resemblance theory, however, needs extensive defence against historical rivals and contemporary objections. I shall therefore now examine the problem of universals in general, propose as a solution to it a more detailed version of the network model already introduced, and use this model to sharpen up the concepts of 'similarity', 'truth' and 'meaning' that have been presupposed in that model.

There have been two classic types of theories of universals, which I shall call respectively the absolute theory and the resemblance theory. (I do not use the term 'objective' for the first theory, as is sometimes done, because as we shall see I do not wish to deny that the resemblance theory is also objective.) The *absolute theory* asserts that $P$ is correctly predicated of an object $a$ in virtue of its objective quality of $P$-ness. If the theory of absolute universals is to provide any answer to the *epistemological* question 'What is it about $a$ that leads us in the absence of other information correctly to predicate $P$ of $a$?', then at least in the case of some predicates it must be further assumed that this quality of $P$-ness is directly recognized in the experience which leads to its prediction of $a$. The epistemological question is distinct from the *ontological* question 'What is it about $a$ that makes it $P$?'. The answer to that question might be in terms of absolute universals whether these are directly recognizable or not; the answer to the epistemological question can only be in terms of absolute universals if at least some of these are directly recognizable.

According to the *resemblance theory*, we predicate $P$ of objects $a$ and $b$ in virtue of a sufficient resemblance between $a$ and $b$ in a certain respect. The 'certain respect' is not $P$-ness, as in the theory of absolute universals, but is just that resemblance itself which has had attention drawn to it in learning the application of $P$. It is not that we see $a$ and recognize that it is 'red', but we see *a*, *b*, ... (post boxes, St George's cross, roses), and learn to associate the conventional term 'red' ('rouge', 'rot', ...) with them in virtue of the resemblance in the respect in which they are in fact seen to resemble each other, and to differ from, say, sky, grass and common salt.

Two common objections have been made to this form of the resemblance theory. First, it is claimed that the theory after all reduces to the absolute theory, because the notion of 'resemblance in some same respect' presupposes that an absolute quality is recognized as this respect. Instead of directly recognizing 'red', and 'red again' in red objects, we recognize *a* and *b* as resembling each other, and *b* and *c* as resembling each other in the same respect, which respect we then call 'red'. It has also been objected that the relation of 'resemblance' itself is a universal, whether used as in 'the members of two pairs resemble in the same respect', or 'the members of each pair resemble each other, but the resemblance in one pair is in a different respect from that in the other'. It is therefore claimed that we are either back with a theory of the absolute universal 'resemblance', or caught in an infinite regress of resemblances of resemblances.

Before considering these difficulties, however, let us first contrast both classic types of theory with the well-known theory which Wittgenstein called 'family resemblances'.[1] This theory seems to be held by some of its adherents to be incompatible with both classic theories. I shall interpret it, however, as a theory which is more general than either of the others in the sense that it can accommodate any factual situation which is accommodated by them, and other situations which cannot, and I shall claim that this interpretation will not initially beg the question between the classic theories, or even the question as to whether there is a question between them.

According to a version of Wittgenstein's theory which I shall call FR, objects may form a class, to the members of which a single descriptive predicate is correctly ascribed in common language, even though it is not the case for every pair of members that they resemble each other in any respect which is the same for each pair. I shall assume that the members of each pair *do* resemble each other in *some* respects, against what seems to be the interpretation of Wittgenstein's theory favoured by some commentators. Later I shall argue in more detail that such resemblances are *necessary* but *not sufficient* to explain the use of universal terms. Meanwhile let us assume that it would be impossible to learn to recognize correct occasions of application of any universal term unless there were some sense in which the two members of pairs of objects resemble each other.

In asserting that FR is more general than either the absolute or the resemblance theory, what I mean is the following. *If*, as in the absolute theory, there is an absolute recognizable *P*-ness residing in all objects correctly

[1] L. Wittgenstein, *Philosophical Investigations*, sec. 66, 67. *Cf.* R. Bambrough, 'Universals and family resemblances', *Proc. Aris. Soc.*, 62 (1961), 207.

predicated by '*P*', then our predicate classes will be special cases of FR classes, such that the objects of every pair in the class resemble each other in all having recognizable *P*-ness. If, on the other hand, as in the resemblance theory, the objects of every pair resemble each other in the respect *P* which is the same for every pair, then our predicate classes will be special cases of FR classes in which the resemblance of each pair is resemblance in the same respect. Thus every case accommodated by the absolute or the resemblance theory is accommodated by FR, and moreover no assumption has been made about the absolute or relative character of *P*-ness or the nature of resemblance. On the other hand, if there are cases of FR which are neither cases of common *P*-ness nor cases of common resemblance in the same respect *P*, then the more general FR theory will be required, and *this* may affect the general acceptability of either or both of the classic theories.

There are such cases. Wittgenstein gave some examples, for instance 'game'; and people who classify varieties of plant species, bacteria, diseases, archeological remains, psychological types, schools of painting, and many other objects, are familiar with many other examples.[1] In particular, recent analyses of scientific concepts which take account of 'changes of meaning' from one theoretical framework or categorization of classes to another, raise the same question in a context which seems to make the succession of scientific theories incommensurable in meaning and hence in logic.[2] It is therefore very important to discover whether the FR theory is logically viable. It seems to be very much more important to do this than to try in the case of each example cited to show that after all common characteristics or common respects of resemblance *can* be elicited if analysis is carried far enough. Perhaps all the presently correct ascriptions of 'game' (necessarily finite in number) *do* have one or more characteristics in common if we peer closely enough, just as any finite random selection of numbers can be described by some, indeed infinitely many, power series. But this fact, if it is a fact, is unilluminating precisely because it is not (in Goodman's term) *projectible*;

---

[1] For applications of the family resemblance (or 'polytypic') concept in philosophy of biology, see M. Beckner, *The Biological Way of Thought* (New York, 1959), 22f. For scientific applications of techniques of classification equivalent to FR, see R. R. Sokal and P. H. A. Sneath, *Principles of Numerical Taxonomy* (San Francisco and London, 1963); *Numerical Taxonomy*, ed. A. J. Cole (London and New York, 1969); and N. Jardine and R. Sibson, *Mathematical Taxonomy* (London and New York, 1970). For a discussion of the correlativity of 'similarity' and 'natural kinds', see W. v. O. Quine, 'Natural kinds', *Ontological Relativity* (New York and London, 1969), 114. I do not, however, accept Quine's contention here that this analysis is sufficient to solve the 'raven' and 'grue' paradoxes. See below, chapters 3 and 7 sec. II.

[2] Especially P. K. Feyerabend, 'Explanation, reduction and empiricism', *Minnesota Studies in the Philosophy of Science*, vol. III, ed. H. Feigl and G. Maxwell (Minneapolis, 1962), 28, and T. S. Kuhn, *The Structure of Scientific Revolutions* (Chicago, 1962). Attempts to resolve the 'meaning-variance' problem in terms of an FR-type theory are made in T. S. Kuhn, 'Second thoughts on paradigms', *The Structure of Scientific Theory*, ed. F. Suppe (Urbana, 1972), and in section V below.

it tells us nothing about the next case that will seem to us to be an example of 'game', any more than the set of power series tells us what the next term of the series is. The next case may just fail to have the characteristics we have laboriously uncovered but may have other characteristics in common with some members of the class, and it is not obvious that by exhaustive prior analysis of just that class we should be able to say which characteristics ought to belong to its next acceptable instance.

It seems that FR has been generally dismissed by older writers on universals, or would have been dismissed had they explicitly discussed it, because it was thought to lead to inconsistencies and circularities. It is therefore necessary first to show that this is not so, and it is fortunately easy to do this, since there are now many methods of practical classification which make essentially the assumptions of FR without falling into circularity or inconsistency. I shall describe the general principles of these by taking a very simple type of case.

## II. The correspondence postulate

Suppose we are given a number of objects $a_1$, ..., $a_n$, ..., and a number of two-valued predicates $(P_1, \bar{P}_1)$, ..., $(P_r, \bar{P}_r)$, ...,[1] which are not necessarily 'primitive or 'simple' predicates, the nature or possibility of which is in any case obscure. We first make a group of assumptions which I call the *correspondence postulate*. Firstly we assume that we can recognize pairwise similarities and differences with respect to $P$, where $P$ is that property or respect of resemblance which we have learned to be correctly assigned the predicate '$P$'. On pain of regress or circularity there must be *some* recognitions that cannot be further *verbalized* in terms of other predicates. (We cannot learn Mandarin from a Mandarin–Mandarin dictionary; there must be some recognizable connection between the words and the world.) Secondly, the recognition involved need not be infallible, but it must be assumed that it is correct on at least some occasions, and over the whole relevant range of experience on most occasions. It does not, however, follow from this assumption that we can judge at the time from the evidence then available which occasions these are. The appeal to 'correctness' here of course commits us to some form of ontology with regard to the reference of universal terms. I hope to show later that this is not objectionable; meanwhile it must be emphasized at this stage that we have no access to the empirical other than the recognition that issues in assignment of predicates such as $P$.

Suppose initially we judge that $P_1$ applies to $a_1$, ..., $a_s$, and $\bar{P}_1$ applies to $a_{s+1}$, ..., $a_n$, and make similar classifications for $P_2$, $P_3$, .... The first thing to notice about such an *initial classification* is that it *loses verbalizable empirical information*. It does this in two ways. Firstly, however large our stock of

[1] I use '$\bar{P}$' for 'not-$P$'.

predicates, as long as it remains finite there will always be further observational respects in terms of which the objects *could* be classified, but for which we have no *names*. Secondly, with regard to the predicates we do have, once the initial classification has been made, we have necessarily lost information about the detailed circumstances of recognition that a given property applies to a given object. The relation of resemblance must ultimately be taken as unverbalizable. For if it were possible to say, for example, that two close shades of red resemble each other in some property $Q$ which is found in each of them, the same question would arise about that in virtue of which $Q$ is predicated of both of them, so that in the last resort we reach a level of language where more or less resemblance is not further verbalizable, although it may be experienced. We may know that several shades of 'flamingo flame' are more or less different, but we may have no further predicates in which to communicate their similarities and differences.[1] Along with the inevitable loss of information, we do of course gain in the possibility of general communication because, as Locke nicely put it, 'it is impossible that every particular thing should have a distinct peculiar name. . . . Men would in vain heap up names of particular things, that would not serve them to communicate their thoughts.'[2] But it should also be noticed that there is no natural limit to the degree of refinement of our talk. If it becomes important to distinguish the shades of flamingo flame, this can be done by introducing new predicates and teaching and learning them by recognition of similarities as every predicate is originally learnt.

These last remarks help to explain a possible interpretation of the FR theory in which it is implied that it is not only incorrect to look for a common set of necessary and sufficient properties in objects that are classified together, such as 'games', but also that it is misleading to represent this fact by means of properties repeated in even two objects, as in the representation

$$ABC, \quad BCDE, \quad CDE, \quad DEA.$$

It is true that no one property is common to all objects of this class, but it may be held that FR also implies that, for example, '*A*' itself is not properly regarded as common to the first and fourth objects. The family resemblance concept, it may be argued, applies not only to complex objects classified together, but also to whatever single properties they are that lead to classification of the objects together. I suggest that this interpretation of FR is no more than a slightly obscure way of stating the loss of information I have described above. Expressed as loss of information it becomes clear that any explicit classification of objects in terms of properties starts, as it were, in the middle of a hierarchy of recognition processes. Some take place below the

---

[1] Lest it be thought that the shades could be measured on a continuum, it should be noticed that empirical measures can be given only within finite intervals, and however small these are, shades lying within them cannot be explicitly distinguished.

[2] J. Locke, *Essay Concerning the Human Understanding*, book III, chap. III, 2, 3.

level of verbalization, for example by judging that there is some sufficient resemblance between the first and fourth objects (though we can't further analyse what) that makes it proper to ascribe the same predicate that we call '*A*' to them. Above this level, recognition is assisted and classification effected by explicitly setting out under what conditions of similarity and difference of the properties ascribed to them objects should be classified together. The classes resulting from this explicit process may or may not also be intuitively present to recognition. Goodman has given the example of the van Meergeren forgeries,[1] which could not be 'seen' to be distinct from Vermeers, even by the most experienced critics, until there were enough of them to enable their own peculiar characteristics to be explicitly stated, after which any art historian could immediately recognize the next van Meergeren as a forgery. The essential point is that *we are always in this intermediate position* between direct recognition and self-conscious classification, because on the one hand we have no direct access to any 'primitive' elements of recognition in terms of which to describe, and so what we recognize may always be certain complexes of these, and on the other hand it is always possible that we may be able to analyse these complexes into relatively more primitive elements.

The same sort of considerations may be adduced in reply to the objection that appeal to resemblances, if not itself an appeal to the absolute universal 'resemblance', generates an infinite regress of resemblances of resemblances. . . . This regress, however, is not infinite. Just as we cannot and need not ultimately *say* in what the primitive resemblance of properties of two objects consists, so we cannot and need not say in what the resemblance of resemblances consists. It may always be possible to extend the explicit account where necessary, but that it cannot be extended indefinitely is no weakness of the theory. Explicit description has ultimately to give way to causal processes and conditioning in the learning of language.[2]

As we saw in the previous chapter (section I), several writers have challenged the view that some such relation of physical resemblance must ultimately be regarded as a matter of causal recognition. One possible source of misunderstanding can be dispelled at once by emphasizing that reference to a 'causal process' here does not of course imply that this process is ultimately open to scientific investigation and consequent formulation of causal laws. To formulate a causal law we need to have some empirical access to all physical events related by the law. However, *this* causal process is precisely the one which is presupposed by any such empirical access, because it is the process without which we cannot recognize and describe *any* physical events at all.

---

[1] N. Goodman, *Languages of Art* (London, 1969), 110.

[2] A perceptive account of universals by a philosophical analyst in which this causal basis is recognized (though not quite explicitly) is D. F. Pears, 'Universals', *Logic and Language* (2), ed. A. Flew (Oxford, 1959), 63: 'If a word is explained ostensively, then however difficult this process may be it really is explained ostensively. It is no good trying to combine the concreteness of ostensive definition with the clarity of verbal definition.'

It may help to make these points a little clearer if we express the correspondence postulate and the process of making initial classifications in terms of a *machine analogy*. Consider an artificial intelligence machine programmed to learn certain regularities in the environment in which it is situated. Part of its programme will consist of a set of terms in machine language which represent within the machine some of the differentiable properties of the environment. These terms correspond to the descriptive predicates of a natural language. The physical construction of the machine must be such that a given stimulus from the environment triggers off a response in the machine in the form of a coding of that stimulus into a report in machine language asserting the presence of certain properties which produce the stimulus. The relation between the properties and their coded report is a cause–effect relation, and as far as the initial classification is concerned it must be assumed both that the cause–effect mechanism works correctly on most occasions (although a flexible machine will allow for a small proportion of errors) and that when working correctly the correlation of cause and effect is constant, that is, the same or a sufficiently similar property triggers off a report containing the same term of machine language.

Now when we are ourselves responsible for programming a machine to learn in a definite kind of environment to which we also have access (for example, a handwritten-character recognition machine), the problems of satisfying the correspondence postulate are merely technical and not epistemological. But if we use this epistemologically transparent situation as an analogue for our own situation of being learning organisms situated in an environment, the status of the environment and its properties, and the nature of the assumed cause–effect relation and of the 'correctness' of the initial classification, all become epistemologically opaque, precisely because in *this* situation we have lost independent access to the environment. Talk of the environment, the causal process and correctness cannot now even be regarded as a refutable empirical *hypothesis*, because any observational descriptions at all are bound to be consistent with it.

I shall return to these epistemological issues later, but meanwhile it is necessary to supplement the correspondence postulate and the initial classification by equally essential considerations of theoretical coherence. In doing so the machine analogy just introduced will be further developed in the hope of illuminating the concepts of truth and meaning of scientific theories.

## III. Coherence conditions

The initial classification by primary recognitions as so far described is still very crude. It is likely to be highly complex and intractable, but it is also likely to suggest some immediate simplifications of itself. For example, we may be interested in finding universal lawlike generalizations among the data, but we may find that some of these potential laws, say 'All $P_1$ are $P_2$', do not quite fit the initial classification, because while *most* $P_1$ are $P_2$, there is a small proportion of $P_1$ that are not $P_2$. Our desire for an economical and coherent system of laws and theories among the predicates may involve more elaborate considerations than universal generalization: it may include requirements of symmetry, analogy, conformity with certain types of model and so on, and of course the richer the predicate language used in the initial classification (relations as well as properties, many-valued or continuously variable predicates, etc.), the more complex may be this set of *coherence conditions* in terms of which we wish to represent the classification. Without some such organizing conditions it is clear that a world described by even a small number of predicates in all apparently observed combinations is likely to become quickly unmanageable. Not even the most rigorous inductivist would suggest that no further processing of the initial classification should take place.

What is the source of the coherence conditions? This is perhaps one of the most interesting and fruitful problems exercising a variety of disciplines at present. It can be approached by empirical investigations such as taxonomy, cognitive psychology, artificial intelligence, decision-making and structural linguistics. Or it may be approached from the history of science, by studying the conditions that have been in fact imposed, sometimes tacitly, on the structures of actual scientific theories. It may be approached by more normative studies such as confirmation and decision theory, or the logics of simplicity or classification. Recent experience of such normative studies discouragingly suggests, however, that the kind of constraints upon classification systems thrown up by purely logical or formal considerations are not nearly powerful enough to account for the comparatively economical and manageable character of most of our cognitive systems, or the comparatively limited mechanisms of data processing and storage that we have available, either in heads or in hardware. It is likely that the more stringent coherence conditions required come both from relatively *a priori* and perhaps culturally conditioned metaphysical principles, and also from physical constraints (for example, linguistic deep structures) which may have been selected during the evolution of learning organisms.

This, however, is not a problem I want to consider further here, for what concerns the epistemological question more closely is the relation between this set of coherence conditions, whatever they are, and what I have called the initial classification. At the point where it is found that a certain theoretical structure is the *best theory* in that it satisfies the conditions and best fits the initial classification, what can be done about the remaining misfits? Apart from just living with them, there seem to be three possibilities.

(1) Interpret the misfit as a mistaken observation, either after a recheck has come up with a different and more amenable result, or without a recheck,

if that is impossible, when the anomaly is judged unimportant relative to the rest of the whole coherent system.

(2) Even supposing no mistake of recognition was made, that is, if on re-investigation 'it looks the same as was reported before', it is possible that without much disturbance to the predicate scheme we can reclassify certain of the objects to fit the theory better. In the elementary type of predicate classification just described, recognition of an object as $P$ may have been a matter of degree of $P$-ness, or of degrees of resemblance of the object to other objects recognized as $P$. Degrees of resemblance are not transitive, and thresholds of degrees may be modified for peripheral objects without seriously changing the conditions under which something is recognized as $P$. In a complex classification such a change of threshold may also affect other, previously non-anomalous, data, and may result in quite radical changes of the whole classification. Some such shift of threshold in recognition of the class 'Vermeer' took place when doubt was cast on the first forgery, carrying with it out of the class many other objects previously regarded as non-anomalous. One may even imagine that in the ensuing reclassification other forgeries than van Meergeren might have been for the first time recognized. This response to misfits may be regarded as a correction of the assignment of $P$ in a particular situation in the light of information about other empirical situations. It need not imply, however, any change in the conditions under which $P$ is recognized to be correctly applied on most occasions. In the machine analogy, the conditions under which $P$ is reported would not in this case be reprogrammed, either externally by a programmer or internally by any 'monitoring' component of the machine itself.

(3) The third possible response to misfits is potentially more radical. If reinvestigation does not reveal that small changes of threshold are sufficient to remove an anomaly, it may be decided to 'change the meaning' of $P_1$ or $P_2$ in order to produce a better fit with the theory. What exactly is meant by 'meaning-change' here will be examined in the next section; in the machine analogy this may be regarded provisionally as a reprogramming of the physical conditions under which $P$ is recognized to be correctly applied on most occasions. The change may be effected by distinguishing two predicates $P_1$ and $P_1'$, discriminated by whether they are associated with $P_2$ or not, or by extending the application of $P_2$ to all $P_1$, thus saving a law similar to 'All $P_1$ are $P_2$', though with different extensions of the predicates.

Put thus crudely, the proposal sounds like an abandonment of empirical constraints in favour of a kind of conventionalism of theory, but there are obviously possibilities of comparison of different theoretical systems and different responses of types (2) and (3) into which empirical considerations will enter. Even in case (3), redefinition together with further empirical examination may reveal that the $P_1$s are indeed distinguishable into two classes in virtue of *another* recognizable property $Q$ which had not previously been brought into the classification, and that indeed all objects which are $P_1 \& Q$ are $P_2$, thus restoring the original generalization over a more closely specified range of objects. And in judging the empirical content of these proposals, it must not be forgotten that the notion of coherence of the classification and of the related theory rests on the assumption that *most* of the initial classification is to be retained in the theory, so that modifications cannot be made indiscriminately. The essential point, however, is that it cannot be known *a priori* and independently of the coherence conditions and subsequent evidence *which* part of the initial classification will be retained, and it is this that differentiates the present account from traditional accounts of observation statements.

(4) We may also conceive that the coherence conditions themselves may, usually as a last resort, be modified in the light of success or failure of the sequence of best theories in accounting for the available observation statements and in making successful predictions. There seem to be a number of examples of this kind of modification in the history of science: abandonment of the postulate of circular motions of the heavenly bodies; rejection of the notion that some theoretical postulates, such as Euclidean geometry or universal determinism, can be known *a priori*; adoption and later rejection of the mechanical philosophy as a necessary condition of scientific explanations; the postulate of reducibility of organic processes to physicochemical theories.

This network model of universals can now be used to expose some myths and illuminate some problems of classic epistemology.

## IV. Some epistemological consequences

### 1. *Empirically privileged statements*

The myth that there is a set of privileged observation statements with timeless meaning and truth-value has often been exposed. But it may be thought that privileged statements are reintroduced in the network model by apparently giving temporal and epistemological priority to the initial classification. This appearance, however, is misleading in two ways. First, it is true that if theoretical systematization retains any pretensions to be empirical, it must be assumed that *most* of the initial classification yields true predications in terms of the dispositions to ascribe predicates learnt in the current descriptive language. But it does not follow that any *given* ascription of a predicate is true, nor that it will retain the same meaning in subsequent development of theory and language. Whether a given ascription is true and will retain its meaning can be determined only relative to the systematization subsequently adopted. Neurath's raft metaphor (quoted by Quine[1]) is the appropriate one here—we may remain afloat on the sea of empirical facts by replacing the

---

[1] W. v. O. Quine, *Word and Object*, 3.

planks one by one, but not all at once, and there is no particular set of planks that must be retained intact throughout the whole operation.

The second way in which the account so far has been misleading is in introducing a temporal relation between initial classification and subsequent theoretical processing. It is much more likely that the two processes go on simultaneously, though not necessarily consciously. That is to say, recognition of something as *P* may already be partly dependent on noticing what else goes with *P* in this case and in related cases, and what the organization of experience has previously suggested. There is no moment at which we are innocent of such coherence conditions and their applications, even though they may not be explicit. This fact, to examine and document which is the concern of the psychology of perception and learning, can be accommodated immediately into the general account just given, for, unlike other epistemologies, this account does not depend on there being some set of basic observation statements which can be known *a priori* to be timelessly true and fixed in meaning or even relatively so fixed. Whether the processing of data by coherence takes place tacitly at the level of recognition or explicitly at a conscious level makes no difference to the general principles of the account, though it may of course make a difference to the type of investigation that has to be carried out to determine what the coherence conditions are.

## 2. *Paradigms of universal terms*

The network model also exposes the myth that for correct use there must be necessary and sufficient conditions for, or paradigm cases of, universal terms. On the contrary, the network model incorporates the FR insight that whether an object belongs to a given class or not depends not only on its possession of specific properties, or on its resemblance to a paradigm, but also on the whole complex of its resemblance relationships with other objects, and on considerations derived from the coherence conditions.

In this respect the present account differs fundamentally from that of Körner,[1] where he does indeed express more adequately than most classic views the uncertainty that sometimes attaches to ascription of a given property by admitting a third class of *neutral* candidates, but where he yet assumes that these are neutral only *relative to* certain paradigm cases taken as conclusively positive or negative. On the present account, on the other hand, there is a sense in which *all* candidates for a given property are neutral, since there may be reasons connected with the coherence of the rest of the network for assigning positive or negative status to any object with respect to that property, and these reasons cannot be known by just observing that object, for the decision may change with changing information about *other* objects and properties.

[1] S. Körner, *Experience and Theory*, 26.

In classic accounts of the theory of universals it has frequently been held that to abandon the notion of paradigms or 'exemplars' for each property-class leads to vicious circularity. Take for example H. H. Price's account of the resemblance theory. He argues that, according to this theory, for a new object to be placed correctly in class $P$ it must resemble the other members of class $P$ sufficiently closely, and indeed that there must be some members of $P$ (which he calls 'exemplars'[1]) such that any other member resembles them at least as closely as they resemble each other. He admits that there may be alternative sets of exemplars, but still seems to imply that these alternative sets must all generate the same class, for if not, he maintains, there is nothing to 'hold the class together'. Is Price correct in holding that there must be exemplars for the resemblance theory to work?

It does not follow from the present account that there are no exemplars; only that they are not necessarily timeless. It may or may not be the case that the initial classification is elicited in some such way as Price requires, but it certainly does not follow that subsequent assignments of $P$ to these or other objects need to coincide with the initial assignments. There is no set of exemplars in the initial classification that may not in some circumstances be cast out of the reference class of $P$. There is no circularity involved, since what 'holds the class together' is just as much pressure from outside as cohesion within. An amoeba passes through the ocean continually changing its shape and changing its internal make-up of molecules, but it remains in some sense the 'same entity'. Similarly a $P$-class that is continually changing its membership defines in this sense the 'same property' $P$, although what constitutes 'same property' can only be determined by looking at its history and its continuous interaction with its environment.

### 3. *Truth as correspondence*

In spite of the importance of coherence conditions in determining estimates of truth and correctness of application, however, the concept of truth involved in the network model is fundamentally a *correspondence* concept. The initial and subsequent classifications and the predictive statements are expressed wholly within the learned descriptive language, and there is no access to the world other than what produces the classifications. There is thus no question of an independent external check of a different kind as to the true reference of any part of a classification. The checks involve only (fallible) classifications and the theoretical coherence conditions. However, the correspondence postulate is the basic presupposition that most of a classification corresponds to the world, in terms of whatever conventional language descriptive of the world has currently been learned. This presupposition is independent of the coherence conditions, although those conditions will determine from time to

---

[1] H. H. Price, *Thinking and Experience* (London, 1953), 20.

time *which* part of a classification is the best estimate of truth relative to the best theory, and change in coherence conditions may dictate change of conventional language. Coherence conditions, however, always pre-suppose that some empirical input is present to be classified; they cannot independently determine the classification, or even affect whatever truth-relation in fact subsists between the empirical world and a given language, so long as that language as learned by the perceiver remains the same. That the truth-*value* of an observation statement is relative to coherence conditions is a matter of epistemology, but the *concept* of truth that is presupposed is a matter of ontology, that is, of a relation between existents. Truth is *a relation between the state of the world that produces empirical stimuli and the observation statements expressed in current descriptive language.*[1]

Classical correspondence theories of truth have had great difficulty in explaining the relation implied here between the empirical world and its description in statements. I believe, however, that the difficulties can be traced to four assumptions made by these theories which are unnecessary in the present account. Firstly, it is assumed that some account must be given of the truth-relation between the world and its description which guarantees that at least *some* specific descriptions are infallible and therefore incorrigible. We have seen how to do without that assumption.

Secondly, it is assumed, more generally, that some philosophical account must be given of the nature of the truth-relation in terms which are somehow more primitive or more accessible than itself. But in the present account it is assumed that the relation is itself primitive. If it is to be described at all, it will be in terms of the physiology and psychology of intersubjective language learning, for the relation is concerned with how particular verbal utterances are associated with particular situations in ways acceptable to the speaker's language community. However, although such an account of language learning enables us to understand how human learners interact with their world, it does not of course solve what was thought to be the philosophical problem, namely: What *is* the relation of 'correctness' or 'correspondence' between statements and the world? It does not solve this problem principally because empirical accounts of learning processes are also scientific, and pre-suppose in the acceptance of their data a correspondence relation of precisely the kind that is in question. I conclude that the relation of correspondence is a philosophical primitive, inherent in any claim to knowledge that is externally constrained, and though it is subject to explication and analysis in

---

[1] Tarski's correspondence theory of truth (rightly designated by Popper as 'absolute or objective', *Conjectures and Refutations*, 223f) may be adopted here, provided it is recognized that the rubric ' "Snow is white" is true if and only if snow is white' is trivial without some epistemological account of the conditions under which truth is asserted, and their relation to the assumed ontology of the external world.

C

some such way as described here, it is ultimately not reducible to any other type of relation.

Thirdly, it has often been suggested that just because one relatum of the relation of correspondence is not accessible in a way expressible in language, and that no direct account can therefore be given of it, the notion of a 'relation' of correspondence is idling, and could be replaced by merely designating certain statements of the theory as those having a certain claim to or probability of empirical truth, where the grounds for this claim are not described because they are ultimately indescribable. The suggestion has some attraction, and could be adopted here without affecting the general thesis. However, the problem would remain of designating those statements which are to be ascribed empirical truth in this way. The designation would have to be made by appealing to the unverbalizable information we have when we believe ourselves to be in a position of responding with conventionally learned descriptions to the constraint of external empirical stimuli. It may be argued, with some phenomenologists, that we do then experience ourselves in a situation of relationship with the world as other than ourselves. Or to put this somewhat obscure way of speaking in other terms as suggested above, we may regard the experience as illuminated by making an analogy with situations of watching other people, or other organisms, or even data-processing machines, in interaction with their world, where our experience in these cases includes both the perceiver and the world. In such cases we can in an ordinary sense speak of a relation of correspondence between the world and the organisms' response to it. Much philosophical ink has been spilt in showing that the analogy between these cases and our own 'solipsist predicament' is not perfect.[1] But analogies are by definition not perfect identities; their function is to illuminate a new way of speaking, in this case a new use of 'relation' and 'correspondence', which directs attention to the fact (or, if preferred, the non-solipsist's theory) that models of organism–world interaction provide useful parallels to the ultimate situation of the human perceiver in the world.

Fourthly, classic correspondence theories believed themselves obliged to give some independent account of or justification for the existence of the external world. This has been simply presupposed in the present account, and I see no compelling reason to accept any such obligation. The programme of building the world from a minimum of primary immediately accessible components has proved notably unsuccessful, and must be regarded as misconceived. It is somewhat analogous to the Baconian programme of building science inductively from infallible simple notions. Just as science is better conceived as involving models and theories which are not wholly dependent on data, so epistemology is better conceived in terms of an imaginative model

---

[1] For an examination of the error involved in this type of philosophical objection, which presupposes an impoverished theory of analogy, see D. Bloor, 'The dialectics of metaphor', *Inquiry*, **14** (1971), 430.

of world–learner interaction, where there is no obligation to prove the existence of the world, or of the interaction, by some argument independent of the consistency of the model with what we know from, as it were, the inside of the learning process. To the further objection that the world–learner model is not analogous to a scientific theory or model because there is no conceivable situation in which it could be falsified, one might reply along Quinean lines that similar unfalsifiability afflicts all sufficiently general conceptual frameworks for science. Choice between frameworks depends ultimately on intersubjectively preferred criteria of 'simplicity'; choice between the objective world–learner model and its alternatives also depends ultimately on an intersubjectively accepted judgment which is in part a judgment of preference or value.[1]

It may be suggested[2] that the present account does not need a concept of correspondence *truth* or *correctness*, but only of *consensus* among language users in their use of descriptive terms in classifying objects. This would not of course eliminate the need for something like the correspondence postulate, but the postulate would now be expressed in terms of a relation between language users rather than between any given language user and the world. The suggestion has the advantage that in describing the application of predicates by societies whose theories we do not share, we are not obliged to hold that these applications are 'correct' or 'true' of the world on most occasions. For example, if a villager tells us that his sick cow is 'bewitched', he may be using a predicate correctly, relative to the consensus in his language community, but it seems odd to have to accept that he is using it *truthfully*, even relative to his language community.

On balance, however, it is preferable to maintain the language of truth-as-correspondence, even though this entails that some descriptive statements in archaic or otherwise unacceptable languages, which seem to be patently false, have to be held to be probably true relative to those languages. The truth-as-correspondence of a descriptive statement, as defined here, has to do with empirical expectations aroused not only by the statement itself, but by other statements which are related to it via an acceptable, coherent and comprehensive theory. Now in the first place the correspondence postulate does not entail that a majority of applications of each single predicate is correct, but only that a majority of *all* applications of predicates is correct. We shall see in a later chapter how this can be represented by probability distributions over statements, and how other statements of the theory may reduce the posterior probability of correctness of particular predicates, so long as most of the others are assumed correctly applied. This might be the situation with, for example, 'phlogiston', where it would be quite consistent with the correspondence postulate to say that, with respect to all the relevant evidence eventually

---

[1] *Cf.* Introduction, above, p. 8.

[2] This suggestion has been made in private correspondence by D. Bloor.

collected, the probability of ever applying 'phlogiston' correctly becomes very small. It is possible, in other words, for a whole language community to be mistaken every time certain of its predicates are applied, and the community may itself come to accept this in the light of all their evidence expressed in their language.

We must also, however, take account of languages which are more radically different from ours than the language of phlogiston was, of which we would be reluctant to accept that a majority of *all* the predicates are correctly applied. Consider a language community whose normal empirical expectations either cover a restricted range of phenomena compared to ours, or cover a range which is comprehensive but quite different from those we are mainly concerned with in our community. For example, their expectations may be largely concerned with a lifelong preparation for the mode of a man's death, or about the hypnotic powers of some men over other men and animals and plants. It will follow that many and perhaps most of their locutions will contain terms such as 'saved' or 'bewitched', which are meaningless or false in our language but are accepted as correct on most occasions of descriptive use in theirs, and which do indeed indicate many empirical expectations which are subsequently fulfilled to *their* satisfaction. If we then ask, as we might about 'bewitched', 'But surely it is possible for a whole language community to be mistaken every time statements positively ascribing these predicates to objects are uttered?', we betray a confusion about the truth conditions of different languages. If a natural descriptive language works as a mode of communication about the empirical expectations that are of interest to that community, it is inconceivable that more than half their utterances are false *relative to their own acceptable* theories. They may be false relative to our theories, but this is not because each individual statement taken as a statement of our language is judged false as a description of the world, but because all the conditions which make our theories acceptable to us, including comprehensiveness and satisfaction of expectations over phenomena which interest us, have either changed the meaning of the terms of the archaic language, or have rendered them meaningless.[1]

---

[1] R. Rorty reports an unpublished argument by D. Davidson (in 'The world well lost', *J. Phil.*, **69** (1972), 652) to the effect that (1) for anything that could be recognizable by us as a language, most of the beliefs expressed in it must be true. Davidson draws this conclusion from a prior argument to the effect that (2) radically alternative conceptual frameworks are impossible. Thus it follows not only that most beliefs are true relative to their own language, as I am suggesting here, but that most beliefs are of the same truths, because languages, or conceptual frameworks, cannot be 'radically alternative'. According to Rorty, Davidson means by 'radically alternative' 'untranslatable into our language in such a way that all or most of the foreigner's beliefs are true', from which, of course, (1) follows (although note that (2) does not follow from (1)). I do not want to argue here either for or against the existence of radically alternative conceptual frameworks in this sense (although I am not convinced of their impossibility). In Davidson's terms, if we know or guess the meaning changes from the foreigner's language to ours, translation is possible, and I am happy to accept Davidson's conclusion that with translation their truths are (or could be) our truths. It should not

The question of meaning will be taken up again in the next section. But meanwhile it may finally be remarked about the proposal to replace 'truth' by 'consensus' that this latter term has at least as many objectionable overtones as does 'truth'. In suppressing reference to the *world* and emphasizing relations between speakers, it conjures up a threat of newspeak in which a consensus with respect to apparently descriptive statements might be manipulated for purposes other than communication of empirical expectations. Again, it suggests that epistemological problems about the relation of language to the world are evaded, but this suggestion is mistaken, for two reasons. First, as I have already pointed out, a correspondence postulate about the use of a given term by different language users and on different occasions is still required. Second, the model of human communication in which there are, as it were, other independent but disembodied language users, is no less subject to epistemological difficulty than the naturalistic model of language users communicating with each other about, and via, situations in an independent external world. If the consensus model is developed adequately, it needs a conception of the environment to avoid the threat of newspeak, and then what counts as consensus will become identical with what counts as truth in the naturalistic model. The differences between the models are more verbal than substantial, but the model using the notion of truth is less open to the currently dangerous misinterpretation of mistaking brainwashing for objectivity.

## V. Meaning-change

Something must now be said about the concept of meaning-change presupposed by the network model of universals. In the discussions that have gone on about the theses of meaning-variance and the theory-ladenness of all descriptive terms, appeal has been made to the classic distinction between the meaning of terms as *reference* or *extension*, and their meaning as *sense*.[1]

---

be forgotten, however, that *such translation may need an expansion of our own interests and even of our own experience*, for example from a purely naturalistic into a religious framework, for it does not follow in Rorty's account that it is *now* the case that 'most of our beliefs' are identical with, or translatable into, 'most of their beliefs'. For recent discussion of 'charitable interpretations' among anthropologists, see articles collected in *Rationality*, ed. B. R. Wilson (Oxford, 1970), and *Sociological Theory and Philosophical Analysis*, ed. D. Emmet and A. MacIntyre (London, 1970).

[1] See, for example, the attempts to interpret meaning-stability in terms of stability of extension only in A. Fine, 'Consistency, derivability, and scientific change', *J. Phil.*, **64** (1967), 231, and I. Scheffler, *Science and Subjectivity* (Indianapolis, 1967), and my own reply to Fine in 'Fine's criteria of meaning change', *J. Phil.*, **65** (1968), 46. P. Achinstein, on the other hand, adopts an interpretation of meaning-stability in terms of 'semantically relevant conditions', which seems to come nearer to the classic account of meaning as *sense*, although he does at the same time consider that semantically relevant conditions can be known independently of theory, and hence approaches what I call below 'intensional reference'. *Cf. Concepts of Science*, especially 6, 101.

Consideration of the network model indicates that this distinction is by no means adequate or unambiguous. Consider first *extension*. The network model, at any given stage of collection of evidence, produces a best theory which has the effect of discriminating objects into classes specified by predicates in such a way as to fit best the theory and all the available evidence. But these classes are not adequately described as extensional, for they are not defined merely by the objects contained in them. They involve also what I shall call *intensional reference*, that is, they depend on recognitions of similarities and differences in producing the initial classification in a given language.

Intensional reference is the relation which subsists between a descriptive predicate in a given language and a property of an object when the statement ascribing that predicate to that object is true. Since in any given classification we know only that *most* of the statements are true, but not which these are, it follows that we know only that most predicates in the classification have intensional reference, but not which. And since which statements *we take* to be true at any given time depend on the current best theory and the coherence conditions, it follows that what *we take* to be intensional references will also be modified by theory and coherence conditions in the light of the rest of the evidence, and will change when they change. What we take to be the extensions of the classes defined by these intensional references may or may not change, for it may or may not happen that the change of conditions of classification affects the actual objects that are classified under a given predicate. The criteria for a Vermeer might have been sharpened up or otherwise modified without there actually being any forgeries or other objects to exclude from the class of Vermeers. Thus, same extension does not entail same intensional reference, but same intensional reference does entail same extension, since that any object has an intensional property is a sufficient condition for placing that object in the extensional class corresponding to that property.

The relation 'same intensional reference, same extension, but not conversely' is reminiscent of the traditional account of meaning as *sense*; we must therefore enquire into the relation between sense and intensional reference. The concept of sense is notoriously ambiguous. It is sometimes used to denote simply recognition of similarity, as when 'same star appears near the sun at dawn' gives the sense of 'morning star' by ostensively directing attention to the object referred to. More commonly, to avoid obscure dependence on the notion of ostension, 'sense' is used in connection with the *definition* of a term, or with the set of synonymous terms which contains it. Thus 'cybernetic' means (has the sense of) 'concerned with self-regulating mechanisms', 'vixen' means 'female fox', 'gorse' (in English) is synonymous with 'furze'. There is here an objectionable implied appeal to some basic terms required to provide the definitions, and even if there were such terms, it is in any case doubtful whether nearly enough other terms would be strictly definable by them, or would be members of sets of strict synonyms, to make this definitional analysis of sense at all adequate for most descriptive predicates.

In order to specify the sense of a predicate more closely without depending on definitions or synonyms, it is necessary to add more and more descriptive phrases to the specification in the hope of capturing the sense uniquely. The 'morning star' is 'a bright planet seen in the eastern sky around dawn, which turns out to be the same planet as that called Venus and sometimes described as the "evening star" because it is seen in the western sky . . .'. Thus arises the suggestion that the sense of $P$ should be understood as being given by the *context* a term has in a theoretical system. It is possible to understand in this way much talk in current philosophy of science about the 'meaning' of theoretical predicates: for example, the 'meaning' of 'neutron' in a particular theory is said to be given implicitly by the postulate system of that theory. When a term enters into different and sometimes conflicting definitional and factual sentences in different theories, then it is said to have different meaning or sense in these theories. Thus 'mass' has one sense in Newtonian mechanics, and a different sense in relativity mechanics.

This sense of 'change of sense' is, however, almost trivial. In making sense relative to theoretical context, we cannot in principle stop short of taking in the whole set of sentences constituting the theory. For suppose we tried restricting the context-meaning of term $P$ just to 'the set of sentences $S$ to which $P$ belongs in the theory'. In order to know the meaning of these sentences we would have to know the contexts of all the other terms in these sentences, which would generally involve sentences other than those in $S$, and ultimately all the sentences of the theory. But then meaning as context would necessarily and trivially change whenever a theory changed, however little. If meaning-change raises any definite and interesting problems it cannot be specified in this way by change of total theoretical context.

The only useful and unambiguous specification of 'meaning-change' in the network model concerns the *causal* conditions of ascription of predicates, or what I have called the *intensional reference*. The 'meaning of $P$' changes when some or all of the empirical situations to which we have conventionally learned by recognition of similarities and differences to assign the predicate $P$, are deliberately ascribed the predicate $\bar{P}$, according to some rules derived from the coherence conditions. If the change of ascription can be said to be due to a mistake in recognition which is rectified on next examination of the same situation, as when a Pieter de Hooch is mistaken at first sight for a Vermeer, we shall not say that the intensional reference of 'Vermeer' has changed. But if the threshold of degrees of similarity delimiting the class 'Vermeer' is changed, or if some objects previously in the class are deliberately excluded for theoretical reasons, such as the desirability of making distinctions between paintings by a master and copies by his immediate pupils, then the meaning of 'Vermeer' ascribed to a painting has changed, possibly in extension and clearly in intension. The most important aspect of meaning-change is therefore concerned more closely with reference, albeit intensional reference, than with sense or context. If the physical conditions under which recognition of a property as $P$ is correct are unchanged, the meaning of $P$ is unchanged, no matter what changes of theory may be dictated by all the evidence and the coherence conditions.

This account of 'same meaning' is the correlate of the correspondence account of truth, for the correspondence postulate for descriptive sentences applies only while the conventional language remains the same, that is, while recognitions of similarity remain the same. And just as the conditions for truth cannot be described as it were externally and independently of the coherence of sentences with all other sentences similarly produced, so 'the meaning of $P$' cannot be specified other than by saying that it remains the same in the same language. We know what it is like for the conditions of conventional application of a predicate to change; hence we can identify at least some occasions of meaning-change, even though it is impossible to *state* explicitly what is the causal relation between learned language and the world that constitutes the 'meaning of $P$'. It follows that the claim, in any particular case, that intensional reference remains the same unless there is a *specifiable* change in the physical conditions under which the term is applied, is always a conjecture based on apparent intersubjective agreement of the language community. We shall see in the next chapter that, among other insights, the grue paradox sharply exposes the possibility that such intersubjective agreement may not indicate sameness of what different speakers of the language community take to be the intensional reference of, for example, 'green'.[1]

Two questions relevant to the meaning-variance thesis remain. First, is it correct to say, in this sense of meaning-change, that the meaning of, for example, 'mass', has changed from Newton's theory to Einstein's? Mere change of theory, or change of the set of sentences containing the term 'mass' in the two theories, is not enough, for this is concerned with context-meaning and not with reference. But if, as is often the case, such change of context-meaning is accompanied by change in the conditions under which the term is ascribed and withheld in empirical situations, then there is meaning-change in the present sense. 'Mass' is withheld from light in Newton's theory, and ascribed to light in Einstein's; 'simultaneous' is ascribed to certain pairs of events in Newton's theory, and withheld from those events in Einstein's. In these respects the meaning of 'mass' and 'simultaneity' have changed, for in a correspondence theory of truth it cannot be the case that 'Events $A$ and $B$ are simultaneous' is both true and false unless some of its terms, in this case at least 'simultaneous', have changed meaning. But the fact that 'the measure

---

[1] See especially p. 88 below.

of mass increases with velocity relative to the observer' is true *in Einstein's theory* and false *in Newton's* does not in itself indicate a change of meaning, because it does not imply that the term 'mass' is to be applied to different objects or in different situations in the two theories. ('Mass' is here taken of course to be the determinable property, not a particular determinate numerical measure of that property.)

The second remark concerns the substitutability of the token 'mass' from theory to theory. Is it a token of the same type, substitutable without change of truth-value? Since truth is correspondence and not 'truth in a given theory', and meaning is reference and not 'meaning in theoretical context', strictly speaking a term is substitutable if and only if its intensional reference is the same. But the demand for sameness of intensional reference as a necessary condition for logical substitutability is too strict. For since 'mass' in Newton's and 'mass' in Einstein's theories do not have the same intensional reference (because they have different extensions), this demand would make 'mass' strictly equivocal and unsubstitutable between the two theories. But we *do* want a statement such as 'mass increases with velocity' to be a statement of Einstein's theory which is *contradicted* in Newton. This can be allowed for by specifying more closely the concept of 'intersection' of theories introduced in the last chapter, as follows. There are many objects which are within the reference of 'mass' in both Newton's and Einstein's theories; where 'mass' is used of these objects denote it by 'mass$_i$'. Then there are many statements of the two theories that are logically comparable; some are identical, some consistent, some contradictory. Denote by 'mass$_n$' the old predicate 'mass' when applied to objects in Newton's theory but not in Einstein's (if any), and by 'mass$_e$' the predicate when used of objects in Einstein's theory but not in Newton's. Then Newton's theory will contain a statement ' "mass$_i$" and "mass$_n$" are the same property', whereas Einstein's theory will have no use for 'mass$_n$', and conversely Einstein's theory will contain a statement ' "mass$_i$" and "mass$_e$" are the same property', whereas Newton's theory will have no use for 'mass$_e$'. It then follows, as required, that some statements containing 'mass' are logically comparable in the two theories and some are logically incommensurable. Notice that this solution of the problem of meaning-variance depends both on taking 'meaning' as intensional reference, and on having two theories available in terms of which to specify the referential intersection. There is no 'essential meaning' of a term in any *one* theory that *must* survive into subsequent theories.

A further example may help to clarify the relations between extension, sense and intensional reference. In the course of a discussion of 'spatial congruence' relative to different metrics, Grünbaum remarks

> . . . the case of congruence calls for a generalization of the classical account of the relation between the intensional and extensional components of the meaning of a term. According to that account, the intension of a term determines its

extension uniquely. But the fact that 'being spatially congruent' means sustaining the relation of spatial equality does not suffice at all to determine its extension uniquely in the class of spatial intervals.[1]

The reason it does not is that different pairs of intervals will be said to be congruent relative to different metrics:

> But since the use of 'spatially congruent' in conjunction with *any one* of the metrics $ds^2 = g_{ik}dx^i dx^k$ does mean sustaining the spatial equality relation, I shall refer to this fact by saying that 'congruent' has the same 'non-classical intension' in any of these uses.

Grünbaum is right in detecting the use of *three* different concepts of meaning here in place of the traditional two. 'Intension' in his usage ('intensional reference' in mine) does indeed change whenever extension changes, since if, for example, 'congruence' applies to the interval between two given objects in a Euclidean geometry, it generally does not apply to the same interval in a non-Euclidean geometry. But *in the context of* each of the two different metrics, 'congruence' may be said to have the same 'sense', or 'intension' in one of its classic uses, or 'non-classical intension' in Grünbaum's neologism.

## VI. Goodman's strictures on similarity

The network model depends so crucially on the intensional concept of similarity that it is advisable to forestall some further objections to this concept by examining it in relation to somewhat different concepts of similarity that seem to underlie the challenge of 'Seven strictures on similarity' put forth by Nelson Goodman.[2] Goodman has seven complaints about 'similarity', namely that

(1) it 'does not make the difference between representations and descriptions, distinguish any symbols as peculiarly "iconic", or account for the grading of pictures as more or less realistic or naturalistic';

(2) it 'does not pick out inscriptions that are "tokens of a common type", or replicas of each other';

(3) it 'does not provide the grounds for accounting two occurrences performances of the same work, or repetitions of the same behaviour or experiment';

(4) it 'does not explain metaphor or metaphorical truth';

(5) it 'does not account for our predictive, or more generally, our inductive practice';

---

[1] A. Grünbaum, 'Reply to Hilary Putnam's "An examination of Grünbaum's philosophy of geometry"', *Boston Studies in the Philosophy of Science*, vol. V, ed. R. S. Cohen and M. W. Wartofsky (Dordrecht, 1968), 45.

[2] 'Seven strictures on similarity', *Experience and Theory*, ed. L. Foster and J. W. Swanson (London, 1970), 19.

(6) '[as a relation] between particulars [it] does not suffice to define qualities';
(7) it 'cannot be equated with, or measured in terms of, possession of common characteristics'.

If all this were indeed the case with the concept of similarity we have been working with, we would be in an unhappy situation. However, it does seem that the notion or notions of similarity assumed by Goodman are not our notion, and particularly in the following four respects.

(*a*) Similarity in the present account is a *primitive* symmetrical, intransitive relation between objects, holding in varying degrees in respect of particular properties. It is a relation given in the causal interaction of the perceiver and the world. It follows that it is not possible to *state* further conditions for the relation to hold. Goodman, on the other hand, requires that we should be able 'to say how two inscriptions must be alike to be replicas of one another' (p. 21), and asks 'what resemblance must the objects a term metaphorically applies to bear to the objects it literally applies to?' (p. 23), and concludes from our inability to answer such questions explicitly that 'to say that all *a*'s are alike in being *a*'s, amounts simply to saying that all *a*'s are *a*'s', making the relation of similarity superfluous. But it can be seen that this conclusion does not follow as soon as it is remembered that similarity is intransitive. '*a*'s are alike in being *a*'s' is construed in our account as '*a*'s bear a sufficient similarity to each other and distinctness from other objects to warrant their being placed in the same class in the current state of all the evidence'. This patently does *not* amount simply to saying that 'all *a*'s are *a*'s'.

There is a sense in which Goodman also recognizes the primitive character of the relation between things often said to be similar (he even suggests that in this sense the relation is 'indispensable' (p. 27)), but he takes this primitive relation to be more simple than it is. Similarity on the present account is primitive but at the same time complex, since it comes in varying degrees and relates pairs of objects in respect of different property-dimensions. Thus the similarity relations recognized among objects form a complex totality which may be analysed into more primitive similarity relations.

(*b*) On the network account judgments of similarity are partly dependent on theoretical considerations, that is, on the coherence conditions, and on the character of the rest of the evidence connected in a lawlike system. Judgments of similarity cannot be *wholly* dependent on these conditions, however, since if they were there would be no lawlike system with which to compare them. The influence of 'theory' is essentially the influence of the whole set of similarity relations given in experience. Goodman, on the other hand, suggests that theory 'creates' or 'governs' judgments of similarity.

> The fact that a term applies . . . to certain objects may itself *constitute* rather than arise from a particular similarity among those objects (p. 23).

> I suspect that rather than similarity providing any guidelines for inductive practice, inductive practice may provide the basis for some canons of similarity (p. 24).

> Laboratory results *create* rather than reflect a measure of sensory similarity . . . they tend to *govern* ordinary judgments at least as much as to be governed by them (p. 29).

> We cannot repeat an experiment and look for a covering theory; we must have at least a partial theory before we know whether we have a repetition of the experiment (p. 22).

All these propositions have a certain air of tentativeness about them: '*may* constitute', 'I suspect', '*tend* to govern', '*partial* theory'. In other words, with regard to the effect of theory, Goodman could be construed in a fashion that is almost consistent with what has been argued here, somewhat as follows. There is a primitive relation of similarity (else how could we even know what counts as another *application* of a given theory, kind of behaviour, performance, etc.?) which Goodman effectively equates with simple generic identity of individuals, and it forms the basis of theory in terms of which more elaborate judgments and specifications of similarity may be made. The only point of conflict is that by equating the primitive relation with generic identity, Goodman eliminates the possibility of correction even of these primitive recognitions by subsequent theory. In the network account the relation is primitive but not simple, and the possibility of correction without circularity is allowed for.

Again, Goodman remarks that comparative judgments of similarity often require selection of relevant properties, weighing of their relative importance, and variation in both relevance and importance. This can also be taken care of in the present account. Relevance and importance will be functions of the theoretical system accepted up to date. The colours of flowers, for example, are not important in the classification of plants, although they are obviously recognizable similarities, because they have not been found to be usefully correlated with other properties in an economical and comprehensive theoretical classification. Theory may cause us to neglect some similarities of apparently and perceptually important properties and to rely on others, such as numbers of stamens, which may not even be perceived at first glance with the naked eye.

If we have little or no information about any theoretical system, however, relevant similarities may be determined just physiologically. Goodman gives the example of extrapolating a data curve which is so far linear. If we extend the straight line, all the points, both observed and projected, will be similar in being on that straight line. If we give the curve a right-angled bend, all the points will be similar in being on *that* curve. Which is the relevant similarity? Experiments have been reported which indicate that in such cases similarity is just determined physiologically—we automatically extrapolate in the straight line.[1] There can be no rational *justification* for this, unless we appeal to natural selection in a race of organisms in an environment in which most important variations have been linear. If we wish to accommodate Goodman's insistence on the primacy of theory, we may even regard such recognition of similarity as incipient theory, where the 'theory' is a physiologically innate determination of perception. In the light of the various levels of conscious reflection on which theoretical or coherence conditions operate, this would not be such a misleading characterization either. Some innate determinations we are born with; some we acquire by explicit learning, as in driving a car; others are thrust upon us by conscious application of theory to *overcome* our natural projections, as in recognizing that carnations are not relevantly similar to roses, or whales to fish, for purposes of theoretical science.

(*c*) In the network account, similarity is not intended to provide a simple reductive basis or explanation for universals, theories, inductive practice, etc. Similarity is a complex determinable, determined by degrees and property-dimensions. Its very complexity as a relation enables us to analyse and explain complex judgments regarding many objects and many properties. Goodman, on the other hand, seems to hold that the only rationale for introducing the relation of similarity would be as a reductive explanation, and since he concludes that the only reductive basis is generic identity (which of course is far from simple as a basis, since it requires as many primitives as there are property-classes), similarity is superfluous for him even in this role— ' "is similar to" functions as little more than a blank to be filled' (p. 27). In the present account this dictum must be construed as 'filled by determinate degree and property-dimension'. But Goodman would presumably be unhappy with this construal, for two reasons. First it makes similarity complex, but this objection amounts to no more than philosophical prejudice in favour of simple reductive bases, and as far as we can see our experience affords no evidence that such a basis exists. Secondly, to speak of the *respect* of similarity seems to introduce not only a complex similarity relation, but also further primitives, namely all the properties. This is not the case, however, for as we have seen, the resemblance theory of universals does not require *exemplars* of each property, only many-dimensional primitive respects of similarity, which themselves permit the construction (below or above the level of consciousness and explicit statement) of properties and their exemplars.

(*d*) Finally, the network account is irreducibly intensional, since our experience demands intensional properties. Goodman rejects the attempt to identify and distinguish intensional properties for measuring similarity as a 'notoriously slippery matter'. But it is just this 'identification and distinction' which we have seen in the last resort to be a primitive physiological matter, and not subject to further explicit analysis. We can agree with Goodman that

[1] R. L. Gregory, *Visual Perception* (Oxford, 1972).

similarity is not to be eliminated by counting or measuring intensional properties, but for different reasons from his. In his account this suggestion is an abortive attempt to eliminate similarity, first by definition in terms of extensional classes of objects, which as he rightly argues leads to a vacuous similarity relation in which all pairs of objects are equally simple, and then by rejecting intensional properties as no more economical than the many-dimensional relation of similarity itself. But in our account we are not looking for economy but accurate analysis, in which similarity is not *eliminated* by intensional properties but *recognized* as having intensional property-dimensions.

Similar remarks apply to Goodman's proof that similarity of objects in virtue of shared properties cannot be used to give an extensional definition of properties. For example, the three pairs of three objects *a*, *b*, *c* may be equally similar in sharing at least one property, for instance $PQ(a)$, $QR(b)$, $RP(c)$, but this does not serve to define any one property as an extensional similarity class. Our concept of similarity here, however, is not required to provide any such definition, for it is already assumed that we know that these pairs of objects are similar in three different respects, and indeed the property $Q$, for example, could be defined on this view of similarity as that which *a* and *b* have in common, and not *c*. Similarity is a relation between objects in respect of a certain property, not a relation between objects as such.

It may be concluded that Goodman's strictures do not capture the concept of similarity required by the network model, and are not inconsistent with that concept.

## VII. Absolute universals again

Let us now return for a moment to the theory of absolute universals and ask whether it can be maintained in any sense in the light of the process of initial classification and modification by coherence conditions that has been described. It might be suggested that the theory of absolute universals has to face the problem of modifications in any case, for this is the problem of *errors* or *illusions* or recognition. The mirage 'looks like' the oasis, but no one who believes in absolute universals would regard himself as committed to the view that they therefore both share in absolute oasishood. However, the situation in the network model aggravates the problem. For it has been argued that $P$ may come to be regarded as incorrectly predicated of *a* in cases where there is neither radical change of meaning of $P$ nor illusion, but in virtue of further evidence and resulting theory changes. That whales were for long described as fish, or all species of mice as members of the same family, is not usually regarded as the result of error of recognition or illusion, nor was the description of the two 'companion stars' of Saturn that were later recognized as part of Saturn's rings. Or rather, if these are said to be illusions, then all descriptions given in terms of discarded theories must be said to be illusory, and in this sense 'illusion' comes to mean a description corrected not only by further evidence but also by later theories unthought of when the description was given. What seems difficult for the theory of absolute universals to accommodate without radical modification is the idea that there can be recognition of a quality of $P$-ness in $a$ in what are now regarded as normal and typical circumstances, which can later be discarded in favour of predication of $a$ by not-$P$.

A persistent believer in absolute universals (or, indeed, in absolute resemblances as being themselves universals) might seek to counter this argument with some such view as the following: there *are* absolute universals, but recognition of them is never infallible, even in apparently paradigm situations. Indeed, he might continue, we shall not know what the (or a) true set of universals is until we have the (or a) true and comprehensive science of the whole cosmos. Whatever its merits, such a position is now an avowedly *ontological* one, and it is not only quite consistent with the network model, but even might be held to be presupposed by that model in its assumption that there is always a high proportion of direct recognitions that are correct predications in the current state of the language, that is, that some recognitions of objective resemblance are correct. This ontological position can, however, only yield an answer to the original epistemological question by being developed into a network theory of FR type, not by its classical epistemological counterparts.

In the history of philosophy the problems of universals and of natural laws are closely connected. Aristotle's 'stabilisation of the universal in the mind'[1] as a result of reflection on experience, is his account both of how we come to predicate a new object correctly as 'swan', and also of how we know 'All swans are white', for 'swanness' is a complex universal incorporating 'whiteness'. And just as Hume declined to find in the relation of two interacting objects a third entity in the form of a causal tie, so he declined to find between two objects which resemble each other a third entity which is the universal in virtue of which they resemble each other.[2] The account of causality, or lawlike relations, which depends on regularities of co-presence, co-absence and covariance, may thus be seen as parallel to an account of qualities as classes defined by their similarities and differences. Directly experienced spatial and temporal relations between objects required for causality are then seen as parallel to directly experienced resemblances required for the definition of reference classes.

These parallels can be extended by noticing that the network account of what it is to predicate a universal correctly corresponds closely to a generally accepted answer to the question 'Given a regularity view of laws, how do we

[1] Aristotle, *Posterior Analytics*, 100a.
[2] D. Hume, *Treatise of Human Nature*, I, part I, 7.

distinguish natural laws from accidental generalizations?' For example, it may happen that all professors in Cambridge were born on a Tuesday, but we should hardly suppose that this is a natural law with predictive force, nor that counterfactual inferences can be drawn from it such as 'If *A* had been appointed to the chair of Etruscan (which he wasn't) he would have been born on a Tuesday'. The generally accepted solution of this problem is to appeal to the theoretical context within which the regularities are asserted.[1] If they are regularities of the kind we expect, given the rest of our knowledge, and in particular if they are entailed or made highly probable by acceptable theories, then they have the character of laws, and predictions and counter-factual inferences may with confidence be drawn from them. If on the other hand we have good reason from the rest of our evidence for thinking that a particular regularity has no place in our theoretical system, we regard it as accidental, as in the case just given. It is a feature of this account, however, that no *fundamental* distinction is made between laws and accidental generali-zations, because the very evidence we appeal to in support of the lawlike character of a regularity itself consists of other such regularities, about which the same question as to their lawlikeness could arise. The distinction made is only between regularities which cohere in a system of regularities, and those which do not.

The parallel with a network account of universals is very close. Here too it might be held (and has indeed been held by Popper, as we have seen)[2] that there are far more resemblances than are significant. Any two objects could be said to resemble each other in some respect, and might give rise to a predi-cate class. How do we know which of these classes are significant? According to the resemblance theory as put forward by, say, Price, it is not easy to answer this question, for the theory appeals to nothing more than the situation of resemblance itself. In this it is on a level with a regularity theory of laws which confines attention to the co-occurrences of the events themselves without examining their context in the rest of the evidence. In the network model, on the other hand, the answer to the question is obvious, and is the same for identification both of significant laws and of significant resemblances. Regularities are laws when predictions from them are also supported by the rest of the evidence; resemblances define significant predicate classes when these predicates enter laws and when supported predictions can consequently be made about further properties of members of a class. But significant classes do not indicate 'real essences' any more than lawlike regularities indicate 'necessary connections'.

In the light of the closeness of the parallel between them, it may be wondered why the problem of predicate classification has not recently pre-

---

[1] See, for example, R. B. Braithwaite, *Scientific Explanation* (Cambridge, 1953), chap. 9, and E. Nagel, *The Structure of Science* (New York, 1961), chap. 4.

[2] *Logic of Scientific Discovery*, appendix *x.

sented itself with such force as the problem of lawlikeness. It is easy to suggest reasons, but not easy to be confident that they go to the heart of the matter. It may be that we are less often presented with problematic classes than with problematic regularities, at least at the primitive level. New predicates rarely have to be introduced into the language to refer to newly significant resemblances, and this may be because such resemblances can in any case be described by appealing to newly discovered laws. That the planet Venus is just a big bright stone was expressed not by labelling previously overlooked resemblances by some new predicate, but by formulating the laws satisfied by stones and planets. A related possibility is that the sciences which have been considered in most detail by philosophers are those in which problems of the law-system are much more striking than problems of classification. Physics and chemistry passed the classification stage long ago, so it is claimed, and indeed at present their basic classification system is well entrenched. It is not so long since biological classification reached this point, however, and there are areas, for example in microbiology, where it has not yet been reached. The same applies to even younger sciences such as archeology, sociology and linguistics. Philosophical analysis of these sciences is forced to take account of problems of classification, and is therefore quickly led to the issues lying behind a theory of universals.

# CHAPTER THREE

# The Grue Paradox

## I. Principles of solution

The notion of similarity is fundamental for the present account of scientific inference, as the basis both of classifying universals and of organizing them into systems of laws and theories. We have already noticed and replied to some objections to taking similarity as thus primitive, including some objections due to Goodman. Goodman's most cogent argument in this context, however, is his discussion of the 'grue' paradox, which calls in question the intensional assumptions involved in recognizing any similarity and difference relations or any potential natural kinds.[1] If Goodman were right in the conclusions he draws from his paradox, no account of the inductive inferences based on a network model of universals, such as I shall subsequently develop, would be tenable. So before going any further it is necessary to examine Goodman's claims for his paradox in some detail.

Consider the predicate 'grue' which is defined in terms of the two predicates 'green' and 'blue', which are normally considered to be primitive in our observation language, as follows:

Grue applies to all things examined before $T$ just in case they are green but to other things just in case they are blue

and a corresponding definition of 'bleen' in terms of 'green' and 'blue':

Bleen applies to all things examined before $T$ just in case they are blue but to other things just in case they are green.

Now consider the two hypotheses 'All emeralds are green', 'All emeralds are grue'. Each is supported by the same evidence before $T$, for all emeralds observed have in fact been green before $T$ and therefore grue. By induc-

---

[1] N. Goodman, *Fact, Fiction, and Forecast* (Indianapolis, 2nd edition, 1965), chap. 3. Unlike several recent writers (W. v. O. Quine, *Ontological Relativity* (New York and London, 1969), 114; L. J. Cohen, *The Implications of Induction* (London, 1970), 99f), I take the 'grue' paradox to be intrinsically more fundamental than the so-called 'raven' paradox, although both appear to be concerned with similarities or natural kinds. I shall show later that the raven paradox can be adequately dealt with by presupposing the concept of similarity; the grue paradox calls in question the relevance of that concept itself.

tion by simple enumeration of positive instances, both hypotheses are equally supported. But they yield different and contradictory predictions after $T$, and we should have no hesitation in accepting the prediction of the first rather than the second. Therefore in any adequate confirmation theory they ought not to be regarded as equally supported. In Goodman's terminology, we ought to be able to show that 'green' is more *projectible* than 'grue'.

Some attempts to dissolve the puzzle have depended on trying to locate an asymmetry between 'green' and 'grue' in the time or other positional reference that is required in defining 'grue' in terms of 'green' and 'blue'. But a perfectly symmetrical definition of 'green' in terms of 'grue' and 'bleen' is entailed by the former definitions, as follows:

> Green applies to all things examined before $T$ just in case they are grue but to other things just in case they are bleen

and there is a corresponding definition of 'blue'. Thus the time or other positional reference is not sufficient in itself to generate a syntactical asymmetry which would give reason for the projectibility of green and not of grue.

Other attempted solutions have depended on the tacit interpretation of 'is green' as 'looks green', and the objection that a thing cannot 'look grue' to unaided observation at $t$ without knowledge of the date, that is, of whether $t$ is before or after $T$.[1] It should be noticed first that on either an extensional or an intensional interpretation of 'green', there is no reason to suppose that green 'looks' any different to grue before $t$, since both predicates have been learned in exactly the same circumstances. The question is, of course, whether grue must 'look different' before and after $t$, and as we shall see, to suppose that it does, or is expected to, is to trivialize the paradox. In any case, this proposed solution does not do justice to Goodman's explicit rejection of intensional interpretations of 'green', on the grounds that there is no definition of such interpretation except perhaps psychological definitions, which, as we have seen before, would beg the question about intensional interpretations of *other* predicates.

Now it may be thought that since arguments for the primitive recognition of intensions have formed the basis of the network model, a solution in terms of intensions should be sufficient within that model. This, however, is not so, for in the network model it is admitted that what constitutes the intensional reference of a predicate cannot be *stated* (thus far agreeing with Goodman)

---

[1] See the early response along these lines of S. Barker and P. Achinstein, and Goodman's immediate response to them, in 'On the new riddle of induction', *Phil. Rev.*, **69** (1960), 511, 523. The rapidity of these exchanges has not prevented the same suggestion being made in much subsequent literature, without apparent recognition that it does no justice to Goodman's intentions (see note 2 on p. 79 below).

but only indirectly got at via the evidence and coherence conditions of the whole network. Although we have seen how to deal with individual mistakes and linguistic inconveniences in ascribing descriptive predicates, we have not taken account of breakdown in intersubjective agreement as to the use of such predicates. We cannot directly check that the physical conditions of recognition of similarities are or remain the same in all speakers of the language; neither can we be sure that when a predicate such as 'green' is learned, everybody 'catches on' to the *same* objective property in the world. The checks are at best behavioural, and involve the internal relations of the network. It is this characteristic of intersubjective meaning that Goodman strikingly dramatizes in the grue paradox. The argument that follows here is designed to provide a solution which recognizes the inadequacy of appeal to intensional meaning in the absence of the corrective devices potentially present in the theoretical network.

I shall regard the problem, as Goodman does, as the *new* problem of *induction*, where both italicized components are important. The problem is *new* because it does not demand that we show that inductive inferences must be true or even probable. What it does demand is that we describe fully those inductive inferences we do in fact make, and uncover their presuppositions. In this context this means that we must describe the asymmetry of 'green' and 'grue' which makes 'green' projectible and 'grue' not. On the other hand, the problem is about *induction*, and this means that the asymmetry, if it can be discovered, must be of a kind that can be accepted as relevant to the expectation that one or other prediction is true. Thus, clearly, mere verbal difference between 'green' and 'grue' is not a relevant asymmetry, nor are some more subtle asymmetries detected in the literature. I shall take it without further argument that detection of difference of *simplicity* in the predictions would be a relevant asymmetry, so long, of course, as it can be shown in the context that simplicity is well-defined. More will be said later about the relation of simplicity to induction, but it will be sufficient here to appeal to the very general practice of adopting the simpler of two conflicting hypotheses, all else, including degree of support by evidence, being equal. Moreover, prediction of *change* will be taken to be less simple than prediction of no change, on the principle that change requires further explanation, and absence of change does not.

I shall now suggest two principles which I take to be essential to any satisfactory solution of Goodman's puzzle, and which have not always been accepted in discussion of it, but which will play a crucial role in what follows.

(*A*) In order to generate a puzzle about *induction* at all, the language describing the present evidence must not be merely verbally different, but must yield predictions which are both genuinely different, and different in respects which the 'green' and 'grue' speakers (who will be called 'Green' and 'Grue' respectively) can explain to each other and agree to be different. For it must be assumed that the results of a test of their respective predictions after $T$ can be agreed upon, and that if the green and grue hypotheses are assumed the only alternatives, one or other of them but not both will then be agreed to be correct, and the other to be mistaken. That is to say, whatever it is that Green and Grue predict before $T$, it must be the case that after $T$, in either language, the true description of what has in fact happened agrees with either Green's prediction or Grue's but not both, and that this can be known to both Green and Grue.

The puzzle arises from the fact that the evidence is the same; the predictions are objectively different ('objectively' because different in either language); but the predictions appear to be statable in terms which are symmetric in both languages, for example not obviously simpler in one than the other.

(*B*) The second principle to be satisfied by any interesting solution follows from these considerations. It is that the problem should be shown to be soluble in its *strongest* form, and that since the solution is to be sought by finding asymmetries in the predictions of Green and Grue, it should not be set up in such a way as to introduce needless asymmetries into the definition or interpretation of the problematic predicates. This may seem trite, but as we shall see it has been violated by some suggested solutions.

Goodman's own claim is that no relevant asymmetries between the predicates can be found short of the pragmatic difference of actual historical *entrenchment* of predicates in the language. He therefore suggests entrenchment as the criterion of projectibility. I shall argue that Goodman's conclusion is in principle correct, but that examination of possible asymmetries has not been carried nearly far enough. Pressing this line of attack more strongly reveals that the puzzle is a real one, but that we are only rarely in a situation in which it actually arises. Most of the time some relevant asymmetries short of entrenchment *can* be found to justify choice between the competing predictions.

Several different versions of Goodman's original definition are to be found in the literature. Any definition which is to generate the puzzle in conformity with principle (*B*) must not introduce trivial logical asymmetries relevant to the solution of the puzzle, and it must in addition be perfectly clear what predicate is being defined. If these two conditions are satisfied it matters little which version we adopt, for the subsequent discussion will be easily adaptable to fit other acceptable versions. I shall adopt the type of definition originally due to Barker and Achinstein, and recently clarified by Blackburn,[1] which does not involve the notion of 'being examined' that appears in Goodman's original version. Blackburn's amended definition is:

At any time $t$, a thing $x$ is grue at $t$ if and only if
($t < T \supset x$ is green) and ($t > T \supset x$ is blue).

This allows a perfectly symmetrical definition of 'green' in terms of 'grue' and 'bleen' as follows:

At any time $t$, a thing $x$ is green at $t$ if and only if
($t < T \supset x$ is grue) and ($t > T \supset x$ is bleen).

In this interpretation it is made explicit that colour predicates are applied to things *at a given time*, and no assumption is built into the definition about persistence of the colour of a given object. After all, we are perfectly familiar with the idea that green is not projectible of beech leaves after 15 October. The definition is, however, still not nearly clear enough to be starting point of the puzzle, because we have not yet indicated how it satisfies principle (*A*).

Before *T*, Grue has the same evidence as Green, and I shall assume at first, in conformity with (*B*), that they express this evidence in languages which differ only in two descriptive predicates, namely that 'grue' replaces 'green', and 'bleen' replaces 'blue' wherever these predicates occur in describing the evidence. I also assume that both agree that all these predicates are names of colours. In accordance with (*A*) predictions made before *T* about green and blue objects after *T* must be not only different, but understood by both speakers to be different in agreed respects. It is at this point that the definition of 'grue' and its companion definition of 'green' need to be supplemented. What is it that Green understands Grue to be predicting? The obvious answer is 'that emeralds will *change colour* at *T*'. But if this is also what Grue agrees that he is predicting (and remember we are assuming so far that all the *rest* of his language is the same) then symmetry is immediately violated, because Grue is predicting a change where Green predicts none. So in the absence of other evidence Green's prediction is the simpler.[2]

---

[1] S. Barker and P. Achinstein, *op. cit.*, and S. W. Blackburn, 'Goodman's paradox', *Studies in the Philosophy of Science*, ed. N. Rescher (Oxford, 1969), 128. It should be noticed that Goodman does not object to Barker and Achinstein's amendment in his reply following their paper.

[2] Suggested solutions in which this asymmetry is introduced are to be found in Barker and Achinstein, *op. cit.*; K. Small, 'Professor Goodman's puzzle', *Phil. Rev.*, **70** (1961), 544; W. C. Salmon, 'On vindicating induction', *Induction*, ed. H. Kyburg and E. Nagel (Middletown, Conn., 1963), 27; R. J. Butler, 'Messrs Goodman, Green and Grue', *Analytical Philosophy*, ed. R. J. Butler (Oxford, 1965), 181; J. J. Thomson, 'Grue', *J. Phil.*, **63** (1966), 289, and 'More grue', *ibid.*, 528; and Blackburn, *op. cit.* Replies by Goodman to this move immediately follow the Barker and Achinstein paper and Thomson's first paper; J. S. Ullian replies in 'More on "grue" and grue', *Phil. Rev.*, **70** (1961), 386.

To preserve (*B*) we must therefore adopt the opposite assumption that both Green and Grue commit themselves in their respective predictions to 'Emeralds remain the same colour after *T*'. How can this assumption be made consistent with principle (*A*)? Look at it first from Green's point of view. He predicts that emeralds remain green and that they remain the same colour. He understands that Grue predicts that emeralds remain grue and that they remain the same colour. He therefore expects that, after *T*, his prediction will be vindicated and Grue will be forced to admit a mistake. According to Green, Grue will say 'That is very strange, I see that emeralds have become bleen. They have changed colour'. Green knows that he honestly and diligently reports his experiences, and so according to the symmetry required by (*B*) he must assume that Grue's reports are likewise honest and diligent. Green can therefore only conclude that in this situation which his own prediction commits him to expect, Grue will have either

(*a*) misremembered 'what grue looked like' (for, of course, before *T* Green assumes it must have 'looked like' green, even to Grue), or misremembered the 'meaning of grue', so that Grue admits a change of colour where in Green's experience there is none; or

(*b*) inexplicably suffered a physiological change at *T*, so that an object whose colour remains 'objectively' (according to Green's claim) the same, now 'looks different' to Grue.

(There is perhaps no need to distinguish the possibilities (*a*) and (*b*), for what is 'misremembering' other than a physiological change?)

I now request the reader to start with the sentence 'Look at it first from Green's point of view' and check that by mutual replacement of 'Green' by 'Grue', 'green' by 'grue', and 'blue' by 'bleen', the account of the last paragraph is entirely symmetrical for Grue and Green. Both of them understand one another's situations perfectly, and know what will settle the argument after *T*. What they do not of course know is which of the different expectations described will in fact come to pass after *T*. It is extremely important to remember throughout the discussion that the asymmetry must be found, if at all, in predictions made *before T*, although they are predictions about what is expected to happen after *T*. So which prediction is most reasonable, given that both are still symmetrical? The excursion into 'change of colour' has left the original puzzle untouched.

## II. Objective tests of 'grue'

It may be remarked at this point that in the description of Green's expectations there certainly is essential reference to some kind of change, though not to the change of colour of an object. Green expects an inexplicable change in Grue's reports of his experiences. This is entirely symmetrical with Grue's expectation with regard to Green and so cannot yield directly a solution of the puzzle. But it does prompt the question of what exactly has been shown by developing the symmetrical situation involved in the prediction on both sides that there will be no change of colour. Suppose that Green, being, symmetrically with Grue, a troubled and intelligent fellow, pursues this point.

*Green:* We have been assuming up to now that all our predictions other than those about green and blue objects are identical. But if this is so then we haven't shown that our predictions about the colours of objects really differ as is required by principle (*A*). All we have shown is that the colours will 'look different' to one or other of us, according to which of our predictions is fulfilled. It does not follow from this that we are committed to disagreeing about what the colours of objects actually are. If I am right, I shall assume you have suffered an inexplicable breakdown, and conversely if you are right. But your (or my) breakdown of memory or perception can be corrected by objective tests of colour. Only if there is a real difference between us here will the puzzle be a serious one. Now I will try to show that if your prediction really differs from mine about objective colour, it is bound to be more complex than mine.

*Grue:* How so?

*Green:* We must first agree upon a test for objective colour which will not depend upon what objects 'look like'. Can we agree that *wavelength* is a property related to colour which we both know how to measure, and both agree that we shall obtain the same result as each other from such a measurement whenever it is carried out? If so, the difference in our predictions required by (*A*) consists in you predicting that this object will change in the wavelength of light it reflects, while I predict it will not change.

*Grue:* Wavelength as you have explained it to me in our previous conversations is indeed a property upon which both of us can agree at any time, and although I think it is a strange property to consider, in order to get the puzzle going in conformity with principle (*A*), I will agree that our predictions about it differ in the way you state.

*Green:* Good. Then it is perfectly clear that your prediction is not symmetrical with mine: it is more complex in predicting a change while mine does not. Simplicity does not of course guarantee success, but it provides a reason, and a reason generally acceptable to scientists, for distinguishing your prediction from mine, and, all other things being equal, for preferring mine.

*Grue:* Not at all. Your claim of asymmetry does not follow. There are other properties about whose measures we could agree at all times, and whose values you predict will change at *T*, while I do not.

*Green:* I suppose you mean other artificial properties like 'grue'? For example 'wengthlave', which I will define for you as the property of having the same value as the wavelength of green if the time is before *T* and as the wavelength of blue if the time is after *T*?

*Grue:* No, certainly not, because I can see as well as you can that predictions about *this* property cannot be stated symmetrically by me. Since we always agree about measures of wavelength, your proposed 'wengthlave' of grue has a different value before $T$ and after $T$, whereas the wavelength of green has the same value before and after. No, the properties I am thinking of are not constructed *ad hoc* from your language, they are ordinary simple properties in my language, although of course I can define them for you in your language, otherwise we shall not be able to agree upon a crucial test for our different predictions. For example, you are predicting a change in the kwell-measure of this object.

*Green:* What on earth is that?

*Grue:* It is very simple, although if you must have it translated into your language, of course it appears complex, just as 'grue' did. You would describe the kwell-measure as the number of electrons in this object if the time is before $T$, and the number of its neutrons if the time is after $T$.

*Green:* But of course I expect this number to change at $T$. Why should it not? And in any case, what has this got to do with the colour of the object?

*Grue:* Oh, that would be a long story. Perhaps if we could meet at the same time tomorrow I will expound my theory to you.

Grue, of course, is bluffing. There is at the moment no alternative theory to Green's about colour which is statable symmetrically with Green's and which yields objectively different predictions after $T$. To that extent Goodman's puzzle is spurious. Even so, this dialogue draws attention to two important features of the puzzle which are frequently overlooked.

First, it usually seems to be assumed that in its most interesting form the puzzle concerns 'purely qualitative properties' and not metric properties like wavelength, or even order properties like 'next to yellow in the spectrum', which might equally well have been used in the above argument. But if only purely qualitative properties are involved, it is easy to slide into the assumption that 'looks' of such properties are sufficient tests of their presence, and that 'looks' do not need to be checked and possibly corrected by some application to objects of predicates *upon which Green and Grue can agree*. If this assumption is made, the puzzle becomes trivial. No doubt it would be very disturbing to science if what people claimed to see when looking at green objects inexplicably changed, but it need not be disturbing to the physics of colours, so long as some predicates are still applied to objects in a manner about which there is agreement. It may be objected that the distinction between 'looks' and objective properties is a spurious one, because apparent changes in another man's physiology could themselves in principle be checked by objective tests of his brain states, and that these tests could also be agreed upon by Green and Grue. But this objection only strengthens the argument, for it admits that it is not in the last resort people's reports of 'looks' which decides what are to be taken as the objective colours of objects. In physics colours are usually checked by metric or order properties, but they may be checked by other qualitative properties, for example that chemical properties show the object to be a sample of $Mn(SO_4)_3$ crystals, and $Mn(SO_4)_3$ crystals are taken as the standard of objective green. The essential point is not whether the predicates to be projected are purely qualitative or not, but that they should be defined in accordance with principle $(A)$. Principle $(A)$ introduces a difference into the predictions of Green and Grue, and because all objective differences (unlike 'looks') are related in a network of physical laws with other objective differences, the differences must introduce asymmetries unless wholesale modifications of Green's theory are assumed.[1]

It may now be suggested that not all objective qualitative properties are associated in this way in a network of laws, and that this is what is assumed in Goodman's statement of the puzzle. Perhaps his reference to a highly structured property like colour is accidental and unnecessary. Suppose then we replace colour predicates by predicates which enter no theory, such as 'psycho-kinetic transmitter' $(PKT)$. Construct a grue-like predicate '$TKP$' of object $x$ such that at $t$, a thing $x$ is $TKP$ at $t$ iff $(t < T \supset PKT(x))$ and $(t > T \supset \sim PKT(x))$. But with such an 'unattached' predicate as $PKT$, about which we know very little, it is quite clear that not only would we be hard put to it to say what would count as evidence or crucial test in order to satisfy principle $(A)$, but also that we should not know what to say about the legitimacy of the rival hypotheses '$(x)(PKT(x))$' and '$(x)(TKP(x))$'. Goodman puzzles arise only in the context of predicates embedded in the language about which we have inductive intuitions. To be so embedded is to be related with other predicates in laws and theories which can be used to reinforce and sometimes correct applications of particular predicates.

Secondly, it has often been claimed that such a wholesale difference in Grue's theory as is required to preserve symmetry can always be constructed by *ad hoc* and artificial introduction of more grue-like predicates into Green's theory. This, however, is mistaken. It has already been argued that if the theory contains a predicate like 'wavelength' of which the numerical measure is agreed to be an objective test, there is no way of restoring symmetry without substantial modification of other aspects of Green's theory. It may be thought, however, that the possibility of trivial 'gruification' of Green's theory has not

---

[1] This aspect of the puzzle is pointed out but not pursued by C. G. Hempel, 'Inductive inconsistencies', *Aspects of Scientific Explanation*, 70f; J. J. Thomson, reply to Goodman, *J. Phil.*, **63** (1966), 533; J. Hullett and R. Schwartz, 'Grue: Some remarks', *J. Phil.*, **64** (1967), 269; and H. Kahane, 'Goodman's entrenchment theory', *Phil. Sci.*, **32** (1965), 377. It is also tacitly presupposed by those who appeal to possible introduction of asymmetry via the initial probabilities of a confirmation theory, e.g. R. C. Jeffrey, 'Goodman's query', *J. Phil.*, **63** (1966), 281; and H. Smokler, 'Goodman's paradox and the problem of rules of acceptance', *Am. Phil. Quart.*, **3** (1966), 71.

been pursued far enough. It may be suggested, for example,[1] that such a measure can be trivially produced by constructing a grue-like $\gamma$-meter, which indicates the same value (the $\gamma$-value) on its scale, say $\gamma_1$, when measuring the wavelength of green before $T$, and the wavelength of blue after $T$. Then although Grue's prediction implies a change of wavelength at $T$, it implies a constant $\gamma$-value through $T$, whereas Green's prediction implies constant wavelength and changing $\gamma$-value. Thus symmetry seems to be preserved. But if we now consider how the $\gamma$-meter is defined and constructed, we see that the only way we *know* of constructing it is by making a meter which measures wavelength, and then putting in a correcting device which switches the needle's readings of green and blue wavelengths at $T$. Thus the description of the $\gamma$-meter, and consequently the concept of the $\gamma$-value, is less simple than that of the wavelength meter and the concept of wavelength value. The only way for Grue to avoid this asymmetry is to have an alternative theory in terms of which to construct and describe the $\gamma$-meter as simply as Green describes his wavelength meter within his theory. This suggested trivial modification of Green's system to preserve symmetry therefore fails, and we are led back to the requirement of a radically different theory which Grue has not revealed to us, and of which we know nothing.

The general form of such attempts to save symmetry by trivial gruification of Green's theory seems to go as follows: Every time Grue seems committed to prediction of a change at $T$ which Green is not committed to, construct a grue-like predicate which does not change value in Grue's prediction, but changes in Green's. This can always be done by trivial definition using $T$ and the predicates of Green's theory. However, in order at some point to satisfy principle $(A)$, there must be an end to this process at which at least one determinate value of a predicate expressible in both languages cannot be so treated, namely that which Green and Grue agree they will agree upon when it is tested for after $T$. A predicate whose values thus count as a test as required by $(A)$ is in the peculiar position of being necessarily the same predicate in both languages, because it is what is measured (or indicated if the predicate is not metric) by the same test set-up. (Of course, Green and Grue may call it by different *names*, but these will name coextensive predicates by definition, and hence be only verbally different.) To mark the objective character of such test predicates let us call them $O$-predicates. One $O$-predicate, say $O_1$, will be one which Green expects to be constant with green, and Grue expects to change, because we know of many predicates whose values remain constant with green which might be candidates for the test predicate. The question is,

[1] I owe this suggestion to a discussion at the Philosophy Colloquium of the University of Pennsylvania, and particularly to contributions by Professors Cornman and Grunstra. Since the first publication of this paper E. Sober has investigated the possibility of systematically 'gruifying' every predicate in the language while retaining the conditions of the paradox, and has concluded that this cannot be done merely by a constructive algorithm.

can Grue preserve symmetry by constructing out of Green's predicates another $O$-predicate (say $O_2$) satisfying all of the following necessary conditions: ($i$) its value remains constant for grue objects at $T$, ($ii$) Green and Grue can agree on its value before and after $T$, ($iii$) its place in Grue's and Green's systems is symmetrical with $O_1$'s place in Green's and Grue's systems respectively? The answer is that no predicate constructed merely *ad hoc* to satisfy ($i$) and ($ii$) can satisfy ($iii$), for the following general reason: $O_2$, since it is an objective predicate by ($ii$), is also in Green's language, and in Green's predictions it must be expected to change value at $T$ for green objects. But there is no such predicate already available in Green's system if this is our physics of colour, and any predicate constructed *ad hoc* by Grue therefore cannot have as simple a relation to Grue's and Green's systems as $O_1$ has to Green's and Grue's systems, just because it was constructed *ad hoc* out of Green's system, whereas $O_1$ was not so constructed out of Grue's system. Notice that although it was claimed by Grue in the dialogue that 'wavelength' was *not* initially available in his system, he agreed that it had been *explained* to him in an objective manner by Green, and this was not by definitions in terms of Grue's predicates but by some method such as showing him a wavelength meter. The problematic $O_2$, on the other hand, *cannot* be explained to us by Grue unless he has a radically different theory from our physics.

This is not a knock-down argument against the possibility of some local and relatively closed regions of physics where *some* grue-like transformations may yield alternative, symmetrical sub-theories, and genuinely different predictions. But if there are such cases it is unlikely that the ambiguity of prediction would offend our inductive sensibilities. They would just be cases of alternative sub-theories with equal inductive support, where we do not know which prediction to consider most reasonable. What the above argument does show is that with regard to *colour*-predicates, and no doubt to any other set of alternative hypotheses where we have equally strong intuitions, our inductive expectations can be explicated by the absence of a grue-like alternative symmetrical theory. We must conclude that Grue cannot provide this by trivial construction from Green's theory.

## III. Meaning variance and entrenchment

Unless Grue has some such long and non-trivial story available Goodman's puzzle in practice is spurious. However, genuine examples of the puzzle may perhaps be found in the context of clashes between radically different fundamental theories such as have been discussed in connection with the 'meaning variance' dispute. Whether there have ever been any pairs of such fundamental theories which could be stated symmetrically as required by the puzzle is doubtful, but is in any case a historical question. What Goodman has shown—and it is a fundamentally important insight—is that if there were such pairs of conflicting theories, they would be confirmationally incommensurable. Each theory would determine its own list of primitive predicates, and hence the initial probability distribution obtained by symmetry of primitive predicates, and consequent comparison of the posterior confirmation even on the basis of the same evidence, would be worthless. Consider again the situation of Green and Grue, and suppose now that Grue has his radically different theory, but that he has been brought up in the same language community as Green, and has therefore learnt all his descriptive terms, including 'green' and 'blue', in the same way as Green. In other words, whenever Grue describes already examined evidence he uses the words 'green' and 'blue' whenever Green does. But, of course, he 'means' by 'green' what we previously defined him to 'mean' by 'grue', because all his predictions and his confirmation theory correspond to the grue-theory. We might well say this is a case of radical meaning variance between the two theories, because terms used in common in ordinary descriptive talk yield quite different inferences, and hence are clearly being used 'in different senses' depending on which theory is presupposed. In so far as the puzzle about meaning variance is a puzzle about incommensurability of theories with respect to confirmation and testing, it is identical with the non-trivial form of Goodman's puzzle.[1]

It must be concluded that when Goodman's puzzle is genuine it is insoluble, but its 'solution' in practice lies in the fact that our inductions rarely involve such fundamentally conflicting theories, and when they do it is even rarer to find the theories perfectly symmetrical in all relevant respects. If the true context of the puzzle is to be found in fundamental theory conflict, it seems that Goodman's own pragmatic solution in terms of 'entrenchment' of predicates is misleading. His suggestion is that predicates are inductively projectible in proportion to the number of times they, or predicates extensionally equivalent to them, have actually been projected in generalizations and predictions in the past. By 'extensionally equivalent predicates' must be meant, of course, predicates giving the same predictions, not predicates extensionally equivalent only with respect to past descriptions, because this would not distinguish 'green' from 'grue'. Goodman denies that this makes entrenchment equivalent to familiarity, or that it forbids introduction of new predicates, because newly introduced predicates may be extensionally equivalent to old ones.[2] He develops some further properties of entrenchment

---

[1] Although if the meaning variance problem is taken also as a problem about incommensurability of meaning or non-translatability (*cf.* Feyerabend, 'Explanation, reduction, and empiricism'; Kuhn, *The Structure of Scientific Revolutions*; Quine, *Word and Object*), it is 'deeper' than Goodman's puzzle. Green and Grue have no *translation* problems, only inductive ones.

[2] '. . . we must continue to be on guard against throwing out all that is new along with all that is bad' (*Fact, Fiction and Forecast*, 97).

to deal with newly introduced specific predicates and hypotheses which inherit entrenchment from old hypotheses ('overhypotheses') of which they are special cases. There appears to be some doubt whether his detailed criteria for elimination of less entrenched predicates and hypotheses in this wider sense can be made to work,[1] but even if they can, it seems that appeal to entrenchment involves general assumptions about the historical sequence of theories which are not necessarily acceptable.

Consider once more the situation of Green and Grue, and suppose that they do indeed have equally supported comprehensive theories which give genuinely different predictions about whose crucial tests they can agree. We may assume that Green's predicates and hypotheses (that is, ours) are in fact better entrenched according to all Goodman's criteria, because if Grue's theory is radically different from Green's there is no guarantee that it inherits from Green's theory any considerable entrenchment even of 'overhypotheses'. (Did Newton's theory inherit any such entrenchment from Aristotelian physics?) Now it does not at all follow that Green's theory should be preferred to Grue's merely because it is better entrenched, because other considerations should certainly be given priority over entrenchment. It is in fact inconceivable that such a pair of theories could ever be said to be symmetrical in all the respects required for a non-trivial statement of Goodman's puzzle. Thus appeal to entrenchment should never be necessary, and to put emphasis upon it without investigating the full theoretical ramifications of the puzzle itself is to betray an unacceptably restricted and conservative view of the progress of science.

Finally, two points need to be made about this proposed solution of the paradox. The first is an objection to the proposal on the grounds that it appeals to a system of *physics*, which cannot enter the intuitions of non-scientific persons, and yet it is quite clear that these people would unhesitatingly opt for the green hypothesis. However, appeal is made to physics only because physical tests are suggested by the need to find predictions related to colour in Goodman's example, which are independent of problematic 'perceptions' of green and grue, and because it is well accepted in physics that terms are interrelated in a network of laws and that alternative sets of tests are available for them. But in the light of the network model of universals this is seen to be true in principle of *all* descriptive terms whether in science or ordinary language. The reason why the layman intuitively projects green is that no alternative system of classification in which grue becomes a primitive has ever been suggested in his education. There is at present no such workable alternative system. Goodman appeals to this fact himself in claiming that the solution of his paradox lies in the greater 'entrenchment' of green in the language, but he does not carry the point far enough. That green is entrenched

---

[1] H. Kahane, 'Goodman's entrenchment theory'; P. Teller, 'Goodman's theory of projection', *Brit. J. Phil. Sci.*, **20** (1969), 219.

is not an isolated or accidental feature of the language, but follows from the fact that the term is entrenched in a network of common expectations or low-level laws, and that there is currently no alternative network of such laws. If grue had been entrenched in a workable system instead of green, it might have been a primitive of the language, and 'All emeralds are grue' the reasonable induction. 'Deciduous' is a term in our network which need *not* be explained in terms of more primitive colour-words—it might be learned by someone only visiting the temperate regions in summer, who recognized permanent features of vegetation other than colour which are associated in our network with 'deciduous'. Alternatively, grue might have been *equally* entrenched with green, and then we would be in genuine uncertainty over what are the reasonable inductions to make about the terms green and grue, just as we now are about, for example, 'peace-loving'—does it imply expectations of non-violence or of militancy?

This brings us to a second point about Goodman's paradox. It may be said in immediate response to the last example that what we are uncertain about in regard to 'peace-loving' is not facts about the world but the *meaning* of the term. What the grue example shows, however, is that this distinction cannot easily be made, and this is exactly what we would expect if the network account of universals is correct. For facts believed about the world are captured in the total network of correspondence and coherence conditions, and may be reflected in the case of some universals in terms of their 'meanings', that is, in terms of what dispositions to recognize the universal it has been found convenient to learn. What grue shows is that these dispositions may be present in different ways even in a language that refers to the same facts and apparently has common meanings as between speakers. For suppose the grue speakers in fact used the vocable 'green' for 'grue'. Then no observation before $T$ would reveal that they are different from green speakers, unless they specify the predictions they make about those observations after $T$ which are different in ways not involving the term 'green' from predictions made by green speakers. But no one ever reveals *all* his expectations. Hence I am never in a position conclusively to know whether my fellow English speakers are using words with the same 'meaning' that I am. The notion of 'same meaning' is at best a behavioural approximation in well-tried contexts, where even the notions of 'natural kinds' and 'observability' can be given content. But when the dynamic characteristics of a natural language are taken into account, there is no guarantee that a logic which demands substitutability of identical terms without change of meaning will be applicable. The indeterminacy of truth and meaning values which infects the network model is far from being an objection to this model. On the contrary, it is a requirement of any model that aims to reflect a natural language, and the grue paradox is important precisely in so far as it exhibits this indeterminacy.

# CHAPTER FOUR

# The Logic of Induction as Explication

## I. Hume's legacy

The deductive model of scientific theory structure, with its overtones of the logician's or mathematician's standards of validity, has served well a philosophy that thought it had learned from Hume that there is no logic except deductive logic, and that saw science as a static, hierarchic ordering of statements resting on a firm base of observational truth. This model, however, is highly inappropriate to the dynamic conception of a theoretical network such as has been described in the last three chapters. For firstly, it implies an ordering in which the relative distinction between theoretical and observation statements coincides with the order of logical deducibility. But to construe the distinction in this way becomes very strained in some examples. Much experimental effort in microphysics, for example, is directed towards 'reading off' the values of fundamental constants from observational set-ups—for example, charges of particles from their tracks. In the deductive reconstruction, however, the ascription of a value to the electron charge comes 'high up' at the theoretical end of the deductive ordering. Conversely, some deductive consequences of theoretical systems are less 'observable' than their premises— for example, the behaviour of a complex synthetic molecule deduced from the wave functions of its constituents. Secondly, the deductive model implies a natural starting point of the *modus tollens* falsifying argument, and this starting point must be assumed given. But if all statements of the system are in principle corrigible, then no such starting point is given. Thirdly, the deductive model implies that valid logical inferences are directed only from theoretical premises to observational conclusions, and are deductive; but this misrepresents both the pragmatic nature of the distinction between theory and observation, and also the fact that many inferences in science are inductive and analogical in character, and move from observation to theory, particular to general, and particular to particular.

Even in what has been taken to be typically deductivist literature, the need has been recognized for some account of a theory of inductive confirmation of theories and predictions, to supplement the deductivist account. For example, Hempel and Oppenheim's 'Studies in the logic of

D

explanation'[1] is the *locus classicus* for the 'covering-law' or deductive model of explanation for non-statistical hypotheses. But Hempel had earlier written 'Studies in the logic of confirmation',[2] in which he asserted

> Unquestionably scientific hypotheses do have a predictive function; but the way in which they perform this function, the manner in which they establish logical connections between observation reports, is logically more complex than a deductive inference.

More recently Hempel reinforced this assertion by noticing that the deductive model of explanation itself stands in need of a theory of confirmation. In 'Aspects of scientific explanation' he replaces his earlier requirement that an explanans must be *true* (which we can never know) with the requirement that it be *'more or less strongly supported or confirmed* by a given body of evidence'.[3] But the notion of confirmation is not further analysed in this paper, and without such an analysis it cannot be claimed that necessary and sufficient conditions for a set of postulates to constitute a potential explanation have been given.

Another example comes from a discussion of empirical meaning by Wesley Salmon, who has consistently maintained the need for a theory of induction and related this need to typical problems in the deductivist analysis of science. Salmon suggests that the verifiability criterion of meaning could be revived for scientific theories if it were explained in terms of inductive as well as deductive relations.

> . . . we might say, roughly, that a statement . . . is empirically verifiable if and only if it is either an observation statement or the conclusion of a correct inductive or deductive argument from verifiable premises.[4]

Again, however, the 'rules of inductive inference' required have yet to be specified in terms general enough to deal with complex scientific theories.

Thus it appears that many of the difficulties which beset the deductive analysis of theories, whether concerned with meaning, prediction or explanation, can be traced to the absence of a satisfactory confirmation theory yielding inferences in the inductive as well as the deductive sense. On the other hand, Hume's legacy puts a question mark against the suggestion that we can have a logic of scientific inference in which induction supplements deduction. Hume argued that we cannot show that there are cogent logical reasons, in the sense of deductions from indubitable premises, for accepting inductive inferences. He also considered and rejected the possibility of show-

---

[1] First published 1948, reprinted in *Aspects of Scientific Explanation* (New York and London, 1965), 245.

[2] First published 1945, reprinted in *Aspects*, 3. The quotation is from p. 28.

[3] *Aspects*, 338; Hempel's italics.

[4] 'Verifiability and logic', *Mind, Matter, and Method*, ed. P. K. Feyerabend and G. Maxwell (Minneapolis, 1966), 360.

ing that inductive inferences are *probably* rather than deductively true, given the evidence.[1] For however we interpret the notion of 'probability' in '*p* is probable given *q*', some non-probabilistic premise is required to be known before this claim can be made, and this premise itself will not be deducible from the evidence, and will therefore need independent inductive justification. For example, given that the great majority of living Italians are dark-haired, it is probable that the first child born in Milan on 1 January 1980 will be dark-haired, but in order to make this inference we need the premise 'If the incidence of a character in a sample is high, the probability of its occurrence in another instance of the same population is high', and to make the inference logically demonstrative, this premise must be true and not merely probable, on pain of infinite regress.

The problem of induction that Hume stated and found intractable has been called by Feigl the problem of *validation*, in contrast to a later approach, initiated by Pierce, which Feigl calls *vindication*.[2] Pierce, and subsequently Reichenbach, suggested that Hume's problem should be replaced by the attempt to show that if it is possible for *any* method of predicting future occurrences to be successful, then the particular methods we call inductive must be successful. Reichenbach considers in particular the problem of arriving at the frequency of *A*s in a possibly infinite population from finite samples.[3] If the population is in fact infinite and no limiting frequency of *A*s exists, then no method will enable us to arrive at such a limit. If on the other hand the population is finite or infinite *with* a limiting frequency of *A*s, let us describe as inductive those methods which give us *convergent* estimates of the frequency as the sample size increases. Such methods will certainly be successful in the long run, although of course there is no guarantee that only these will be uniquely successful, for one might have equal success from breathing deeply and taking the frequency to be the first proper fraction one thinks of. Let us call any rule which *ensures* success in the long run a *convergent rule*. One such rule would be to take the limiting frequency to be the same as the frequency $m/n$ of *A*s in the sample of *n* obtained up to date. This is what Carnap calls the *straight rule*. Unfortunately, however, there is an indefinite number of other rules which also ensure convergence, and predictions in the short run will be greatly affected by the particular convergent rule we adopt. Indeed, for every possible prediction of the relative frequency in a finite future sample, there is some convergent rule which directs us to make that prediction. So the decision to adopt some one or other of the convergent rules gives no assurance at all about the accuracy of predictions

[1] *Treatise*, I, part III, 6, 12.

[2] H. Feigl, 'De principiis non disputandum . . . ?', *Philosophical Analysis*, ed. M. Black (Englewood Cliffs, N.J., 1950), 116.

[3] H. Reichenbach, *Experience and Prediction* (Chicago, 1938), 339f, and *The Theory of Probability* (Berkeley, 1949), chap. 11. For a detailed account and defence of Reichenbach's theory, see W. Salmon, *The Foundations of Scientific Inference* (Pittsburgh, 1966), 52f, 97f.

in the short run, which are after all what we are primarily concerned with in a finite lifetime.

Are there, however, further acceptable criteria for choosing between convergent rules? Reichenbach relied upon 'simplicity', holding that the straight rule should be preferred because it is simpler than its rivals. But this does not give any justification for that particular rule unless it can also be shown that simple rules are more likely to lead to true conclusions. It is very difficult to see how this could be shown without relying on still further principles that themselves need inductive justification. The same applies to any other criterion that might be thought of for choosing between convergent rules. Moreover, even if the straight rule is chosen, the fact that it converges to the true limiting frequency in the long run does not exclude the possibility that its value oscillates widely in the short run, and if the population is infinite, or not known to be finite, there is no way of knowing how close we are to the limit after observing a given sample, no matter how large that sample may be. Convergent rules do not therefore seem to provide a solution to the vindication version of the problem of induction in the sense Reichenbach intended, for such a solution would certainly require good reasons for adopting a unique member of the set of convergent rules. The straight rule is, however, an important contribution to confirmation theory, and we shall return to it.

Since little success seems to have attended attempts to solve the problems either of validation or of vindication, some have drawn the sceptical conclusion that the problem of induction is in principle insoluble, or is a pseudo-problem. An influential example of the latter view is Strawson's argument[1] to the effect that there is no problem about the rationality of induction, any more than about the rationality of deduction. In circumstances where an inductive conclusion is required, it just *is* rational to follow our normally accepted inductive rules. We do not ask of a deductive inference such as 'If $2 + x = 4$, then $x = 2$' any justification other than the rules of arithmetic which are used in the inference. Similarly, it is suggested, we should not ask of an inductive inference any justification other than the rules of induction themselves. To conclude from observation of a large number of black ravens and no non-black ravens that all or most ravens are black *is* to infer rationally; no further external justification is possible or necessary.

The comparison between deduction and induction is instructive, and is itself sufficient to show that this attempt to dissolve the problem of induction is inadequate as it stands. In the first place, the rules of induction are not actually well formulated, as those of, say, propositional logic or arithmetic are. Moreover there is no agreement about how they should be formulated. Are we, for example, to take eliminative induction more seriously than enumerative induction? Is the straight rule adequate as a predictive method in some

---

[1] P. Strawson, *Introduction to Logical Theory* (London, 1952), chap. 9.

statistical problems, and if so, which? All these and many other questions are matters of controversy in inductive logic and statistical inference. With regard to inductive logic we are in a position which would only be paralleled in deductive logic if we imagine the multiplication table and the rules of syllogism to be unformulated and controversial, and logical and arithmetical arguments to be carried on according to a variety of unanalysed and mutually inconsistent procedures.

Again, it is certainly misleading to claim that *no* kind of justification is sought for the rules of deductive logic. The history of foundational studies since Frege, Russell and Hilbert is itself a refutation of this claim. That it is admittedly impossible to avoid using deductive methods in the discussion of deductive methods is beside the point. It may be that discussion of inductive methods will involve inductive methods, but this does not necessarily make the construction of an inductive logic valueless, any more than mathematical logic is valueless. There are more ways than one of taking the 'problem of induction', and to take it as demanding logical analysis of the same general kind as that already developed for the 'problem of deduction' is one which remains untouched by Strawson's attempted dissolution.

Another well-known attempt to dismiss the problem of induction is Popper's.[1] For Popper there is no logic save deductive logic, and therefore no *logical*, or even philosophical, problem of induction, since Popper accepts Hume's conclusion that induction cannot be reduced to deduction. There may be a remaining *psychological* problem, concerned with how people in fact pass from evidence to theory or from evidence to prediction, but this is a subject for the science of psychology, not the philosophy of science. According to Popper, the logician can give no other account of inductive inference than to assimilate it to conjectures or guesswork. Evidence *causes* us to have certain general expectations upon which we may act. We can have no *reasons* for these expectations; the best we can do when they have been conjectured is to subject them to 'severe tests' which will falsify them by the deductive rule of *modus tollens* if they are false. But universal generalizations can never be shown to be true or probable, even if they successfully pass every test we can devise for them, and therefore they remain conjectures upon which we cannot, as a matter of logic, place any reliance at all. We can compare one generalization with another only in respect of its *past* success in standing up to tests, and this gives no rational indication of *future* success, much less of universal truth.

Popper does indeed introduce a concept of confirmation or, as he prefers to call it, *corroboration*. This is a property of theories which have the following characteristics: they are highly falsifiable and not *ad hoc*, and they have survived serious and risky tests designed to refute them if they are false.

---

[1] *Logic of Scientific Discovery*, chap. 1 and *passim*.

Corroboration, if measurable at all, would not be a probability measure, but would rather depend on the *inverse* of probability, since the corroboration of a theory varies directly with its content, and its content inversely with its probability. Corroboration is in any case not a measure of the reliability or approach-to-truth of theories. Regarding the possibility of defining a measure of corroboration, Popper says

> I do not believe that it is possible to give a completely satisfactory definition. My reason is that a theory which has been tested with great ingenuity and with the sincere attempt to refute it will have a higher degree of confirmation than one which has been tested with laxity; and I do not think that we can completely formalize what we mean by an ingenious and sincere test.[1]

Popper also introduced a concept of 'verisimilitude', which he calls a measure of 'approach to truth'. In his article 'Truth, rationality, and the growth of knowledge', he introduces the notion with the remark

> Looking at the progress of scientific knowledge, many people have been moved to say that even though we do not know how near or how far we are from the truth, we can, and often do, *approach more and more closely to the truth*.[2]

He goes on to argue that such talk is perfectly respectable if we understand 'approach to the truth', or 'degree of correspondence to truth', in terms of the *truth-content* of a theory. This indeed sounds like a claim to deal with the problem of induction. But it immediately becomes clear that such is not Popper's intention. After defining the truth-content $Ct_T(a)$ of a theory $a$ as the class of its true logical consequences, and the falsity-content $Ct_F(a)$ of $a$ as the class of its false consequences, he states the criteria for the notion of approach to truth or *verisimilitude* $Vs(a)$ as

(a)　$Vs(a)$ should increase if $Ct_T(a)$ increases while $Ct_F(a)$ does not, and
(b)　$Vs(a)$ should increase if $Ct_F(a)$ decreases while $Ct_T(a)$ does not.

But this is admittedly a *metalogical* definition—as a putative measure of verisimilitude it makes no mention of *evidence*, that is, of how many consequences of $a$ have in fact been found to be true. Popper makes a point of distinguishing the question 'What do you *intend to say* if you say that the theory $t_2$ has a higher degree of verisimilitude than the theory $t_1$?' from the question 'How do you *know* that the theory $t_2$ has a higher degree of verisimilitude than the theory $t_1$?', and he goes on, 'We have so far answered only the first of these questions'.[3] Indeed, Popper draws the correct conclusion from his definition of verisimilitude when he says 'our preferences [relative appraisals of verisimilitude] need not change . . . if we eventually refute the

---

[1] *Conjectures and Refutations*, 287–8. See also *ibid.*, 36, 280f, and *Logic of Scientific Discovery*, chap. 10 and appendix *ix.
[2] *Conjectures and Refutations*, 231; see also 391f.
[3] *Ibid.*, 234 (my italics).

better of the two theories',[1] and he gives the example of the preferability of Newton's dynamics, though refuted, over Kepler's and Galileo's theories. That $t_2$ has a higher verisimilitude than $t_1$ does not even exclude the possibility that $t_2$ may be known to be false, while $t_1$ is, as far as is known, true. The concept of verisimilitude is clearly not intended as an explication of what I have called confirmation or degrees of belief in a theory, and it is quite irrelevant to the epistemological problem of induction.

Objections can be made to Popper's view on the grounds that it is impossible even to state it without making some inductive assumptions. For example, it is not clear that the notion of a 'severe test' is free of such assumptions. Does this mean 'tests of the same kind that have toppled many generalizations in the past', which are therefore likely to find out the weak spots of this generalization in the future? Or does it mean 'tests which we should expect on the basis of past experience to refute this particular generalization'? In either case there is certainly an appeal to induction. Again, one past falsification of a generalization does not imply that the generalization is false of *future* instances. To assume that it will be falsified again in similar circumstances is to make an inductive assumption, and without this assumption there is no reason why we should not continue to rely upon all falsified generalizations.

Again, Popper argues that the application of any general term to an object carries implications that are themselves universal generalizations.[2] In other words, there are strictly speaking no singular statements, because when I assert, for example, 'there is a glass of water in my hand', I am tacitly asserting along with this an indefinite number of further claims: 'If I drop this on the floor it will break', 'the contents of the glass have mostly the same chemical composition as all other samples of water', and so on. Thus the application of general terms shares with all other generalizations in Popper's account the character of conjecture. Our dispositions which trigger off descriptive terms in response to certain stimuli are discussable only in psychological and never in logical or philosophical contexts.

This account of general terms is not unlike that of the network model, and could be interpreted in a way consistent with that model by supposing that the burden of Popper's argument against induction is to point out the ultimate unanalysability of the recognitions of similarity upon which ascriptions of descriptive terms rest. If this is a fair interpretation, then indeed we have no logical or explicit inductive grounds for primitive ascriptions of general

---

[1] *Ibid.*, 235. In a subsequent paper ('A theorem on truth-content', *Mind, Matter and Method*, ed. P. K. Feyerabend and G. Maxwell (Minneapolis, 1966), 343), Popper argues that total logical content can be used as a rough measure of actual truth content. This has the advantage that the logical content of a theory is accessible whereas its actual truth content is not. But this development leaves the present comments on verisimilitude unaffected, because it remains, and is intended to remain, a wholly *a priori* notion.

[2] *Logic of Scientific Discovery*, 94.

terms. But this situation is not happily described as 'conjecture'; it is rather a matter of physical disposition, something which, having learned a language, we can't help. We do not *conjecture* that the daffodil we have just seen is yellow. And it is certainly misleading to assimilate the situation of primitive recognition to that of putting forward generalizations and theories, for these may involve formulatable rules, not only in inferring the theories themselves but also, as we have seen, in judging and possibly modifying the immediate results of recognition. As with Strawson's dismissal of induction, Popper's is not so much wrong as directed at the wrong target. The problem of stating and systematizing the rules according to which we make conjectures remains. If there were no such rules, not only would agreements about scientific procedure between different scientists be impossible, since they would all constantly make different 'conjectures' about even the simplest generalization, but also much intersubjective description would break down, because we would have no rules to tell us how to go beyond, and sometimes to correct, the primitive ascription of predicates for which we have acquired physical dispositions.

## II. A more modest programme

I conclude that neither Strawson nor Popper has made out a case for abandoning the 'problem of induction'. The discussion so far suggests that inductive assumptions are all-pervasive in human discourse, not only in scientific inferences from evidence to theories and predictions but also in the very use of general descriptive terms in stating the evidence itself. But we must be more specific about what constitutes an 'inductive assumption' and about what is the most profitable way of understanding the 'problem of induction', bearing in mind the failure of all attempts to solve previous versions of it.

By an inductive assumption or an inductive inference I shall mean an assertion or argument about the unobserved which has two essential features.

(i) It provides some rule or set of rules for passing from statements about evidence to generalizations or singular predictions about the as yet unobserved.

(ii) It carries some claim about the truth of such generalizations or predictions; that is to say, to put it imprecisely at present, there is some rational expectation that the conclusion of the inference corresponds to true statements about the unobserved.

Clearly the second feature is required as well as the first in order to generate any 'problem', for a mere set of rules for passing from the observed to the unobserved makes no claim and serves no function. A function other than the inductive function might of course be given to it. Plato, for instance, might be said to have used the observed motions of the planets to suggest a problem in pure mathematics, rather than to make empirically true general statements about planets, but this would merely mean that Plato was not concerned with inductive inference in the sense intended here.

As for the most profitable way of taking the problem of induction, I shall regard it as an example of what Carnap calls the *explication* of concepts, of which other examples are the explication of deductive arguments in formal logic, and the explication of various concepts of 'probability' by means of axiomatization and interpretation. The process of explication starts with some vaguely formulated concept whose rules of use are implicitly embedded in some area of ordinary language or in a semi-technical process like arithmetic. We attempt a rigorous formulation of these rules, perhaps in a formalized axiom system, as for instance in Peano's axioms for arithmetic. We may find in the process that our original understanding of the concepts was not only vague but also ambiguous or inconvenient, or even internally inconsistent. A successful explication is one that so adjusts the formalization to the original intuitive concept as to capture as much as possible of the latter, while at the same time changing it where desirable to simplify, sort out ambiguities and correct contradictions. Examples are: the relations of syllogistic and propositional logic to inference in ordinary language; formalized geometry to physical space-measures; and Kolmogoroff's probability postulates to the probability talk of ordinary discourse, gambling and statistics. In many such areas the resulting explication may not be unique, and may remain a matter of dispute among logicians and among the 'ordinary users' of the concept concerned.

I am going to take the problem of induction in its more modest version to be the problem of *explicating intuitive inductive rules*. Thus the problem falls into two parts.

(i) To formulate a set of rules which capture as far as possible the implicit rules which govern our inductive behaviour.

(ii) To formalize these in an economical postulate system.

As has already been remarked, in the first task we are at a more primitive stage even than Peano and Frege in the formalization of arithmetic, because there are as yet no agreed inductive rules to form the subject matter of the explication. But it should not be assumed that absolutely definitive formulation of these must be arrived at before the second task can be begun, for we expect to find, in a successful explication, that light is thrown on the intuitive understanding of induction as well as upon its systematization. Indeed, we shall later see reason to question some of the maxims that have traditionally appeared in accounts of scientific induction. Meanwhile it is sufficient to remark that any explication of inductive rules ought to take account of such traditional formulations as Mill's methods and other analyses of enumerative and eliminative induction, analogical argument, elementary statistical inference, the use of simplicity criteria in induction, and the problem of inference to and from laws and theories as expressed in the deductive or 'covering law' model. Some of these formulations have also been explicitly designed by their authors to solve the problem of induction in a stronger sense, namely to show that the postulates arrived at could be justified independently of the inductive processes themselves. I shall assume, however, that a sufficient, and perhaps the only possible, justification of a set of postulates of inductive inference would be that they form a 'good explication' of the intuitive inductive rules. Justification in this sense resides in the interaction of postulates and rules, and not in any external support for the postulates independently of the rules.

As a preliminary list of the intuitive rules which should be taken into account by any adequate theory of inductive confirmation, consider the following.

(*a*) *Enumerative induction:* a generalization should in general be increasingly supported or confirmed by observation of an increasing number of its positive instances and no negative instances.

(*b*) *Variety of positive instances* in the evidence should have more effect in increasing confirmation than uniformity of instances.

(*c*) *Analogical argument:* a high degree of similarity between two objects or systems of objects gives high confirmation to the prediction of their similarity in further, as yet unobserved respects.

(*d*) *Eliminative induction:* confirmation of a hypothesis can be increased by observational refutation of an increasing number of its competing hypotheses, or more generally by reduction of the confirmation of its competitors on given evidence.

(*e*) *Converse entailment:* a hypothesis is deductively refutable by observation of the falsity of one of its logical consequences; conversely, the confirmation of a hypothesis should be increased by observation of the truth of one of its logical consequences.

(*f*) *Initially unexpected evidence:* if a logical consequence of a hypothesis is initially unexpected before the hypothesis was proposed, and is then observed to be true, it has more effect in raising the confirmation of that hypothesis than does observation of an initially expected consequence.

(*g*) *Predictive inference:* if a hypothesis is confirmed on given evidence, some further and as yet unobserved consequences of the hypothesis ought to be predictable with high confirmation.

(*h*) *Simplicity:* The simpler of two hypotheses supported by given evidence ought to be regarded as more highly confirmed by that evidence.

To propose such a list is to raise a host of issues in the logic of induction. Some of the problems that arise concern how a 'positive instance' of a generalization should be understood; whether hypotheses are ever deductively refuted; how 'simplicity' should be measured; what it is for an observation to be 'initially unexpected'; and so on. We might also add to the inductive rules requirements of 'power', 'high empirical content' and 'utility', which raise further problems about the correct interpretation of these desirable features of hypotheses and about their relation to inductive characteristics. The list is enough to indicate that the task of formulating a systematic inductive logic which does justice to all these problems has hardly begun.

## III. Probabilistic confirmation

The second requirement for an explication of induction is a postulate system in which to systematize the inductive rules. The logical system that is *prima facie* the most adaptable and best adapted to represent the scientific theory–observation network is undoubtedly probability theory. This offers the possibility of explicating inferences of the inductive type from observation to theory and from observation to prediction, and also makes it possible to loosen the deductive relations between theory and observation as these are presupposed in the deductive model, so that the notions of 'looseness of fit' and 'best theory' relative to uncertain evidence may be represented by probability functions.

Attempts to develop a probabilistic theory of scientific inference along these lines have met with many difficulties, and there have been claims to have conclusively refuted even the possibility of such a theory. Most of these claims, principally those of Popper and L. J. Cohen,[1] rest on two questionable assumptions: first, that no probability theory can assign non-zero probability to any universal generalizations, whatever the evidence, and second, that science essentially involves assertions of universal generalizations and theories in infinite domains, so that if any scientific statements are given probability in an adequate confirmation theory, then these must be. I shall investigate both these assumptions in subsequent chapters, and find them both false. Meanwhile it remains the case that probability still does not have any serious rivals in its potential comprehensiveness as a logic of science. Moreover, as I hope to show, not all its potentialities have yet been fully exploited.

To locate the present account on the map of logically possible probabilistic confirmation theories, it is necessary to distinguish it from various other approaches and from some preconceptions about what is being attacked when the whole project of a probabilistic theory is allegedly refuted. Firstly, I shall not follow Carnap in understanding probabilistic confirmation theory as a system of logical or analytic truths regarding the measure of degree of confirmation. It is more correctly described, as Carnap himself implicitly describes

---

[1] *Ibid.*, 33, and L. J. Cohen, *The Implications of Induction*, chap. 1. Popper does allow that 'degrees of corroboration' are representable by functions of probabilities, but Cohen claims to have shown that there are no functions of probabilities that will explicate the induction of causal laws. This claim has been controverted in H. E. Kyburg's review, *J. Phil.*, **69** (1972), 106. For critical comparisons of Popper's and Carnap's positions on probabilistic confirmation, see I. Lakatos, 'Changes in the problem of inductive logic', *The Problem of Inductive Logic*, ed. I. Lakatos (Amsterdam, 1968), 315; and A. C. Michalos, *The Popper–Carnap Controversy* (The Hague, 1971).

it in his more careful formulations,[1] as a system for the explication of intuitively accepted inductive and theoretical inferences. As with all explications, this does not imply that intuition is accepted blindly. It implies that, on the whole, what we take to be valid inductions are explicated as such, but it does not exclude the possibility that some intuitive types of inference may turn out to be inconsistent or useless or misleadingly specified. *A fortiori*, no claim is made for an explicatory confirmation theory that it 'solves the problem of induction', since all Hume's problems will break out again with respect to the postulates of the confirmation theory. But as I have argued, it by no means follows that the process of explication is worthless, any more than the foundations of mathematics are worthless once it is accepted that arithmetic cannot be reduced to propositional logic or otherwise 'justified'.

Secondly, it may be thought that the general problem of support for and choice between scientific hypotheses on the basis of evidence should be identified with the problem of choice between hypotheses about the statistical properties of a population on the basis of samples drawn from it. This suggestion has the advantage that it can make use of well-developed techniques and concepts of the theory of games and of statistical decision theory, as well as more conventional statistical inference. The suggestion is essentially to interpret confirmation theory as *part of* the normal inference procedures carried out in statistical theories such as those of classical thermodynamics, quantum theory and genetics, instead of interpreting confirmation theory as explicatory of those inferences. That this would be too limited a view is sufficiently indicated by the fact that many theories in science are not statistical in form, and the methods developed in statistical inference and decision theory are trivial and unilluminating when applied to theoretical postulates of universal form. This objection is not removed by the claim that, since the most fundamental theory at present available to us, namely quantum theory, is statistical, all scientific theories must therefore be at bottom statistical. For one thing this claim begs the question of whether all theories *are* reducible in principle to fundamental physics, and further, even if they are, this reduction has not yet been carried out and meanwhile we have a problem about scientific theories which are expressed non-statistically.[2]

I shall not interpret confirmation theory in this way as identical with statistical inference within science, but shall follow Carnap to the limited extent that confirmation theory will be a metalogic of science, having a status comparable to that of a logic of entailment relative to scientific theories understood as deductive systems. The degree to which one statement renders

---

[1] R. Carnap, especially in *Logical Foundations of Probability* (Chicago, 1950), appendix, and *The Continuum of Inductive Methods* (Chicago, 1951).

[2] For a development of statistical decision theory as a general account of theory choice in science, and for a contrary view to that presented in the last sentence, see Braithwaite, *Scientific Explanation*, 115f, 196f, and 'The role of values in scientific inference', *Induction*, ed. H. E. Kyburg and E. Nagel (Middletown, Conn., 1963), 182.

another *probable* in this theory can indeed be viewed as a measure of 'partial entailment'. Just as a clear distinction can be made between the logic of deductive systems and the interpretation of the entailment relation, on the one hand, and the postulates and theorems of a scientific deductive theory on the other, so a distinction will be made here between the probability logic of the confirmation theory and interpretations of probability on the one hand, and the scientific inferences to be explicated, whether in statistical or non-statistical theories, on the other. This, however, will not preclude use of some of the techniques and concepts of statistical inference in confirmation theory. Indeed, I shall examine in the next chapter proposals to model confirmation theory on various forms of statistical inference by assimilating science to 'games against nature' or 'decision making under uncertainty', and in particular some arguments to the effect that a confirmation theory *must* be probabilistic in form, which are derived from this assimilation. I shall not adopt this view, for reasons that I give in the next chapter, but rather show that a probability theory using some of the concepts of statistical inference is *sufficient* to represent important features of scientific inference, not that it is a necessary condition of explication, nor that it is unique.

To put the project of a probabilistic confirmation theory for science in a more positive light, there are three types of condition it should satisfy.

(1) It should explicate and systematize inductive methods actually used in science, without grossly distorting any of them. This will generally imply clarifications of confused inductive intuitions and of the relations between them, and in so far as such clarification is provided, confirmation theory will escape the charge of being an *ad hoc* rehearsal of intuitions which gives out no more than was put in.

(2) A confirmation theory should suggest fruitful new problems of inductive inference, and new insights into the structure of science, which may even issue in revision of the intuitively accepted methods themselves. For example, problems about criteria for confirming generalizations have led to fruitful discussions, and it will be suggested below that the probabilistic approach to these problems leads to a revised view of the function of generalizations and theories in scientific inference which is of general philosophical importance.

(3) A confirmation theory may indicate that biological learning and inference are at least isomorphic with the model of inference provided by the theory, since the existence of the model will at least show that there is a self-consistent way of satisfying certain inductive and epistemological requirements. This suggestion, however, should at present be taken only in a heuristic sense. It should be interpreted neither as giving licence to confirmation theory to dictate the results of psychological and physiological research, nor as issuing in a 'discovery machine' for future scientific progress. In its present stage of development, confirmation theory can have only the more

modest aim of providing one logically possible model of scientific inference. Even this will leave much to contingencies, such as the basic language presupposed and the presuppositions or coherence conditions of learning, which, as has already been suggested, cannot be expected to be wholly normative. There are in fact likely to be at least as many contingencies in a confirmation theory as there are in respect of actual human learning; hence no confirmation theory can be expected, at least at the present stage of research, to describe a more efficient discovery machine than the human brain itself.

# CHAPTER FIVE

# *Personalist Probability*

## I. Axioms and interpretation

In order to develop a probabilistic theory of confirmation we shall need first a convenient formulation of the probability axioms. We are concerned with expressions of hypotheses, evidence and predictions satisfying the axioms of propositional logic. These propositions will be denoted by lower-case letters such as $h$, $g$, $f$, $e$, sometimes with suffixes, to indicate that some of the terms in a probability expression will generally be interpreted as hypotheses or laws (as $h$, $g$, $f$), and some as evidence or observable predictions (as $e$). All the axioms and theorems to be discussed will, however, be valid for any substitution of $h$-type letters for $e$-type letters, and conversely, unless something is specifically indicated to the contrary. Probability will be understood as a measure over a pair of propositions; thus, $p(h/e)$ is to be read as 'the probability of $h$ conditionally upon $e$', or 'the probability of $h$ given $e$'. The *initial probability* of $h$ will be written $p(h)$, but this is to be understood consistently with what has just been said as the probability of $h$ given either the tautology, or given previous background evidence which does not enter into the current calculations.

The *probability axioms* are as follows.

(PA1) $0 \leq p(h/e) \leq 1$.

(PA2) *Equivalence:* If $h$ and $e$ are logically equivalent to $h'$ and $e'$ respectively, then $p(h/e) = p(h'/e')$.

(PA3) *Entailment:* If $e$ logically implies $h$, then $p(h/e) = 1$.

(PA4) *Addition:* If $e$ logically implies $\sim(h\&h')$, then

$$p(h \vee h'/e) = p(h/e) + p(h'/e).$$

(PA5) *Multiplication:* If $e$, $e'$ are consistent and not self-contradictory,

$$p(h\&e'/e) = p(e'/e)p(h/e'\&e).$$

Logical implication or entailment will be written $\rightarrow$.

Some immediate corollaries of the axioms will be useful.

(PC1) From the addition axiom it follows immediately that if on evidence $e$ we have a set $S = \{h_1, h_2, \ldots, h_n\}$ of mutually exclusive and exhaustive hypotheses (of which $\{h, \sim h\}$ is the simplest special case), then their probabilities on $e$ must sum to $1$ for all $e$. For the disjunction $h_1 \vee h_2 \ldots \vee h_n$ must be true given $e$, and hence $p(h_1 \vee h_2 \ldots \vee h_n/e) = 1$, and by the addition axiom

$$p(h_1/e) + p(h_2/e) \ldots + p(h_n/e) = 1.$$

(PC2) From entailment and addition, if $e$ *falsifies* $h$, that is, entails that $\sim h$ is true, we have

$$p(h/e) = 1 - p(\sim h/e) = 0.$$

(PC3) If $h_1$ and $h_2$ are independent, that is, if the truth of one has no relevance to the truth or falsity of the other, then $p(h_1/h_2 \& e) = p(h_1/e)$, and $p(h_2/h_1 \& e) = p(h_2/e)$. The multiplication axiom then gives

$$p(h_1 \& h_2/e) = p(h_1/e)p(h_2/e)$$

which can be taken as the definition of the *probabilistic independence* of $h_1$ and $h_2$.

(PC4) The equivalence and multiplication axioms also yield the following result: if $h_1$ entails $h_2$, then the probability of $h_1$ cannot be greater than that of $h_2$, given any $e$. That is

$$\text{If} \quad h_1 \to h_2, \text{ then } p(h_1/e) \leqq p(h_2/e).$$

(PC5) A rearrangement of the multiplication axiom itself is often wrongly called *Bayes' Theorem*—wrongly because, historically, Thomas Bayes proved a different expression from the multiplication axiom which should properly go by his name. Consider the multiplication axiom expressed as a relation between the initial probability of $h$ and its *posterior probability* on evidence $e$, $p(h/e)$. Provided $p(e)$ is non-zero, we have

$$p(h/e) = p(h \& e)/p(e)$$
$$= p(h)p(e/h)/p(e).$$

This is what is often called Bayes' theorem, and I shall follow this historical misusage here.

In some statistical applications, $p(e/h)$ is called the *likelihood* of $h$ on $e$. So we have, informally

$$\text{Posterior probability of } h = \frac{(\text{Initial probability of } h) \text{ times } (\text{likelihood of } h \text{ on } e)}{\text{Initial probability of } e}.$$

In deductive situations, where $h \to e$, using axiom (PA3), this expression reduces to

$$\text{Posterior probability of } h = \frac{\text{Initial probability of } h}{\text{Initial probability of } e}.$$

In general, if the initial probabilities of $h$ and $e$ remain fixed, Bayes' theorem permits us to calculate a transformation (the *Bayesian transformation*) from an initial belief distribution to a posterior distribution when evidence $e$ is observed. We shall see later, however, that it is not always convenient to assume that the initial probabilities remain fixed, and we shall consider the possibility of non-Bayesian rules of transformation.

(PC6) It is useful to notice as a corollary of Bayes' theorem that, provided neither $p(e/h)$ nor $p(e)$ are zero, if $p(h)$ is zero, so is $p(h/e)$ for every $e$, and by replacing $h$ by $\sim h$ it also follows that if $p(h) = 1$, $p(\sim h) = 0$, and hence $p(h/e) = 1$ for all $e$.

(PC7) Consider now the expression $p(e)$. Taking $S$ as the set of exhaustive and exclusive hypotheses as before, since $h_1 \vee h_2 \ldots \vee h_n$ is logically true for this set, we have the logical equivalence

$$e \equiv (h_1 \vee h_2 \ldots \vee h_n) \& e$$

Therefore, by the equivalence axiom

$$p(e) = p((h_1 \vee h_2 \ldots \vee h_n) \& e)$$

And by the addition and multiplication axioms

$$p(e) = \sum_{s=1}^{n} p(h_s)p(e/h_s)$$

or, the initial probability of $e$ is the sum over all $h$ of the product of initial probability and likelihood of each $h$.

(PC8) Bayes' theorem strictly speaking is obtained from (PC5) and (PC7).

$$p(h_r/e) = \frac{p(h_r)p(e/h_r)}{\sum_{s=1}^{n} p(h_s)p(e/h_s)}$$

for all $h_r$ $(r = 1, 2, \ldots, n)$.

The *interpretation* of the probability function to be adopted here is that of a measure of *rational degrees of belief*. This is the so-called *personalist* interpretation, which is to be distinguished on the one hand from objective (logical, statistical or propensity) probabilities, and on the other hand from merely subjective degrees of belief.

Let us distinguish it first from objective probabilities. Personalist probability is a measure of *beliefs* in propositions, and not a property of the propositions themselves. That is to say, it will not be necessary to assume that a hypothesis or other proposition descriptive of the world has either a logical probability or a propensity of being true, or a relative frequency of truth. 'The coin will fall heads' may be said to have an objective probability of truth in virtue either of an indifference distribution among the logically possible alternatives, or of its relative frequency of truth in past tosses, and either of these objective probabilities may be expressed in terms of a propensity of the coin to fall heads. Carnap's interpretation of probability in his first full-scale work on confirmation theory was a logical interpretation, assuming a principle of indifference over states of the world described in a standard language. But we have no grounds for assuming that the world can usefully be described in terms of atomic, equally weighted states even in one scientific language, let alone perennially and independently of the changes of scientific language itself. Every suggested probability assignment such as Carnap's must therefore prove itself by its consequences in explicating intuitive inductive beliefs, and not by *a priori* claims about sets of possibilities in the world. I shall examine the adequacy of his actual probability assignments for state descriptions later, but I shall interpret them as suggestions for measuring degrees of scientists' reasonable beliefs, rather than as having objective status with respect to the world.

On the other hand, the probability of propositions is not to be interpreted in a frequency sense either. It is not that the proposition has turned out true or false on a given number of occasions in the past, or that it is predicted to do so in the future. A proposition is either true or false; there is no sense in the concept of the relative frequency of its truth, although there may of course be sense in speaking of the relative frequencies with which a propositional function of given type is true of given instances. But this sense is not applicable to universal scientific laws which are asserted as true over a given domain of individuals, for if they are false once, they are false, and no truth-frequency can be ascribed to them. Reichenbach has indeed suggested that the probability of hypotheses be interpreted as the relative frequency of truth of hypotheses *of a given type* (such as inverse-square laws).[1] But this suggestion is also inapplicable, for in the first place, we never know of a hypothesis that it is true, however long it has been part of acceptable science; secondly it is by no means clear how to define a 'type' of hypothesis—is it a question of mathematical or linguistic form, or of subject matter, or what? And thirdly, and perhaps most importantly, the truth of a hypothesis in one domain (for example, inverse-square gravitation) may not be at all a reasonable index to the truth of a similar hypothesis in a quite different domain (for example, short range nuclear forces). To determine whether it does or does not have reasonable inferential relevance requires an investigation of the properties of the subject matter, not of the formal or linguistic properties of the hypothesis as such. And investigation of the properties of the subject matter is not confirmation theory, but science itself.

Personalist probability measures a *degree of belief* that a hypothesis is true, but not in a purely subjective sense.[2] The proposal to measure degrees of belief by probability was originally made in the context of games of chance, where it is clearly not sufficient that one should regard belief as a purely

---

[1] *The Theory of Probability*, 434f.

[2] For careful analyses of subjective and personalist probabilities, see the collection of essays in *Studies in Subjective Probability*, ed. H. E. Kyburg and H. E. Smokler (New York, 1964).

individual, temperamental or psychological matter, since there are objective facts (the possible outcomes of a game, for example) to which any rational man should conform his beliefs. So the concept of a *rational* degree of belief was developed, as that belief that is appropriate to the playing of games of chance whose detailed outcomes are uncertain. Specifically, it is assumed that no rational man would accept betting odds according to which he will always lose (or, as a stronger requirement, never do better than evens) *whatever* the chance outcome of the game. In order to bring to bear upon a man's subjective beliefs some objective norms to which all rational men should assent, the personalist proposes (following the original suggestion of F. P. Ramsey)[1] to interpret 'degree of belief' in a prediction behaviourally in terms of the maximum betting odds a man would be prepared to accept on the occurrence of the predicted event, provided no considerations other than his belief influence his attitude to the bet (for example, being honest and having no money, and therefore being unwilling to take *any* risk of loss however strong his belief). Then it is held that a rational man will not allow his betting system to be such that an opponent who knows no more than he does and can choose what game to play, can make a book against him which will result in his loss whatever the outcome of the bet (a *Dutch book*). A betting function is called *coherent* if it assigns real numbers in the closed interval (0, 1) to the possible outcome of the bet in such a way that if betting odds are laid on the occurrence of the outcomes in the ratio of the numbers, a Dutch book cannot be made by the opponent. It can then be shown that a coherent betting function is a probability measure over the possible outcomes, and conversely that any probability measure corresponds to some coherent betting function.

If the condition of coherence is strengthened by requiring that the odds are not such that a bettor must certainly either come out even or lose for every possible outcome (*strict coherence*), then it can be shown that the necessary and sufficient conditions are obtained by replacing the axiom (PA3) by (PA3′).

(PA3′) If and only if $e \rightarrow h$, then $p(h/e) = 1$.

With this replacement, the axioms define a *regular* probability measure.

The personalists' case for a probability theory of decision-making under uncertainty therefore rests on two proposed equivalences

(*A*) Degree of rational belief ≡ coherent betting behaviour
(*B*) Coherent betting function ≡ probability measure.

With these specifications of coherence and probability, equivalence (*B*) for betting functions has been uncontroversially justified by theorems due to Kemeny and Shimony.[2] But the probability axioms place only very weak

[1] F. P. Ramsey, *Foundations of Mathematics*, ed. R. B. Braithwaite (London, 1931), chap. 7.
[2] J. G. Kemeny, 'Fair bets and inductive probabilities', *J. Symb. Logic*, **20** (1955), 263; A. Shimony, 'Coherence and the axioms of confirmation', *J. Symb. Logic*, **20** (1955), 1.

constraints upon the measures that can be assigned to propositions, since except in the circumstances mentioned in (PA3) and (PA3'), particular probability measures are not specified. In fact, if $\{h_s\}$ is a set of mutually exclusive and exhaustive hypotheses, given $e$, then any set of measures $p(h_s/e)$ in the interval (0, 1) such that $\sum_s p(h_s/e) = 1$ is consistent with the axioms.

Justifying equivalence $(A)$ is a much more difficult matter. The suggestion that degrees of belief should be explicated as betting behaviour originally arose from a behaviourist or empiricist desire to unpack the apparently subjective notion of belief into terms of observable behaviour. The highest odds a man is prepared to offer on an unknown outcome is taken to represent his subjective degree of belief in that outcome. Adding the objective requirement of coherence, we obtain the notion of *rational* betting behaviour as an index of degrees of *rational* belief. 'Rationality' is still a very weak notion here—a man is rational in this sense if he does not deliberately court loss; it is not necessary that he use any other criteria of comparison between his beliefs. He may assign a probability 5/6 to six coming up at the next throw, or to the constitutional election of an anarchist government in Britain this year; so long as the rest of his beliefs cohere with these assignments he is rational in the personalist sense.

## II. Bayesian methods

Since any assignment satisfying the probability axioms satisfies the requirements of coherent betting and personal probability, these concepts are clearly not sufficient to determine useful probability assignments to hypotheses. We shall later have to consider how stronger conditions of scientific rationality of an intersubjective character might be imposed upon the probability framework in order to explicate particular kinds of inference which are generally assumed valid. But meanwhile let us consider some strong claims made in the personalist context, at the opposite extreme to the anti-probabilistic views of some other philosophers of science, namely that *all* decision-making must necessarily be representable by probability functions, and in particular that rational scientific inference can be known *a priori* to be in this sense probabilistic.

There is both truth and falsity in these claims. First of all, consider a weakness. It is not difficult to drive a wedge between the conditions under which games of chance are played and those under which scientific decisions are made, in spite of the natural but somewhat misleading equation of scientific decision-making with 'games against nature'. Suppose X and Y are tossing a die. X is of an optimistic turn of mind and offers even odds for each fall of the die. He might, of course, win on one or more tosses. But if the rules of the game are such that Y can propose any combination of bets on the same toss at the odds accepted by X, Y has only to place a sum $m$ against each of the six possible falls of the die to be sure of winning an amount

$$(5/6 \times m/2) - (1/6 \times m/2) = m/3$$

whatever the outcome. In the context of fair games the odds were intuitively unreasonable as well as incoherent. But what has this situation to do with 'games against nature'? Here we suppose the scientist asking himself what is his degree of belief in a hypothesis on the basis of evidence by considering how much he would bet on its truth. (We will suppose for the moment that the hypothesis is a singular prediction in order to avoid the difficulty that no bet on a universal hypothesis in an infinite domain can be conclusively settled in a finite time if the hypothesis is true.) Must his betting rates at modest stakes faithfully reflect his degrees of belief, and even if it is assumed that they must and that his degrees of belief are reasonable in an intuitive sense, must the rates be coherent? The argument from ideal games certainly does not prove either of these assumptions. The hypothesis in question is not the outcome of a closely specified situation which can be made to repeat itself in all relevant respects an indefinite number of times—it may in fact be a once-for-all prediction in conditions never again even approximately repeated. Nature is not an ingenious opponent who seeks out just those bets which will result in certain loss. And there is no clear relation between reasonable expectation of truth and the willingness to gamble on financial reward. The point does not need labouring that there is no close analogy between games and games against nature.

In a second form of the argument we may concede most of the disanalogies referred to, but redescribe the gambling situation as one in which an addicted gambler plays many different games for short and long periods picking his odds at random, and playing against a variety of opponents, both subtle and disinterested. Then, by a generalization of the first argument, although nothing is certain, his *reasonable expectation* on coherent betting odds is to come out approximately even in the long run. This form of the argument is obviously not conclusive for coherence. It refers to expectations, not certainties, and it may be intuitively reasonable on some occasions to assume expectations will not happen, and consequently to bet incoherently. We may be in a highly unexpected corner of the world, even in the long, but necessarily finite, run.

Further props can be knocked away from the equivalence of reasonable belief and coherent betting by removing the restriction to singular predictions that was implicit in the notion of an 'outcome'. Many philosophers argue that scientific laws and theories are in their essence universal in scope over infinite domains. We shall later find reason to question this, or at least to question its relevance to the logic of confirmation, but if it is accepted, then degrees of belief in laws cannot be represented by betting rates in a game in which there is a fair chance of settling the bet. Nor could the rates in any case be strictly coherent. For if we assume that a law can be falsified but never shown to be true, any bet upon its truth may either be lost or may not be settled, but it can never be won. The only odds on its truth a reasonable man would freely choose under these circumstances would surely be o : 1, for since in any case he cannot win anything, any positive sum he places on its truth may be lost, and at best he will simply not have to pay. A zero degree of belief would thereby be assigned to all laws alike, that is, the same degree that is assigned to statements known to be false. We might suggest forcing the scientist to adopt non-zero coherent odds, however small, in order to elicit his *relative* degrees of belief in different laws, but the concept of a forced bet only weakens still further the analogy between reasonable belief and betting behaviour in ideal games.

Some of the disanalogies between games and scientific decisions are removed in the general theory of statistical decision.[1] In this theory we consider finite decision problems of the following kind: given two or more conflicting hypotheses, one of which must be true, and some specified test-procedure for them, which is the best hypothesis to choose? Assume first that the likelihood of each hypothesis is defined for each possible outcome of the test. A *decision strategy* is a general instruction, for each possible test outcome, about which hypothesis to choose, and a set of decision strategies is thereby generated for all possible outcomes and all possible hypothesis choices. Decision theory is the general investigation of the properties of decision strategies and the reasons for adopting them.

At first sight it may seem obvious that the best strategy to adopt is a *maximum likelihood* strategy, that is, one which directs us to choose as a result of each possible test-outcome $e$ that hypothesis $h$ which has greatest likelihood $p(e/h)$ on that outcome. This implies that the outcome we have observed was the most probable if $h$ is in fact true, and consequently that if we choose to accept some other hypothesis $h'$ as true, the outcome we have observed was *not* the most probable given the truth of $h'$. Not to choose $h$ therefore seems, it is argued, gratuitously to expect that what we have observed was not the most probable outcome. This argument is highly inconclusive, but the maximum likelihood strategy has been favoured on other grounds as well. Fisher and other positivistically inclined statisticians have argued that likelihoods can be specified uncontroversially in relation to statistical hypotheses, whereas inverse arguments to posterior probabilities cannot, since they involve assignments of initial probabilities whose status is doubtful and whose values are in any case generally unknown.

---

[1] For introductory accounts, see N. Chernoff and L. E. Moses, *Elementary Decision Theory* (New York, 1959); R. B. Braithwaite, *Scientific Explanation* (Cambridge, 1953), chap. 7. See also R. D. Luce and H. Raiffa, *Games and Decisions* (New York, 1958), and R. M. Thrall, C. H. Coombs and R. L. Davis, *Decision Processes* (New York, 1954), especially J. Milnor, 'Games against nature', chap. 4.

It is true of course that a statistical hypothesis such as '$q$ per cent of crows are black' does in itself tacitly specify the probability $p(e/h) = q$ of obtaining a given sample of black crows by random selection. But it should be noticed that the argument does not transfer easily from statistical to universal hypotheses. If $h$ is the universal hypothesis 'All crows are black', and $e$ is a random collection of crows which all turn out to be black, then $p(e/h)$ is trivially equal to 1. But the evidence for a hypothesis is not always in the form of a 'random sample', and the evidence for universal hypotheses is not always such that the truth of the evidence can be deduced from the truth of the hypothesis. Consider $h \equiv$ 'The moon originated by being flung off the earth in an explosion while the earth was in a molten state', where $e$ is some evidence about the character of moon rocks which is neither deducible from $h$ nor, because $h$ is not a statistical hypothesis, a random sample of a population described by $h$. In such a case, assessing $p(e/h)$ is just as problematic as assessing the initial probabilities of $h$ and $e$ themselves. In fact, calculation of both posterior probabilities and likelihoods presupposes that not only initial probabilities over all hypotheses and evidence statements are known, but also that initial probabilities of all *conjunctions* of hypotheses and evidence are known, since by the multiplication axiom

$$p(e/h) = p(h\&e)/p(h)$$

We shall return later to the general problem of assigning initial probability distributions over all kinds of scientific statements. But meanwhile let us return to the problem of decision theory in statistical contexts. Even here maximum likelihood is not always the best strategy, and the arguments in its favour obscure important assumptions that are as controversial as adoption of the initial probabilities themselves.

To see this, let us first take the general decision-theoretic viewpoint and consider a strategy-set as a whole. If on the likelihood calculations a certain strategy $T_1$ gives a smaller probability of choosing the true hypothesis for *all* possible test outcomes than another strategy $T_2$, $T_1$ is said to be *dominated* by $T_2$. Any strategy which is dominated by another strategy is *inadmissible*; any which is not is *admissible*, or *rational*.

Decision theory can be expanded by taking into account *utilities* and *initial probabilities* as well as likelihoods. In most decision situation it is only in the artificial case of low stakes that choice of hypothesis should depend only upon maximizing the likelihood of correct choice. In general we want to minimize the acceptable *risk* of false choice, whether this is measured in terms of financial, emotional or some other undesirable loss. Thus our betting behaviour would not generally reflect the full degree of our belief where the consequences of losing are in some way disastrous. Someone might not choose to enter into a marriage even though the probability of its success was somewhat greater than that of its failure. These considerations can be taken into account by defining a *utility function* over the hypotheses between which we are choosing in such a way that the utility of a hypothesis reflects the value to us of the hypothesis being true. Then if we are interested in somehow maximizing expected gain or utility, we shall be concerned with maximizing the arithmetic products of the utilities with the likelihoods of correct choice of hypotheses, rather than with the likelihoods of correct choice themselves. It is not obvious, however, how the notion of 'utility' should be interpreted in relation to pure science, since at least monetary gains and losses do not seem to be directly involved. It might be held that all theories are equally valuable if true, so that utilities come out equal and can be neglected; or the notion of 'epistemic utility' may be adopted, depending on the range or content of a hypothesis, thus incorporating the view that a scientific theory is more valuable the more powerful it is.[1] Since we shall consider later how to take account of range or content in a context of probability of hypotheses, we shall meanwhile take the utilities of all hypotheses as equal, that is to say, we shall neglect them.

Initial probabilities may be introduced into the decision-theoretic framework by noticing that, on a personalist view, if we had some means of knowing that the initial probabilities of the conflicting hypotheses are unequal, and assuming the utilities are equal, we should choose the hypothesis that we most believe to be true, that is, the one that maximizes the degree of rational belief or posterior probability on evidence $e$. But if we merely calculate the likelihoods of two conflicting hypotheses, $h_1$ and $h_2$, we have by Bayes' theorem

$$p(e/h_1) > p(e/h_2)$$

if and only if

$$\frac{p(h_1/e)}{p(h_1)} > \frac{p(h_2/e)}{p(h_2)}$$

and hence the maximum likelihood hypothesis is not necessarily the one with maximum posterior probability in this case unless $p(h_1) \geqq p(h_2)$.

Initial probabilities can be taken into account, however, by simply multiplying the likelihood of correct choice on a given strategy by the initial probability that the hypothesis in question is true. Thus, although initial probabilities did not at first appear explicitly in the formulation of the decision problem, nevertheless if likelihoods of hypotheses on evidence are the only probabilities explicitly admitted, there is mathematical equivalence between the decision problem and calculations based on the assumption of equal initial probabilities.

A strategy in which initial probabilities have been taken explicitly into account is called a *Bayes strategy*, and is equivalent to the direction to

---

[1] Suggested by C. C. Hempel, 'Inductive inconsistencies', *Aspects of Scientific Explanation* (New York, 1965), 76, and developed by I. Levi, *Gambling with Truth* (New York, 1967), chap. 5ff.

maximize posterior probability as calculated by Bayes' theorem. Bayes strategies have an important role in statistical decision theory, since it has been proved under conditions of great generality that every admissible strategy is a Bayes strategy for *some* set of initial probabilities (all positive), and conversely that Bayes strategies for positive initial probabilities are always admissible. Moreover, even if an *in*admissible strategy is adopted, assuming that the initial probabilities are specified, for a finite number of decisions based on finite tests there is always some other, Bayesian, strategy with different initial probabilities, which would produce the same decisions on the same test outcomes. This means that there is a very weak sense in which it is *necessary* that any decision procedure, in science or elsewhere, should be representable by probability functions and Bayes' theorem. But in the absence of any independent specification of what initial probabilities should be assumed, this 'necessary condition' is quite vacuous.

There is a further important respect in which the arguments for coherence and admissibility establish weaker conclusions than might at first sight appear. This is that even if initial probabilities are specified, it is by no means necessary that posterior probabilities relative to given evidence should be calculated by Bayes' theorem. To see why not, consider the position of the decision-maker at a given point of time, say $t_0$, when he has a certain amount of evidence $e_0$ and a probability distribution over hypotheses. Suppose he adopts Bayes strategy $T$. $T$ tells him, for each possible combination of evidence statements he may *subsequently* observe to be true, which hypothesis he should choose, and it does this in accordance with the rule of taking the maximally probable hypothesis for each combination of evidence, which is equivalent to using Bayes' theorem. In the case of a bettor this means that he is prepared at time $t_0$ with coherent odds for any bet his opponent may suggest, conditional upon whatever evidence may be observed. Under these circumstances the bettor cannot be presented with a Dutch book, and the decision-maker does not face the expectation of loss whatever the state of nature. It is very important to stress that all this refers only to *possibilities*, none of which are yet realized at $t_0$. It refers to what the decision-maker should be prepared to choose in hypothetical circumstances *before* more evidence is collected. It describes the static situation at $t_0$ in terms of a certain probability distribution satisfying the probability axioms.

The situation is, however, quite different when more evidence is collected at time $t_1$. Denote this new evidence by $e_1$. There is now a new decision situation in which a probability distribution $p_1(h/e_0\&e_1)$ is defined over all $h$ consistent with $e_0\&e_1$. Admissibility demands only that $p_1$ be a probability measure; it does not demand that $p_1$ be related to $p_0$ by the equations

$$p_1(h/e_0\&e_1) = p_0(h/e_0\&e_1) = \frac{p_0(h/e_0)p_0(e_1/h\&e_0)}{p_0(e_1/e_0)}$$

for all mutually consistent $h_1$, $e_0$, $e_1$. In other words, $p_1$ need not be related to the initial distribution of probabilities $p_0$ in accordance with axiom (PA5). Only if it is so related is the transformation *Bayesian*.[1]

Since the coherence requirement has nothing to say about what particular distribution of initial probabilities should be adopted, it cannot prevent this distribution from being changed with new evidence. Indeed a man may change his betting rates *without* new evidence by just changing his mood, or thinking more deeply, and yet remain coherent. Coherence does not merely leave the question of initial probabilities completely open; it also allows them to be changed for any reason or no reason, in ways that do not satisfy the Bayesian transformation. Consequently we have to envisage two different kinds of situation that may face the decision-maker. He may have to choose a set of hypotheses in various domains all at once at $t_0$, when his actual evidence is $e_0$. This corresponds, for example, to the scientist's situation in predicting the outcome of a complex of future experiments. In such a case arguments for admissibility indicate that he should adopt a Bayesian strategy, and this involves calculating conditional probabilities and maximizing. But on the other hand the decision-maker may be in a situation in which he is allowed to collect evidence sequentially. Then he has not only to select a particular Bayesian strategy $T_0$ at $t_0$, but also to select strategies at $t_1$, $t_2$, ... as evidence $e_1$, $e_2$, ... comes in. These subsequent strategies may be the same as or different from $T_0$. The difference between the situations is that between prediction and learning. The learner, of course, also remains a predictor, because it is assumed that he is not merely soaking information in as time goes on, but learning from it to adapt to the future. And the learner has the predictor's problem of choice of initial probabilities, and has it renewed on each occasion upon which he receives new evidence.

We must therefore conclude that coherence alone does not provide a valid *a priori* argument for a probabilistic confirmation theory. The best that can be said for the decision-theoretic formulation is that it provides a sufficient framework in terms of which any choice-problem and subsequent decision *could* be represented, though possibly at the cost of highly implausible assumptions about utilities, initial probabilities and posterior transformations. But clearly in itself this representation is vacuous as a method of induction; for unless there are some restrictions upon the initial distribution, and some grounds for using a Bayes or some other posterior transformation rule on that distribution, the effect of evidence on the posterior distribution is quite arbitrary. Even if the overall probability distribution is required to be co-

---

[1] The non-derivability of Bayesian transformations from coherence was explicitly pointed out by J. M. Vickers, 'Some remarks on coherence and subjective probability', *Phil. Sci.*, **32** (1965), 32, and I. Hacking, 'Slightly more realistic personal probability', *Phil. Sci.*, **34** (1967), 311. For a detailed discussion see D. H. Mellor, *The Matter of Chance* (Cambridge, 1971), chap. 2.

herent or strictly coherent, that is, so long as it satisfies the probability axioms, a long run of black crows, for example, might be represented as good evidence for the statistical generalization 'Most crows are white', if only the probability of this generalization was initially high enough, or if Bayes' theorem is held to be irrelevant to the posterior transformation of probabilities. Rational decision procedures are not sufficient to capture any interesting *inductive* sense of rationality, for which further constraints must be found.

## III. Convergence of opinion

Some personalist statisticians have attempted to provide a long-run justification of all types of scientific inference by adopting a single constraint on initial distributions, namely that all hypotheses to be confirmed have a finite probability. Appeal is made to a convergence theorem for statistical hypotheses, according to which, under certain conditions that must be examined, the posterior probabilities of such hypotheses tend to 1 or 0 with increasing evidence for true or false hypotheses respectively, irrespective of the numerical values of the initial probabilities of these hypotheses, provided only that the initial probabilities of true hypotheses are not less than some small fixed value. This theorem, if relevant to scientific inference in general, would be a very powerful aid to confirmation theory, for it would discharge us from discussing the details of the initial probability distribution if only this one condition upon it were satisfied.

Attempts to apply the convergence theorem in the context of confirmation theory are, however, subject to several kinds of misunderstanding and overoptimism. Thus, Savage claims to show rather generally how a person 'typically becomes almost certain of the truth' when the amount of his experience increases indefinitely, and Edwards, Lindman and Savage remark in a well-known expository article

> Although your initial opinion about future behaviour of a coin may differ radically from your neighbour's, your opinion and his will ordinarily be so transformed by application of Bayes' theorem to the results of a long sequence of experimental flips as to become nearly indistinguishable.[1]

Assimilation of such results to scientific inference is certainly tempting, if only because it gives access to a large and familiar body of mathematical theory and technique. But personalist formulations of this kind, when applied to science, neglect the very stringent conditions on hypotheses and evidence under which the relevant convergence theorem is true, and therefore omit to

---

[1] L. J. Savage, *The Foundations of Statistics* (New York, 1954), 46; W. Edwards, H. Lindman and L. J. Savage, 'Bayesian statistical inference for psychological research', *Psych. Rev.*, **70** (1963), 197.

take account of the relevance of these conditions to typical cases of scientific inference. The second quotation also gives the misleading impression that Bayesian transformations are required for convergence of opinion. If this were the case it might be held to provide an argument for adoption of Bayesian transformations along the following lines. Inductive intuition suggests that in a satisfactory confirmation theory a true hypothesis should be one whose posterior probability tends to 1 after collection of a large amount of evidence. The convergence theorem says that this is the case under a number of necessary conditions. If these conditions included Bayesian transformations, it would be reasonable to adopt such transformations, since they would be necessary for explicating the inductive intuition. But, as we shall see, Bayesian transformations are *not* among the necessary conditions of the theorem. This indeed makes the theorem rather strong, since it is independent of choice of non-zero initial probabilities at *every* stage of collection of evidence, and not just at the first stage. However, the remaining necessary conditions of the theorem do restrict its application in scientific contexts quite stringently. We shall now examine these conditions in detail.[1]

Consider a finite set $S$ of $N$ mutually exclusive and exhaustive hypotheses $\{h_s\}$. Let $E_r$ be the conjunction of all evidence propositions $e_1, e_2, ..., e_r$ collected before time $t_r$. The convergence theorem proceeds as follows. By means of a law of large numbers it is first shown that, for $h_1$ true, the likelihood ratio $\dfrac{p(E_r/h_1)}{p(E_r/h_s)}$, for all $s \neq 1$, becomes greater than any fixed number $\eta$ however large, with second-order probability which approaches 1 as $r$ increases. Then assuming for the moment the Bayesian transformation (which will be relaxed later), we have from (PC8)

$$p(h_s/E_r) = \frac{p(h_s)p(E_r/h_s)}{\sum_s p(h_s)p(E_r/h_s)} \tag{5.1}$$

It follows at once that if $p(h_s) = 0$, no finite observation can make $p(h_s/E_r)$ non-zero, that is, if $h_s$ is judged *a priori* to be impossible, no evidence can make it probable.

Suppose, however, that $p(h_s) > 0$ for all $s$. Consider the second-order probability $P$ that $p(h_1/E_r) > \alpha$ for any fixed $\alpha$ as near to 1 as we please, if $h_1$ is true. Using the above result for the likelihood ratios, it can be shown that, after a sufficiently large number of observations yielding $E_r$, $P$ approaches 1 as $r$ increases. It follows also that the posterior probabilities of all false hypotheses of the set $S$, given $E_r$, become as near to zero as we please with probability which approaches 1 as $r$ increases.

The significance of the convergence theorem, more informally, is that, with sufficient evidence, the posterior probability of a true hypothesis is

---

[1]See L. J. Savage, *The Foundations of Statistics*, 46–50.

overwhelmingly likely to approach 1, and that of a false hypothesis to approach zero, *whatever the set of non-zero initial probabilities* when Bayesian transformations are assumed. The convergence of the second-order probability ensures not only that there is an $E_r$ such that $p(h_1/E_r) > \alpha$ for fixed $\alpha$, but also that it becomes progressively less probable that $E_r$ is freak evidence which only temporarily favours a false hypothesis. Thus, the posterior probability of a true hypothesis is not only overwhelmingly likely to approach 1, but also to remain near 1, as evidence increases. Conversely, if the posterior probability of a given hypothesis approaches 1 as evidence increases, there is only a small decreasing second-order probability that it is not the true hypothesis. Moreover, if there are non-zero lower limits on the $p(h_s)$, lower limits of these second-order probabilities can be calculated. If, however, no lower limits on the $p(h_s)$ can be stated, nothing can be said about the speed of convergence, and in any finite time *any* given hypothesis of the set will have the highest probability for *some* distribution of initial probabilities. Thus for the theorem to be any use in a finite time (that is, for it to be any use), not only must the initial probabilities be non-zero, but they must also have some fixed finite lower limit, say $\varepsilon$, as near to zero as we please.

Under this condition, the requirement of Bayesian transformation can itself be relaxed. Denote probabilities at time $t_r$ by $p_r$, and assume fixed likelihoods, that is

$$p_0(e_i/h_s) = p_1(e_i/h_s), \ldots = p(e_i/h_s)$$

The limiting property of the likelihood ratios then remains as before, but we cannot now assume that $p_r(h_s) = p_{r-1}(h_s)$. The ratios (5.1), however, now yield

$$p_r(h/E_r) > \alpha' p_r(h)$$

for any fixed $\alpha'$ near 1 and $h_1$ true, with second order probability approaching 1 with increasing $r$. But since $p_r(h_1) \geqq \varepsilon$, all $r$, this means

$$p_r(h_1/E_r) > \alpha$$

with $\alpha$ as near to 1 as we please, and we have the convergence theorem as before, but without the Bayesian assumption of constant initial probabilities. In other words, as long as the lower limit of $p(h_1)$ is fixed, however small it is, the posterior probability provided by sufficient evidence for the true hypothesis will approach maximal probability.

Now let us investigate the conditions of the theorem in more detail and in relation to a set of scientific hypotheses. The conditions are as follows.

(i) $p(h_s) \geqq \varepsilon > 0$, all $s$. The possibility of satisfying this condition for scientific hypotheses will have to be investigated later on; meanwhile we assume it satisfied.

(ii) No two hypotheses $h_s$, $h_s'$ yield the same value of $p(e_i/h_s)$ for all possible

$e_i$. Or, if this condition is not satisfied by different hypotheses $h_s$, $h_s'$, then only the disjunction of $h_s$ and $h_s'$ can be confirmed by any $e_i$.

(iii) Given a particular hypothesis $h_s$, the probability of making the test which yields $e_i$ is independent of the particular order in which it and other evidence are observed. This will be called the *randomness assumption*. It is a substantial assumption in scientific contexts, because it is often known to be false, and usually not known to be true; for it contemplates a situation in which we know nothing about the conditions which differentiate one observation from another. If, for example, the structure of the universe includes the true hypothesis 'A proportion $p$ of aircraft are jets', and we know that we are sampling aircraft, and *this is all we know*, then the probability that this aircraft is a jet is $p$ on any occasion of observation. But the situations in which we wish to use the convergence theorem are unlikely to be so simple, because we are likely to want to confirm $h_s$ on evidence which is known not to be random in this sense. This might be because we are able to sample only over a limited part of the spatial and temporal domain of $h_s$, or only over certain sorts of instances $h_s$ (the 'observable' as opposed to the 'theoretical' instances, for example). More obviously, there is something paradoxical about a requirement of random sampling in the context of scientific experiment, because it is universally accepted that science requires deliberately structured experiments, *not* random observation, and that this requirement is precisely what differentiates science from statistics. The difference of method must be regarded as a *consequence* of the non-applicability of many statistical assumptions in the domain of science, not as its cause. Its cause lies rather in the inaccessibility of the whole domain to random sampling, and in the desire to reach conclusions in as short a run as possible. That this is so is shown by the fact that where the random sampling assumptions seem reasonable or even inescapable, as in social surveys, star counts, genetic investigations and the like, statistical methods form an integral part of scientific procedure. Where these assumptions are not reasonable, convergence is inapplicable and structured experiment is required.

(iv) The $e_i$s are independent given $h_s$. That is, for all $s$

$$p(e_1 \& e_2/h_s) = p(e_1/h_s)p(e_2/h_s).$$

I shall call this the *independence assumption*. Like the randomness assumption, it is not always valid in scientific contexts. It implies that the probability of observing a particular piece of evidence is unaffected by having already observed particular other pieces of evidence, an assumption that is satisfied, for example, by 'random sampling with replacement' of balls out of a bag containing a finite number of balls. But, on the contrary, scientific experiments are often deliberately designed in the light of experimental evidence already obtained. Rather than conducting a large number of independent tests of the same hypothesis, scientific hypotheses are often themselves modified in the light of evidence, and new tests devised for the new hypothesis, with the results of previous evidence assumed. Especially where the hypothesis is universal rather than statistical in form, and may therefore need modification as a result of *falsifying* evidence, a limited number of experiments structured in the light of previous evidence, and directed to progressively modified hypotheses, is a much more efficient procedure than blindly conducting a long series of independent tests of the same hypothesis, such as might be appropriate for statistical hypotheses concerning large populations.

Comparison of 'convergence of opinion' with the conditions of scientific inference therefore yields the following results: where the convergence theorem applies there is indeed convergence of opinion with regard to hypotheses which have non-zero probability, and this convergence is independent of the actual distribution of initial probabilities and of Bayesian transformations. But the conditions of the theorem, especially the randomness and independence assumptions, are not valid for typical examples of scientific inference in limited domains and from controlled experiments.

We have now investigated two attempts to assimilate scientific inference to standard statistical methods. The decision-theoretic approach shows that, in a very weak sense, any confirmation function for scientific hypotheses must be equivalent to a probability function, but the coherence constraints provided by this approach for specification of the probability distributions are inductively vacuous. Convergence of opinion, on the other hand, seeks to provide, over and above coherence, minimal constraints that are sufficient to yield inductive inference, but this method turns out to be inapplicable to most kinds of scientific reasoning. Later we shall see how further constraints can be placed upon the probability distributions in such a way as to provide sufficient (though not necessary) explications of scientific confirmation. Among these constraints will be the adoption of Bayesian transformations, but since we have seen that these are not a necessary condition for either coherence or convergence, we shall now digress to examine some alternative proposals.

## IV. Non-Bayesian transformations

We have seen that the mere fact of coherence provides no constraints on the initial probability distribution of beliefs, or on the transformations from initial to posterior probabilities at any stage of accumulation of evidence. The weakest response to this situation is the view that the succession of probability distributions over hypotheses held by a believer is a purely *physical* or *causal* matter of belief-response to stimuli deriving from the evidence. If *no* normative constraints are put upon belief distributions other than coherence, this is equivalent to adopting an inductively arbitrary, and arbitrarily changing, initial probability distribution as a result of each new piece of evidence, and cannot in itself yield inductively or predictively useful beliefs except by accident. We shall consider later what constraints should be put upon the initial distribution itself, but first it is necessary to discuss arguments for and against Bayesian transformations, assuming that an initial distribution has been specified at $t_0$.

The first point to be made is that no confirmation theory is possible unless there is *some* transformation rule for passing from initial to posterior distributions on reception of evidence. For a confirmation theory is required to specify how beliefs should change with changing evidence, and to answer such practical questions as 'Which experiments *will* most usefully test a given hypothesis?' (and therefore which is the best experiment to do if practicable), and 'Which hypothesis *will* be most acceptable on the basis of specified future evidence?'. Answers to such questions require that a transformation rule be specified on the basis of the *present* distribution, which will yield reasonable future distributions under various sets of circumstances not yet realized. But so long as there is some such rule, it does not yet follow that it must be Bayesian.[1]

One of the most cogent objections to Bayesian transformations is that they presuppose that the evidence statements upon which posterior distributions are conditionalized are *taken to be true*, in the sense that such statements have probability unity after they have been observed. That is, after observation of $e_r$, the initial probability of $e_r$ itself is necessarily replaced by the posterior probability $p(e_r/e_r) = 1$, and this probability remains constant throughout subsequent collection of evidence. A radical departure from Bayesian theory which originates from rejection of this assumption of incorrigible evidence has been made by Richard Jeffrey.[2] Jeffrey considers situations in which we may not be quite sure whether we have observed $e_r$ or not; we may be slightly more sure of its truth than before, but the light may have been bad, and we may wish to leave open the possibility that at some future time we may be led to withdraw $e_r$ from the evidence under pressure of other evidence, or at least to give it probability less than 1. A Bayesian might attempt to accommodate

---

[1] An attempt has been made by Abner Shimony ('Scientific inference', *The Nature and Function of Scientific Theories*, ed. R. G. Colodny (Pittsburgh, 1970), 79) to provide an *a priori* argument for Bayesian transformations by examining a purely analytic derivation of the probability axioms which does not depend on the notion of coherent betting. But Shimony concludes that the analytic assumptions required, weak as they are, strain our intuitions about belief distributions too far to be cogent *a priori*, and he suggests that the justification of Bayesian methods must rather rest on examination of their consequences and the consequences of alternatives. In another analytic investigation, Paul Teller has shown that the Bayesian transformation from a belief distribution at $t_0$ to a distribution at $t_1$ is equivalent to the assumption that if $A \rightarrow E$, and $B \rightarrow E$, and $A$ and $B$ have equal initial probabilities at $t_0$, and $E$ is the new evidence observed at $t_0$, then $A$ and $B$ have equal probabilities at $t_1$ ('Conditionalization, observation and change of preference', forthcoming). This equivalence greatly clarifies the nature of Bayesian transformations, but the equivalent assumption is still not so cogent *a priori* as to constitute an analytic proof of Bayesian transformations.

[2] R. C. Jeffrey, *The Logic of Decision* (New York, 1965), chaps. 11, 12. I have somewhat modified Jeffrey's notation in what follows.

this idea by allowing withdrawals as well as accumulations of evidence in his sequence of posterior distributions, but he cannot deal in this way with 'partly observed' evidence to which it is appropriate to give a probability value between 0 and 1. Or he might attempt to get beneath the uncertain evidence to something more conclusive which can be given probability 1 and can communicate a higher probability still short of 1 to the uncertain evidence. But any such attempt would be reminiscent of the search for more and more basic and incorrigible observation statements, a programme which we have already seen reason to reject.

Jeffrey's suggestion is a radical personalism in which not only is the initial distribution a physical rather than a rational bequest, but also every subsequent distribution shares this physical character. As we go around with our eyes more or less open, our degrees of belief in various propositions are physically caused to change. We usually cannot specify any evidence that can be asserted conclusively; thus the sequence of distributions is best not specified by conditional probabilities, but merely by suffixes indicating the order of the sequence. The distribution $p_r(h_s)$, for example, is the equivalent of the posterior distribution at $t_r$, and this distribution includes such probabilities as $p_r(e_j)$ with values less than 1, which are physically related to the uncertain and inexpressible results of observation. Observation-type propositions are distinguished from hypothesis-type propositions by being those propositions in which the changed probabilities physically *originate*. Changes are disseminated over the distribution for hypothesis-type propositions in such a way as to preserve coherence, according to the set of equations

$$p_r(h_s) = p_r(h_s/e_j)p_r(e_j) + p_r(h_s/\sim e_j)p_r(\sim e_j)$$

where $e_j$ is the only evidence-type proposition whose probability has changed, or according to a corresponding set of equations in the general case where several probabilities of evidence have changed. Jeffrey does require that in the sequence of distributions the conditional probabilities such as $p_r(h_s/e_j)$ remain invariant, so that for these probabilities the $r$-suffix can be dropped, and the above equation becomes

$$p_r(h_s) = p(h_s/e_j)p_r(e_j) + p(h_s/\sim e_j)p_r(\sim e_j) \qquad (5.2)$$

Jeffrey seems to introduce this condition as just a device for distinguishing between those propositions which are directly affected by observation and those which are not.[1] If $p_r(h_s/e_j)$ does not satisfy the stability condition for some $h_s$, then this $h_s$ must be included among those propositions in which changes of probability originate. It might be objected that if $h_s$ is hypothesis-like, it is contrary to the spirit of the physical account (which is an attempt

---

[1] In private correspondence, which he kindly permits me to quote, Jeffrey tells me that he intends the condition just as one which defines an interesting situation, *viz.* that in which the observer judges that he made an observation directly relevant to $e$ and to nothing else.

E

to retain a vestige of empiricism) to suppose that our belief in a *hypothesis* can be directly affected by observation in a way not directly related to the probabilities of evidence for that hypothesis. But even if it is accepted that there must be some rule for dissemination of probability-changes from evidence to hypotheses, why choose the invariance of conditional probability, and not for example the likelihood $pr(e_j/h_s)$, or some other function of the $pr_{-1}$-distribution?

Jeffrey's attempt to deal with the problem of uncertain evidence is courageous and ingenious. Unfortunately it presupposes only coherence constraints upon the sequence of distributions, and could hardly form a sufficient basis of a confirmation theory. First, it should be noticed that in Jeffrey's theory, if (5.2) holds and $pr(e_j) \neq pr_{-1}(e_j) \neq 1$, then the likelihoods of statistical hypotheses with respect to $e$ must be assumed to change. For we have

$$pr(e_j/h_s) = pr(e_j)p(h_s/e_j)/pr(h_s)$$

and by (5.2) this is not equal to

$$pr_{-1}(e_j/h_s) = pr_{-1}(e_j)p(h_s/e_j)/pr_{-1}(h_s)$$

Some personalists and most objectivists have regarded the likelihoods of statistical hypotheses as the most 'objective' of the various probability functions entering Bayes' theorem.[1] Of course, Bayesians themselves replace $pr_{-1}(e_j/h_s)$ by $pr(e_j/h_s\&e_j) = 1$ when $e$ is said to have been observed. But for them there is at all times $t_r$ a constant conditional likelihood $p(e_j/h_s)$ which is dependent only on the statistical hypothesis $h_s$ and a random sample $e_j$, whether or not $e_j$ has been observed at $t_r$. This constant likelihood cannot be expressed by Jeffrey, because all his probabilities, including likelihoods, are in effect conditional upon unexpressed (and inexpressible) evidence. The probability of $e_j$ on $h_s$ is well-defined only if the truth of $h_s$ is the only factor upon which $e_j$ is conditional. Where the value of $pr(e_j/h_s)$ in Jeffrey's theory differs from this 'objective' likelihood, $e_j$ must be construed as a sample conditional not only upon $h_s$ but also upon some other unexpressed evidence, for example that the sample is not random, or has been only partially observed. It must be conceded to Jeffrey that such non-random sampling is nearer to practical scientific experimentation than the ideal situations of statistics, as indeed I have already argued.

A more serious drawback to Jeffrey's proposal from the point of view of confirmation theory, however, is that it allows no control over confirmation relations. It is impossible to calculate what the effect of $e_j$ upon the probability at $t_{r-1}$ of $h_s$ will be when $e_j$ has been 'partially observed' at $t_r$. The conditional probability $p(h_s/e_j)$ bears no regular relation to the new $pr(h_s)$, because the physical observation of $e_j$ has an unpredictable causal effect upon

---

[1] This point has been made against Jeffrey by I. Levi, 'Probability Kinematics', *Brit. J. Phil. Sci.*, **18** (1967), 197.

the whole distribution. But, as we have seen, to have a confirmation theory at all we must be able to answer explicitly some general questions concerning the effect of possible future evidence upon the probability of hypotheses. This might be done by means of Bayes' theorem, or by some other posterior transformation, so long as we have some rule for the transformation which can be applied at $t_{r-1}$ to calculate the effects of evidence at $t_r$. But Jeffrey's proposal provides no rule at all apart from coherence, neither Bayesian nor any other.

Jeffrey's proposal was an attempt to deal with the problem of uncertain evidence. In the next section I shall suggest a different method of dealing with this problem, but meanwhile another serious drawback of Bayesian transformations should be noticed, namely the requirement that posterior probabilities be calculated by taking into account the *total evidence*. This may be done directly by calculating the posterior distribution at $t_r$ conditionally upon the conjunction of all the evidence at $t_r$, or, in less cumbersome fashion, by using the posterior distribution over all statements at $t_{r-1}$ as if it were the initial distribution at $t_r$. The two methods are equivalent only if the initial probability of subsequent *evidence* as well as of all hypotheses is taken to be its posterior probability on all previous evidence, or if the new evidence is always probabilistically independent of the old. If we consider the sequence of scientific theories and accumulating evidence, however, it is much more likely that past evidence will be taken into account only in so far as it affects posterior probabilities of hypotheses, and not for its effect on as yet unforeseen subsequent evidence. The state of belief of the scientific community at any given time is far more dependent on theories that have recently proved successful than upon the detailed evidence which originally led to their acceptance, and which may be forever lost in the back numbers of journals and the vaults of libraries. Moreover, the assumed initial probability of later evidence is more likely to be dependent on these accepted theories than upon such discarded evidence.

Unless the successive items of evidence are probabilistically independent, it also follows that its time order of collection will be relevant to the calculation of posterior distributions. If the total evidence of science since Galileo had been collected backwards, our beliefs calculated by the rule of conditionalizing on previous evidence and then forgetting it would not now be the same as they are. It must, however, be rememberd that the decision what to collect at any given time is itself dependent on the order in which evidence and theories have emerged, so this conclusion is not as paradoxical as it may look at first sight.

Patrick Suppes[1] has given examples of transformation rules in which evidence is 'forgotten' in an analysis of various types of learning in behavioural

---

[1] P. Suppes, 'Concept formation and Bayesian decisions', *Aspects of Inductive Logic*, ed. J. Hintikka and P. Suppes (Amsterdam, 1966), 21, and 'Probabilistic inference and the concept of total evidence', *ibid.*, 49.

psychology, where the constraints upon transformations necessary for learning are inferred in models simulating the actual behaviour of organisms. In a simple case of choice between two possibilities $A$ and $B$ in a sequence of trials with reinforcement for correct choice, the rule may be to guess on each occasion that possibility, say $A$, which turned out correct on the previous trial, irrespective of all evidence before that trial, however overwhelmingly it may have suggested that most trials turn up $B$. The rule does not take account of total evidence, but it does ensure that if one of the frequencies $A$ or $B$ is greater than the other, eventually more guesses will be correct than not, and learning has been partially successful.

Abandonment of the requirement of total evidence may be regarded as a merely practical consequence of the finite memory of all actual learning devices. There are, however, other sorts of case which show that Bayesian transformation cannot even in principle be a general requirement for historical development of scientific theories. These are cases of observation of new properties, and invention of new concepts or rejection of old. If a given concept has not previously been named in the language, then no hypothesis containing it can have explicitly been given any initial probability. To take such a concept into account involves non-Bayesian discontinuity in the sequence of posterior distributions. Conversely, terms may disappear from the scientific language (for example, 'phlogiston'), thus removing some hypotheses by giving them probability identically zero, and causing non-Bayesian readjustment in the probabilities of others.

Much work remains to be done on the question of the relative efficiency of Bayes and other transformation rules in given sorts of circumstances. It may be assumed that which rules are most efficient will be a function both of the complexity of the learning mechanism, and also of the general characteristics of the world in which it is learning. With sufficiently complex learning devices one can envisage a 'monitor' which checks the predictive success attained by a lower level probabilistic inductive device, and changes the initial probabilities until better success rates are attained. Something like this process is suggested by Carnap for choosing a probability distribution out of his continuum of different distributions defined by a parameter $\lambda$ in the real open interval $(0, \infty)$.[1] He shows that, if the universe is on the whole homogeneous, the distribution defined by $\lambda = 0$ will be expected to give greatest success in the long run, while if the universe is grossly heterogeneous, $\lambda$ infinite will give greatest expected success. $\lambda = 0$ is in fact equivalent to the statistical method of maximum likelihood and, in simple cases, to Reichenbach's straight rule, which is the direction to predict that the proportion of properties in the world will continue to be the proportion that has already been observed in samples. The success of this method clearly relies heavily

[1] *The Continuum of Inductive Methods*, sections 18, 22.

on the assumption that the world is relatively homogeneous with respect to the samples, especially when these are small.

But of course the essence of the inductive problem is that we do not know *a priori* whether the universe is on the large scale homogeneous or heterogeneous, whether we are in a typical part of it, or how far the past success of any inductive method will be maintained in the future. If choice of pragmatically efficient initial distributions and transformation rules depends crucially on how well adapted the learning organism is to actual contingent conditions in the world, then it is unlikely that any detailed confirmation theory can be developed *a priori*, that is to say, independently of empirical investigations into how organisms (including ourselves) actually learn. It cannot be expected, therefore, that Bayesian methods will be useful except locally. Our strategy from now on will therefore be to investigate the *sufficiency* rather than the necessity of Bayesian methods as an explication of certain local aspects of scientific induction, as these are represented by generally accepted intuitive inductive rules.

## V. Uncertain evidence

Meanwhile, however, there is one problem related to the possibility of non-Bayesian transformations which requires further discussion. As we have seen in the last section, one of Jeffrey's motives for abandoning Bayesian transformations was the problem of *uncertain evidence*. And in an earlier chapter the same problem has been raised by the network theory of universals, for I have argued that if such a theory is to allow for change of the truth-values and meanings of theoretical statements with changing scientific theories, then it must allow for similar modifications in observation or evidence statements. The question now arises, how can such modifications be incorporated into a probabilistic confirmation theory, where some constraints on the truth-value of evidence must be assumed, in order to use Bayesian or any other definite rule of transformation from initial to posterior probabilities? Indeed, a straightforward application of Bayes' theorem to transformations of belief distributions by evidence would seem to presuppose a distinction, of the kind we have already rejected, between observation statements, which can be known to be conclusively true or false after test, and theoretical statements or hypotheses which remain uncertain unless they can be definitely falsified. In this section I shall suggest a way in which transformations of belief distributions, according to a Bayesian or some other rule, can be reconciled with the fact that evidence previously accepted may have to be modified or withdrawn. It will incidentally follow that even in a deductive system, contrary observations do not necessarily falsify theories, thus incorporating the Duhemian symmetry of confirmation and disconfirmation.

Formally, in a probabilistic confirmation theory the problem of uncertain evidence arises as follows. (I shall assume the transformations are Bayesian for ease of exposition, but the same principles would hold for other types of transformation provided rules for them were definitely specified.) When evidence $e$ is observed, the distribution of probabilities over all other statements of the system whose probabilities are less than 1 is changed to the posterior distribution relative to $e$. The posterior probability of $e$ itself is $p(e/e)$, which necessarily goes to 1 by (PA3). This does not of course imply that $e$ is in any sense necessarily true, for necessary truth would be represented only by $p(e) = 1$ independently of any background evidence. It implies only that $e$ is henceforward accepted as empirically true as a result of observation. But even this is to assume too much. Suppose we are in some doubt about the truth of $e$, while still being unable to specify observations of a more 'basic' sort which express any part of $e$ of which we are certain. In these circumstances $p(e/e) = 1$ does not represent our posterior belief accurately. Moreover, we may appeal to the rest of our beliefs to check our belief in $e$. If I think I see a red-headed bird against the Indiana snow, I shall be more inclined to believe my eyes if I also believe to some degree that there are birds called cardinals in the United States which have red crests, even though I may be certain of none of these propositions. Can a probability theory accommodate such mutual reinforcement of beliefs, or are they for ever caught in an indeterminate circle of probabilities?

The problem was first clearly recognized in the context of a probability theory of knowledge by C. I. Lewis.[1] As a phenomenalist, Lewis held that sense appearances are the only ultimate determinants of the truth of empirical belief, and that, though they 'may be difficult, or even impossible, to formulate, . . . they are matters of which we may be certain'. He then attacks the problem of how statements asserting objective empirical facts may be made probable by postulating what he calls 'terminating judgments', whose certainty is warranted by sense. The formal fact, in a probability theory, that $p(e/e) = 1$ represents Lewis's dictum that 'unless something is certain in terms of experience, then nothing of empirical import is even probable'. This truth of the probability calculus is made palatable in Lewis's fallibilist epistemology by distinguishing between sensory terminating judgments and objective statements of fact, a distinction which is likely to be even less acceptable nowadays than that between observation and theoretical statements. However, Lewis's proposal has the merit that it suggests a separation between the question of what it is that causes a belief distribution to change and which is accepted as certain for purposes of the change, and the question

---

[1] C. I. Lewis, *An Analysis of Knowledge and Valuation* (La Salle, Ill., 1946), 235f. I have made essentially the same suggestion in 'A self-correcting observation language', *Logic, Methodology, and Philosophy of Science*, vol. 3, ed. B. van Rootselaar and J. F. Stahl (Amsterdam, 1968), 297, and in 'Duhem, Quine, and a new empiricism', *Knowledge and Necessity*, Royal Institute of Philosophy Lectures, vol. 3 (London, 1970), 191.

of how relatively observational and theoretical statements in science are probabilistically related to one another. Given some category of statements, other than those descriptive of objective fact, which can function as the given evidence, a mutual readjustment of probabilities of all empirically descriptive statements of the scientific network is possible. The question is, what category of statements should be adopted as the given?

To follow Lewis and adopt phenomenalist judgments is unduly subjective. It raises problems not only about how such judgments are related to facts, but also about how the judgments of different scientists are related to one another. In conformity with the general policy in philosophy of science of understanding the objective in terms of the *inter*subjective, I shall make a different suggestion, which restores the 'given' to the empirical world. The suggestion is that what we are given in science are *observation reports*, which may be spoken, written, taped, drawn or printed, and which make up the data upon which the community of science works. No scientific observation is intersubjectively available until it has in some such way been made public. By an observation report is to be understood a singular sentence describing the record that some state of affairs has been observed, it being assumed that the report was intended for general scientific publication. Thus typical observation reports at specified space–time points would be 'It is reported that "At $(s_1, t_1)$ NaCl dissolved in water" ', or 'It is reported that "At $(s_2, t_2)$ the ammeter stood at 100 milliamps" '. In each case the sentence between double quotes is what would normally be called a singular observation statement, but of course the truth of the report does not entail the truth of the corresponding observation statement, and neither need logically equivalent observation statements be substitutable for each other in reports with preservation of truth-value of the reports. Thus no truth-value of any observation statement is ensured by assertion of its corresponding report, but observation statements may be given of varying degrees of confirmation dependent on what reports have been asserted.

It may be thought that little has been gained by replacing observation statements by observation reports as the accepted evidence. Observation reports are themselves empirical descriptive statements, and it may be objected that all that has been accomplished is a shift of the 'protocol' character previously ascribed to observation statements to observation reports. Why should observation reports, any more than observation statements, be regarded as incorrigible once asserted? The short answer to this is that they need not be regarded as absolutely incorrigible, but only as substantially less subject to correction than the observation statements themselves. In considering how evidence and hypotheses are related in a scientific system, it is in fact much less likely that mistakes are made in describing reports than in asserting observation statements. The historian of science, or the reader of current scientific literature, has no difficulty in assuring himself of what is present in the literature, allowing for misprints, mistranslations and other misfortunes, which are cleared up more easily than mistakes in the observation statements that are reported. Moreover, the assertion that something has been reported in a piece of scientific literature is practically independent of the circumstances under which the corresponding observations are made, and of possible influence from the theoretical context. There are occasions on which we see what theoretical preconceptions suggest that we see, but it is rarer to read words, much less whole sentences, merely because we want to read them.

It may, however, be objected here that the whole onus of the relative incorrigibility of the observation reports is now being put upon the judgment of what it is to be a seriously made report designed for general scientific publication. A great number of putatively descriptive statements are made every day, and these include political propaganda, travellers' tales and horoscopes. Moreover, they have included alchemists' recipes for making gold, reports of spontaneous generation of frogs in dung-heaps, and other reports that passed for scientific observation in earlier stages of science. How are 'serious scientific reports' to be identified? Is not the judgment of what is to count as an observation report just as much tied to particular scientific theories as the observation statements themselves?

The justice of this objection must be admitted. However, as I have already argued, the task of an inductive logic can only begin with certain presuppositions about what it is to be an acceptable inductive inference, and, it must now be explicitly added, with presuppositions about what it is to be a scientific observation. We can only move one step at a time into investigation of the dynamics of theories—that is, from the static interpretation of the deductivist model to inductive evolution from theory to theory; not all the way to an investigation of how the inductive and observational method itself becomes established. There may well be another level of investigation in which the concept of 'observation' itself would be seen to be modified by a more comprehensive network of relationships between scientific, technological, social, theological and other pragmatic factors. If so, study of the more limited feedback loop involving theory change within a particular understanding of what it is to be scientific will not conflict with the more comprehensive study; indeed it may well serve as an essential model for it. Meanwhile we pursue the more local aim of understanding how empirical constraints operate upon the posterior probability distribution in the network, without presupposing that any of the statements of the network itself are observationally privileged.

We are now in a position to set up a self-correcting theory–observation network in probabilistic terms. The basic assumption that has to be made is what I have called the correspondence postulate, namely that the observation statement corresponding to a given observation report is true more often than not. It would generally be appropriate to assume a stronger condition, namely that there is a quite high probability of the observation statement being true. That every observation sentence must at least be assumed to have a probability greater than $\frac{1}{2}$ may be defended as follows. Suppose there are statements ascribing a predicate $P$ to objects, and that each statement has a probability less than $\frac{1}{2}$ of being true. This means that in the long run, on more than half the occasions on which '$Pa_r$' is reported, $Pa_r$ is false, and hence $\bar{P}a_r$ is true. Remembering that we have no privileged intersubjective access to the world over and above the observation reports, it follows that the assumption that all observation sentences containing $P$ have probability less than $\frac{1}{2}$ is merely equivalent to interchanging $P$ and $\bar{P}$ in our conventional language and assuming that sentences containing $\bar{P}$ have probability greater than $\frac{1}{2}$. Hence the assumption that $Pa_r$ has probability less than $\frac{1}{2}$ is empty. Moreover, if the probability is equal to $\frac{1}{2}$ for assignments of both $P$ and $\bar{P}$, then no experience is informative, since it is as likely to be falsely reported as truly. In general the nearer the probability is to $\frac{1}{2}$, the less effect observation has on posterior distributions.

Let $e$ be an observation statement in the scientific system, and $e'$ be the observation report that '$e$'. In order to simplify the notation we will assume that all necessary conditions of observation are given as true, and are probabilistically independent of the outcomes of the observations, so that these given conditions can be dropped from expressions of background knowledge. Then $e'$, $\sim e'$ respectively mean that, given that the conditions for observing the reference of $e$ or $\sim e$ were set up and that the observation was reported, there is a report to the effect that $e$, $\sim e$ respectively was observed. Now since there is some probability greater than $\frac{1}{2}$ that $e'$ is true given $e$, and that $\sim e'$ is true given $\sim e$, put

$$p(e'/e) = 1 - \varepsilon_1, \quad p(\sim e'/\sim e) = 1 - \varepsilon_2$$

where $\varepsilon_1$, $\varepsilon_2$ are less than $\frac{1}{2}$. Then we have

$$p(e') = p(e'/e)p(e) + p(e'/\sim e)p(\sim e) = (1 - \varepsilon_1)p(e) + \varepsilon_2 p(\sim e)$$

$$p(e/e') = \frac{(1 - \varepsilon_1)p(e)}{(1 - \varepsilon_1)p(e) + \varepsilon_2 p(\sim e)} \tag{5.3}$$

and if $h$ is any statement independent of $e'$ given $e$

$$p(h/e') = \frac{(1 - \varepsilon_1)p(h\&e) + \varepsilon_2 p(h\&\sim e)}{(1 - \varepsilon_1)p(e) + \varepsilon_2 p(\sim e)} \tag{5.4}$$

With corresponding assumptions we can calculate the compound effect of two or more observation reports on the probability of $h$. In considering the conjunction of observation statements, we must take into account the effect of multiplication of probable errors, even when these are assumed small. It might be thought that, in this account, the probability of the conjunctions of all the observation reports involved in any scientific theory must be very low, certainly much less than $\frac{1}{2}$, and that therefore the account leads to the counter-intuitive conclusion that we become less and less sure of either evidence or hypotheses the more uncertain evidence we collect. This would indeed be the case if all statements of the system were probabilistically independent of each other. But the conclusion does not follow for certain other initial probability distributions where evidence statements are already positively relevant to each other. For example, an observation report $f'$ may increase the probability of $e$ as well as of $f$, if $p(e/e'\&f') > p(e/e')$, and $f'$ may reinforce $e'$ in confirming $h$, if $p(h/e'\&f') > p(h/e')$.

This probabilistic model has several features which well represent a Duhemian theory–observation network. First, it does distinguish between theoretical and observation statements in the sense that observation statements are those with corresponding observation reports. But this may be regarded as a purely pragmatic distinction. At some points of the network external contact with the empirical world is made, but nothing is presupposed about the 'directness' or non-theoretical character of this contact. Some statements involving theoretical terms may be the subject of observation reports as well as statements more traditionally thought to be 'observable'.

Indifference in the system as to which statements are observational and which theoretical is made possible by the fact that the probability distribution over all statements (as opposed to reports) has similar features. It follows from (5.3) and (5.4) that the probabilities of neither observation nor theoretical statements go to 0 or 1 unless $\varepsilon_1$ or $\varepsilon_2$ is zero. Indeed, severe doubt can be cast on an observation statement $e$ even though its corresponding observation report is in the record, for another observation report $f'$ may reduce the probability of $e$ because the conjunction of $e$ with statements $f$ corresponding to $f'$ has low initial probability. Thus part of the Duhem–Quine thesis is satisfied, namely that no observation statement, even after a test with positive results, can be decisively held to be true, since there is a finite probability that the report is an error. And on the other hand, Quine's corresponding requirement that no theoretical statement $h$ can be conclusively falsified is satisfied in the sense that no observation statement that appears to refute $h$ is itself conclusively true, although of course the probability of $h$ may in some circumstances become vanishingly small.

It follows that in this model, unlike the deductive model, there is a symmetry of confirmation and disconfirmation, in that no statement of the system is either conclusively verified or conclusively falsified. There is also symmetry of deductive and inductive inference, in that both are represented by change of probability values. Deductive inference becomes merely the special case of probabilistic inference in which there is transfer of minimum probability value from premises to conclusion in accordance with (PC4). Finally, so long as we are prepared to assign values greater than $\frac{1}{2}$ to $\varepsilon_1$ and $\varepsilon_2$, there is no vicious circularity of uncertainty of evidence and theory, but mutual readjustment of probability values in the light of all observation reports.

The model therefore seems to represent adequately the effects on a belief distribution of uncertain evidence, provided that the idea of locally acceptable 'observation reports' is adopted, and provided that definite rules of transformation such as (5.3) and (5.4) are specified. It does not, of course, follow that the model would in practice be the most fertile or efficient way of constructing an artificial intelligence device to represent scientific discovery; in fact recent studies indicate that conditional probability is far from being the best basis for such a device.[1] But the model is sufficient to show that the probability representation cannot be excluded as an explication of scientific inference on the grounds that it fails to take account of the problem of uncertain evidence.

[1] For discussion and references, see J. S. Williamsen, 'Induction and artificial intelligence' (unpublished PhD thesis, University of Cambridge, 1970).

# CHAPTER SIX

# Bayesian Confirmation Theory

## I. The positive relevance criterion

In the past there have been many objections to Bayesian theory on the grounds that it cannot in principle explicate this or that piece of inductive practice: confirmation of universal generalizations, preference for theories of great power or simplicity, and so on. But it follows from the decision-theoretic formulation of Bayesian theory already considered that all such objections are beside the point. There will always be *some* distribution of initial probabilities which will provide the required explication in particular cases; the significant question is whether such a distribution appears at all perspicuous and illuminating for general aspects of theory structure, or whether it is entirely implausible and *ad hoc*. It is only in the latter case that cogent objections can be brought against Bayesian theory, and not enough effort has yet been put into development of the theory to permit summary rejection of it on these grounds. Moreover, Bayesian theory can lay claim to some immediate successes which have not come to light in other approaches. In this section I shall list some of these, and go on in the next section to discuss the requirements of confirmation of scientific laws and theories which are more problematic and also more illuminating.

What relation has to hold between data and hypothesis if the data are to count as *evidence for, confirmation of* or *support for* the hypothesis? We have to try to pick out what we regard as the essential features of that relation which may be expressed as 'Assuming the evidence expressed in true propositions, under what conditions do we regard the evidence as increasing the reliability of a hypothesis, or as making the hypothesis more acceptable as a guide to truth?' We have first to decide how to interpret 'confirmation' in terms of probability functions. There are two obvious possibilities here. We may regard a hypothesis $h$ as confirmed by evidence $e$ if and only if the probability of $h$ on $e$ attains at least some fixed value $k$ such that $1 > k \geqq \frac{1}{2}$. Thus we obtain the *k-criterion*

$$e \text{ confirms } h \text{ iff } p(h/e) > k \geqq \tfrac{1}{2}.$$

Roughly speaking, we then believe that $h$ has at least a better than even chance of being true. Such a belief might well dispose us to *act upon* or *accept*

$h$, particularly if $k$ is near $1$, or if a choice between $h$ and $\sim h$ is forced upon us. But it may be questioned whether it is a suitable measure of the confirmation of $h$ *relative to* $e$, for the $k$-criterion may be satisfied even if $e$ has *decreased* the confirmation of $h$ below its prior value, in which case $e$ has relatively *dis*confirmed $h$. Further properties of the $k$-criterion will be noted later, but meanwhile it seems better to adopt a relative measure of confirmation, which I call with Carnap the *positive relevance criterion*,[1] which requires the posterior probability of $h$ on $e$ to be greater than its initial probability.

(PR)        $e$ confirms $h$ iff $p(h/e) > p(h)$

I shall also say that $e$ is *negatively relevant* to $h$, or disconfirms $h$, iff $p(h/e) < p(h)$, and that $e$ is *irrelevant* to $h$ iff $p(h/e) = p(h)$.

There are several useful corollaries of these criteria.

(RC1)  $e$ confirms/is irrelevant to/disconfirms $h$ according as $h$ confirms/is irrelevant to/disconfirms $e$.

For $p(h/e) = p(e/h)p(h)/p(e)$, hence $p(h/e) \gtreqless p(h)$ according as $p(e/h) \gtreqless p(e)$.

(RC2)  If $e$ confirms $h$, then $e$ disconfirms $\sim h$.

For $p(\sim h/e) = 1 - p(h/e) < 1 - p(h) = p(\sim h)$.

(RC3)  If $e$ confirms $h$, then $\sim e$ disconfirms $h$.

For then $h$ confirms $e$, from (RC1); $h$ disconfirms $\sim e$, from (RC2); and hence $\sim e$ disconfirms $h$, from (RC1).

All these properties of positive relevance are intuitively desirable for a relation of confirmation.

Some elementary intuitive conditions of adequacy for inductive inference can also be seen at once to be satisfied by positive relevance, independently of any assignment of initial probabilities. Not all of these conditions are uncontroversial, but in line with the self-correcting explicatory approach adopted here let us first state them and then explore their consequences.

## (i) Equivalence

A condition that seems fairly unproblematic is that logically equivalent expressions should have identical effects in confirming logically equivalent expressions. That is to say, we have the *equivalence condition*

If $g \equiv g'$, and $h \equiv h'$, then if $g$ cfms $h$, $g'$ cfms $h'$.

This is satisfied by both the positive relevance and $k$-criteria of confirmation immediately in virtue of (PA2).

The main criticisms of this condition have been in connection with the so-called 'raven paradoxes'. There are, as we shall see, methods of dealing

---

[1] *Logical Foundations of Probability*, chap. 6.

with these paradoxes other than abandoning the equivalence condition, which would, of course, entail also abandoning a probabilistic theory of confirmation, since the condition follows directly from the probabilistic axioms.[1]

## (ii) *Entailment*

Another condition that seems inescapable for the relation of confirmation is that any entailment of a proposition $h$ must be confirmed by $h$. Thus we have the *entailment condition*

$$\text{If } h \to g, \text{ then } h \text{ cfms } g.$$

This is satisfied immediately for empirical propositions by (PA3):

$$\text{If } h \to g, \text{ then } p(g/h) = 1,$$

hence for any $g$ whose initial probability is less than maximal, $p(g/h) > p(g)$, and $h$ confirms $g$. It is also satisfied by the $k$-criterion, for $p(g/h) = 1 > k$.

## (iii) *Converse entailment*

This is the condition

(CE1) $\qquad\qquad$ If $h \to g$, then $g$ cfms $h$.

It is satisfied by the PR criterion of confirmation where neither $p(h)$ nor $p(g)$ is zero or one, for we have immediately from Bayes' theorem: if $h \to g$,

$$p(h/g) = p(h)/p(g) > p(h).$$

The applicability of this condition to scientific confirmation is more problematic than that of conditions (*i*) and (*ii*). It is generally assumed in the deductive model of theories that, given some initial conditions $i$, if a hypothesis entails an experimental law or singular prediction, and the entailment is accepted as true on the basis of observation, the hypothesis is thereby confirmed. This is indeed the only way in the deductive model in which the reliability of hypotheses can be increased. It is also assumed in so-called 'enumerative' induction that, given the antecedent of a substitution instance, a generalization is confirmed by the consequent of that instance which is entailed. Modifying converse entailment to take account of the initial conditions, these assumptions imply

(CE2) $\qquad$ If $h\&i \to g$, and $\sim(i \to g)$, then $g$ cfms $h$ given $i$.

Or, in terms of positive relevance

$$p(h/g\&i) > p(h/i).$$

---

[1] For a useful critique of this and the other probability axioms in relation to confirmation, see R. Swinburne, *An Introduction to Confirmation Theory* (London, 1973), chaps. 4–6.

This condition is satisfied for empirical propositions if neither $p(h)$ nor $p(g\&i)$ is zero, since $h\&i \rightarrow g\&i$, and

$$p(h/g\&i) = p(h\&i/g\&i) = p(h\&i)/p(g\&i)$$
$$= p(h/i)p(i)/p(g\&i).$$

But $p(i) > p(g\&i)$, hence $p(h/g\&i) > p(h/i)$.

(CE3) If, further, the truth of $h$ is independent of the occurrence of the initial conditions $i$, it also follows that $g\&i$ cfms $h$, since then

$$p(h/g\&i) > p(h/i) = p(h).$$

One objection to the converse entailment condition is that, together with the equivalence condition (i) above, it entails the so-called 'raven paradox' of confirmation first stated by Nicod and discussed by Hempel. The paradox will be examined in a later chapter, and a resolution suggested that involves abandoning neither the equivalence nor converse entailment conditions.

It has been further objected to the converse entailment condition that sometimes a single positive substitution instance of a general law is enough to confirm it, and that scientists do not usually go on accumulating positive instances of the same kind as those already observed, since such instances would not be regarded as adding significantly to the evidence.[1] What does add to the confirmation, it is held, is observation of positive instances in a variety of circumstances, the elimination of rival hypotheses, or the placing of the law in a logical relation with other laws and theories, which are themselves well confirmed. I shall consider elimination and variety immediately, and later we shall see how far a Bayesian theory can account for the greater confirmation expected from incorporation of a law within a general theory. With these supplementations I shall maintain the condition that any observed entailment provides some confirmation for a hypothesis whose initial probability is non-zero, however little that confirmation may be under some circumstances.

A useful consequence of converse entailment may be noted here: that is, that it ensures transitivity of confirmation in a deductive system in the following sense. If evidence $f$ confirms a generalization or low-level theory $g$ in virtue of being entailed by $g$, then $f$ should also confirm a higher-level theory $h$ of which $g$ is an entailment. Since then $h$ entails $f$, this confirmation follows at once from converse entailment by the transitivity of entailment.

(iv) *Initially unexpected evidence*
A further assumption generally made about confirmation is that if evidence $e_1$ is entailed by a hypothesis $h$, and $e_1$ would have been initially unexpected

---

[1] *Cf.* J. S. Mill, *System of Logic*, Book III, iii, 3: 'Why is a single instance, in some cases, sufficient for a complete induction, while in others myriads of concurring instances, without a single exception known or presumed, go such a very little way towards establishing an universal proposition? Whoever can answer this question . . . has solved the problem of Induction.'

without consideration of $h$, then actual observation of $e_1$ does more for the confirmation of $h$ than would initially expected evidence $e_2$ which is also entailed by $h$. The rule of searching for variety of instances can be seen as an example of this consideration, for it is initially more unexpected that a general law, such as 'All crows are black', should be true for a great number of different kinds of instances of crows (of all ages, for example, and observed in very various geographic regions), than that it should be true only for old crows in the south of England. Again, Einstein's general relativity theory entailed that light rays should be detectably bent in the neighbourhood of large gravitating masses. This result was highly unexpected in the context of theories and evidence available before relativity theory, and when the bending was actually observed in the neighbourhood of the sun in the eclipse of 1919, it provided more support for relativity theory than would have been provided by evidence that was also accounted for, and therefore already expected, on older theories.[1]

The requirement of unexpected evidence can be rather easily explicated by probabilistic confirmation as follows:

Interpret 'initially unexpected' to mean 'initially improbable', that is, improbable on the evidence available prior to $e_1$ or $e_2$. We can then compare the effect of the posterior probability of $h$ of observing $e_1$ or $e_2$, where $e_1$ is initially much less probable that $e_2$, and $h \to e_1 \& e_2$. We have

$$p(h/e_1) = p(h)/p(e_1)$$
$$p(h/e_2) = p(h)/p(e_2)$$

But $p(e_1) \ll p(e_2)$, hence $p(h/e_1) \gg p(h/e_2)$. Thus the confirming effect of $e_1$ on $h$ is much greater than that of $e_2$, as required, and in a situation of limited experimental resources, it would be wise to do the experiment to test $e_1$ rather than that to test $e_2$.

(v) *Eliminative induction, variety, and the Bacon–Mill methods*
Bayesian theory provides an immediate explication and generalization of eliminative induction.

Suppose $S \equiv \{h_1, h_2, ..., h_n\}$ is a finite, exhaustive and mutually exclusive hypothesis set. Then since

$$p(h_1/e) + p(h_2/e) + ...p(h_n/e) = 1$$

refutation, or more generally disconfirmation, of any proper subset of $S$ by evidence $e$ increases the sum of the probabilities of the remaining subset.

---

[1] A requirement similar to this seems to be what Popper has in mind in speaking of the 'severity' of tests. In *The Logic of Scientific Discovery* he asserts (in italics) that '*the probability of a statement (given some test statements) simply does not express an appraisal of the severity of the tests a theory has passed, or of the manner in which it has passed these tests*' (394, cf. 267). But in *Conjectures and Refutations*, 390, he suggests a measure of severity proportional to the initial *im*probability of the test, as is also suggested here.

Without further information one cannot conclude that the probability of any particular remaining hypothesis, say $h_1$, is increased, since the evidence may in fact reduce it. The only conclusive result that can be obtained is in the case of elimination of all hypotheses except $h_1$, in which case $p(h_1/e)$ becomes 1, and $h_1$ certain. All consequences of $h_1$ also become certain under these circumstances, since their probabilities must be at least as great as that of $h_1$, by (PC4). This result is, of course, ideally the aim of crucial experiments, in which the alternatives are supposed to be reduced to two, one of which is refuted by the experiment, although it is rare to find circumstances in which no other alternative hypotheses are possible.

The requirement of variety of instances can also be seen as a consequence of eliminative induction. Consider the generalization 'All vegetarians are teetotallers', which would generally be said to be better confirmed by a sample of vegetarians taken from all sections of the population than by a sample taken, for example, exclusively from Methodists. General explication of this intuitive inductive rule turns out to be more difficult than might be expected, no doubt because what seems reasonable in examples of this kind depends heavily upon extra information, such as that many Methodists are teetotallers, and the more complex the background evidence assumed the more unmanageable becomes the explication.[1] We can, however, regard the requirement as a corollary of eliminative induction by supposing that the more varied the population from which samples are taken, the more rival generalizations about teetotallers are refuted if they are false, and hence the greater is the confirmation obtained for those generalizations that remain unrefuted.

Assumptions of this kind lie behind the eliminative inductive methods of Bacon and Mill, where it is initially assumed that the set $S$ consists of universal generalizations expressing uniform cause–effect relations between properties, at least one of which must be true. Suppose the competing hypotheses are $h_r \equiv$ 'All $P_r$ are $Q$' where $r = 1, ..., n$. $Q$ is here a particular kind of ante-cedent situation such as, in Bacon's classic example, 'presence of heat', and the $P$s are a finite number of candidates for uniform correlation with $Q$. The assumption that at least one uniform correlation must be true is the assumption that there is at least one property $P$ with which $Q$ is uniformly causally related, and that all the possibly relevant $P$s have been identified and included in $S$. The uniformly correlated $P$ may be a *reductive explanation* of $Q$, as when Bacon seeks the 'form', or corpuscular mechanism, which both causes phenomenal heat in the perceiver, and is objectively identical with heat as a property of the object. In this case not only 'All $P$ are $Q$', but also 'All $Q$ are $P$', hence: 'Heat *is identical with* the unconstrained motion of the small parts

---

[1] *Cf.* P. Achinstein's attempt to explicate variety purely in terms of enumerative induction, without the eliminative assumptions to be discussed below, 'Variety and analogy in confirmation theory', *Phil. Sci.*, **30** (1963), 207, and Carnap's reply, *ibid.*, 223.

of bodies'.[1] In other cases the uniformly correlated $P$ may be an explanation of *another* property known to be uniformly associated with all $Q$s, as in a classic example recently analysed by L. J. Cohen,[2] namely Frisch's investigations of the homing characteristics of bees. Here the known common property $Q$ of bees is ability to 'recognize' their surroundings and find their way about. How do they do this? The answer is assumed to lie in at least one other common property $P$ which is the 'cause' of $Q$. It is antecedently assumed that if one bee has such a property $P$, then all have the same property, and that if $P$ is causally related with $Q$ in one case, it is causally related in all. There is no possibility of the causal relation sought being a statistical relation referring only to some bees, or of $P$ causing $Q$ in some cases, but not being present with $Q$ in some, otherwise similar, cases. Let us also assume for simplicity in exposition that there is only one $P_r$ that is the cause.

With these assumptions, suppose a given population $\{x_s\}$ of bees has properties $P_1 P_2, \dots, P_n Q$, where the $P$s are such characteristics as 'ability to discriminate colours', 'ability to discriminate shapes', 'ability to discriminate odours', etc. Since it is assumed that the hypotheses $h_1, \dots, h_n$ exhaust the possibilities, we have initially $p(h_1 \vee h_2 \dots \vee h_n) = 1$. Assuming also that all the hypotheses have equal initial probability, and that $p(P_r x_s)$ is the same for all $r$, the evidence $P_1 P_2 \dots P_n Q x_s$ does not discriminate among the hypotheses and hence cannot confirm any of them. There is therefore no point in collecting any more instances of this kind, although they may have suggested the possibly relevant $P$s in the first place. Next, tests are carried out in a *variety* of conditions. Suppose population $\{y_s\}$ has properties $P_1 P_2 \dots \bar{P}_n Q$. This eliminates $h_n$ as the uniform causal relation between the $P$s and $Q$, and hence raises the probability of the disjunction of the remaining $h$s to 1. If tests in varying conditions can be carried out which eliminate all the hypotheses except $h_1$, we have $p(h_1/e) = 1$, where $e$ is the evidence from all these tests, and it follows immediately by (PC4) that the probability of the next $Q$ and all following $Q$s being $P$ is also 1.

If such elimination of all hypotheses but one is not attained by variation of properties, it may be attained by testing instances of *absence* of $Q$, to see whether they are also instances of absence of $P_1, P_2, \dots, P_n$. Any instance of $\bar{Q}$ that is accompanied by $P_r$ entails that $P_r$ is *not* a cause of $Q$ and hence eliminates $h_r$. Notice that just as $P_1 P_2 \dots P_n Q x$ does not confirm any $h$ under the conditions assumed, nor does co-absence of all the properties, i.e. $\bar{P}_1 \bar{P}_2 \dots \bar{P}_n \bar{Q}$; for this satisfies all the $h$s and does not discriminate between them.

Search for instances of absence is the basis of what we would call control

---

[1] For a detailed account of Bacon's theory of induction and of 'forms', see my 'Francis Bacon', *A Critical History of Western Philosophy*, ed. D. J. O'Connor (London, 1964), 141.

[2] L. J. Cohen, 'A note on inductive logic', *J. Phil.*, **70** (1973), 27. Cohen's own support function for eliminative induction is non-probabilistic. Its values coincide with those of a probability function only when hypotheses are compared on the basis of evidence that makes one of them certain. *Cf.* Cohen's *The Implications of Induction*, 28.

experiments, and Mill called the method of difference and Bacon the tables of absence. Thus Bacon rightly concludes, given his premises about the form of heat, that the unique cause of this effect will be isolated by collecting, first, instances of co-presence of heat and motion of small parts of bodies with variety of other properties: friction, gas expansion, flame, exercise, etc.; and, second, instances of co-absence having pairwise similarities in the other properties: bodies at relative rest, absence of expansion, absence of flame, etc.

The circumstances under which these results follow are highly idealized. It would be more realistic to suppose that in practice conditions fall short of the ideal in at least two respects, namely that not all possibly relevant uniform causal factors have been isolated, and that there is some probability, however small, that no uniform causal hypothesis is true. Both these possibilities can be taken account of by including in the hypothesis-set the 'catch-all' hypothesis $h' \equiv \sim(h_1 \vee h_2 \vee ...h_n)$, and giving $h'$ a *small* initial probability. The best that can then be hoped for from a series of experiments such as just described is $p(h_1 \vee h'/e) = 1$. If $e$ contains instances of all the types described which are consistent with $h_1$, it will approach random evidence from the total population, and according to the convergence theorem it will become more and more probable that $h_1$ is true, though its probability never attains exactly unity.

These considerations help to explain two features of inductive inference which are sometimes thought to count against the applicability of probabilistic confirmation. The first is that it is often intuitively felt to be useless to continue observing large numbers of instances of just the same kind. We have now seen that with the assumptions about uniform causation made by eliminative induction, further instances of $P_1P_2...P_nQ$ are useless or nearly useless, since they confirm all uniform hypotheses equally. Evidence should therefore be randomized by variety, and, if appropriate, also by instances of co-absence. Secondly, as Mill points out in the passage already quoted, it is sometimes the case that few or even single instances of a generalization will confirm the generalization very highly. This is understandable when the initial probability that all true generalizations are statistical is very low, and when it is almost certain that there are no viable competing uniform generalizations, for then one suitably chosen instance such as $P_1\bar{P}_2...P_nQ$ may be sufficient to make the probabilities of all but one uniform generalization $h_1$ vanishingly small. To take a different kind of example where equivalent conditions are satisfied, one determination of a physical constant, such as the ratio of charge and mass of the electron, or the liquefying point of helium, is sufficient to make its value within the limits of accuracy of the experiment practically certain.

It has often been suggested that enumerative and eliminative induction are in some sense alternative or even conflicting methods, and that probabilistic confirmation is more appropriate as an explication of enumerative induction. In view of the results just obtained this apparent dichotomy is clearly not fundamental, for Bayesian theory can accommodate both methods. They do, however, part company at a less fundamental level, because in *any* type of confirmation theory eliminative induction is applicable only when it is possible to specify a finite set $S$ of hypotheses which have between them overwhelming initial probability. Enumerative induction, on the other hand, will later turn out to be more relevant to the confirmation of singular predictions than of general laws and theories. But I shall argue that of the two, enumerative induction is the more fundamental, since it alone enables us to identify the set $S$ which is appropriate to any application of eliminative induction.

## II. The transitivity paradox

Some of the easy successes of Bayesian theory have now been outlined. However it is well known that difficulties soon arise, even in relation to the comparatively elementary requirements of enumerative induction. These difficulties concern, first, the possibility of assigning finite initial probabilities to all the universal generalizations that may be confirmed by evidence, for if there are more than a strictly finite number of these, assignment of finite probabilities, however small, will violate the probability axioms. Hence arises the problem of choosing which finite set out of all possible hypotheses to permit to be confirmed by suitable evidence. Secondly, problems arise about what is to count as confirming evidence for a generalization in order to evade Hempel's paradoxes of confirmation. As we shall see, difficulties about lawlike generalizations are multiplied when we come to consider confirmation of scientific theories and their resulting predictions.

I shall not consider in detail the various attempts that have been made to solve these problems, but I shall rather use the problems, in conjunction with Bayesian confirmation theory, to suggest an interpretation of the structure of scientific theory which conflicts to some extent with the generally received deductivist model of science. In this way I hope to indicate the potentially *revisionary* function of Bayesian theory, that is to say, the way a systematic attempt to explicate inductive intuitions in science may interact with and modify these intuitions themselves. The intuitions concerned here are those relating to the *universality* of scientific laws and theories. I shall argue later that it is possible to interpret the whole of a scientist's behaviour with regard to lawlike generalizations without assuming that he has any non-zero degree of belief in their universal truth in infinite domains, and indeed that it is more plausible to suppose that he does not have any such finite belief. One motivation for this argument is the difficulty in a Carnap-type confirmation theory of assigning finite initial probabilities to enough universal generalizations. I now want to provide another motivation for the argument by considering some more general features of Bayesian confirmation theory, which seem to indicate that the function of general theories in science is not what it is assumed to be in the deductivist model, and that theoretical inference, like Mill's eductive inference or instance confirmation in Carnap's theory, can better be regarded as going from particular to particular than via general theories and their entailments.

It is convenient to begin the examination of confirmation relations in scientific theories by a discussion of conditions of adequacy given by Hempel, which have been shown to give rise to paradoxes. We have already seen that the equivalence, entailment, and converse entailment conditions are intuitive criteria of adequacy for scientific inference, and that they are all satisfied by Bayesian theory under certain conditions of non-zero probabilities. These are three of the conditions discussed by Hempel in his 'Studies in the logic of confirmation',[1] in which they are initially presented, along with other suggested conditions, independently of whether confirmation should be explicated by a probabilistic or any other type of axiomatic system. Hempel's paper gave rise to the well-known 'paradoxes of the raven', which will be discussed in the next chapter, but some of the other conditions of adequacy give rise to another paradox, which I shall discriminate as the 'transitivity paradox'. My terminology and notation will here be a little different from Hempel's.

Consider first the general question of how far the confirmation relation can be expected to be *transitive*, that is, should it satisfy the condition 'If $f$ confirms $h$, and $h$ confirms $g$, then $f$ confirms $g$'? In this general form transitivity is obviously not satisfied by the usual notions of confirmation or support. For example, the foundations of a house support the walls which support the roof, and it is true that the foundations support the roof, but although the pivot supports the see-saw, and the see-saw the child at one end, it is not in general true that the pivot supports the child. Again, if a man has disappeared, to find him shot dead confirms the hypothesis that he has been murdered, and that he has been murdered in itself confirms that there is a strangler at large, but that he has been shot does not confirm that there is a strangler.

We have, however, already noted that in the case of the entailment relations $g \rightarrow h$ and $h \rightarrow f$, there is transitivity of confirmation as the converse of transitivity of entailment. There are also other special cases in which just one of the confirmation relations of general transitivity is replaced by an entailment, and in particular one in which transitivity seems both plausible and indispensable. Consider first the condition Hempel calls *special consequence*.

If $f$ cfms $h$, and $h \rightarrow g$, then $f$ cfms $g$.

This condition seems to be required for the following sorts of cases which are typical of scientific inference.

(*a*) Positive instances $f$ of a general law $h$ together with appropriate given

[1] *Aspects of Scientific Explanation*, 13f. See also B. Skyrms, 'Nomological necessity and the paradoxes of confirmation', *Phil. Sci.*, **33** (1966), 230.

antecedents $i$ are taken to confirm a prediction of the next instance $g$ as well as confirming the law itself. In this case $h\&i \rightarrow f\&g$; hence by converse entailment $f$ confirms $h\&i$, and by special consequence $f$ confirms $g$.

(*b*) In a more complex theoretical deductive system the condition might be stated in the form 'If a hypothesis is well confirmed on given evidence, some further as yet unobserved consequences of the hypothesis ought to be predictable with high confirmation'. Consider Newton's theory of gravitation, in which we have the theory confirmed by some of its entailments, for example the laws of planetary orbits and the law of falling bodies. The theory predicts other as yet unobserved phenomena, such as the orbits of the comets or of space satellites. We are not *given* the truth of the theory, only of its original data from planets and falling bodies, and yet we want to say that this evidence confirms predictions about comets and satellites. Special consequence would allow us to transmit confirmation in this way in virtue of the entailment of predictions by the theory. There is also an even stronger intuitive type of theoretical inference. Not only do we require theories to yield strong confirmation to their predictions before the predictions have been tested, but sometimes we use a theory actually to *correct* previously held evidential propositions, if the theory is regarded as sufficiently well confirmed by its other consequences. For example, Galileo's 'observation' of constant acceleration of falling bodies, and Kepler's third law of planetary motion (about the relation of the periodic times of the planets to the parameters of their orbits), were both corrected by Newton's theory in virtue of its supposed confirmation by other evidence. If the confirmation here is regarded as transmitted by converse entailment and special consequence to the corrected laws, it must have a higher degree than was originally ascribed to the pre-theoretical laws on the basis of 'observation', since the corrected laws are preferred to the previously observed laws.

It seems clear that any adequate account of confirmation must be able to explicate our intuitions and practice in cases such as these. Unfortunately, however, it can easily be shown that the conditions of converse entailment and special consequence just described will not do so, because these conditions together yield further unacceptable consequences. To see this in terms of a specially constructed counterexample, put $h$ equivalent to $f\&g$. Then $f$ confirms $h$ by converse entailment. Also $h$ entails $g$, hence $f$ confirms $g$ by special consequence. But $f$ and $g$ may be any propositions whatever, and the example has shown that the two conditions together entail that $f$ confirms $g$. A relation of confirmation which allows any proposition to confirm any proposition is obviously trivial and unacceptable, for it contradicts the tacit condition that confirmation must be a selective relation among propositions. A paradox has arisen by taking together a set of adequacy conditions, all of which seem to be demanded by intuition; hence it is appropriate to call it the *transitivity paradox*.

That converse entailment and special consequence are not always satisfiable together can be made more perspicuous than in the artificial counterexample just given. If $h$ entails $f$, and $h$ consists of a conjunction of premises representable by $h_1 \& h_2$, of which only $h_1$ is necessary to entail $f$, it is clear that $f$ may confirm only $h_1$. If in addition $h_2$ is that conjunct of $h$ which is necessary to entail $g$ (for example, $g$ may be $h_2$ itself), there is no reason why $f$ should confirm either $h_2$ or $g$. What seems to be presupposed in the examples mentioned as cases of the special consequence condition is a more intimate relation between the parts of the hypothesis which entail $f$ and $h$ respectively than mere conjunction. We shall later try to determine what this more intimate relation consists of.

It is important to notice how general and powerful the transitivity paradox is. Nothing has been assumed about the relation of 'confirmation' or 'support' except conditions that seem to arise directly out of commonly accepted types of theoretical argument. As soon as the relation of theory to evidence is construed in deductive fashion the conditions seem natural presuppositions of any relation of confirmation within such a system. In particular, no assumption has yet been made about the relation of *probability* to confirmation. Probability is the best explored measure of the confirmation relation, and many difficulties have been found in it. But that it cannot incorporate the intuitive conditions exemplified here in simple theoretical inferences cannot be held against probabilistic confirmation any more than against its rivals. For the conditions as they stand are inconsistent with any adequate confirmation theory, and this inconsistency may indeed drive us deeper into examination of the structure of theories than the disputes about probabilistic or nonprobabilistic theories of confirmation. We shall however first consider what might be done about the paradox in a Bayesian theory.

Expressed in Bayesian language the paradox is represented as follows. We are interested in the value of $p(e_2/e_1)$, where $e_1$ is an observed or otherwise given consequence of a theory $h$, and $e_2$ is an as yet unobserved or otherwise problematic further consequence of $h$, that is, an untested prediction made by $h$. Now $p(e_2/e_1)$ is a single-valued probability function of its arguments alone, and its value cannot depend on whether or not we are interested in some particular $h$ from which $e_2$ and $e_1$ are deducible. $e_1$ and $e_2$ are in general deducible from a number of different sets of premises in the language, whether we have thought of them or not. Some of them are perhaps potential theories. But whether they turn out to be viable as theories or not, $p(e_1/e_1)$ has its unique value independent of them.

If this result still appears counter-intuitive, we can make it a little more transparent in probabilistic terms. We want to show that $p(e_2/e_1) > p(e_2)$, which is the condition that $e_1$ and $e_2$ are not probabilistically independent of each other, but are related by positive relevance. We may be disposed to argue that if $h \rightarrow e_1 \& e_2$, then we should be able to show both that (1):

$p(e_2/e_1) \geqq p(h/e_1)$, since $h \rightarrow e_2$, and also that (2): $p(h/e_1) > p(e_2)$, since $h$ can be imagined to be increasingly confirmed by a greater and greater amount of evidence $e_1$ until its posterior confirmation is greater than the initial confirmation of $e_2$. Then it would follow immediately that $p(e_2/e_1) > p(e_2)$ as desired. But this intuitive argument is fallacious in general. (1) entails $p(e_1 \& e_2) \geqq p(h \& e_1) = p(h)$, since $h \rightarrow e_1$. (2) entails $p(h) > p(e_2)p(e_1)$. Hence both conditions cannot be satisfied together unless initially $p(e_1 \& e_2) > p(e_1)p(e_2)$, which is equivalent to the result which was to be proved, namely $p(e_2/e_1) > p(e_2)$.

Before considering various attempts to resolve the transitivity paradox, we may notice three other versions of general transitivity of confirmation which are not so counter-intuitive as to rate as paradoxes, but which are nevertheless seductive enough to be adopted in some discussions of confirmation, and are nevertheless fallacious for the positive relevance criterion.

## First fallacy

Consider what Hempel calls the *converse consequence* condition

If $f$ cfms $g$, and $h \rightarrow g$, then $f$ cfms $h$

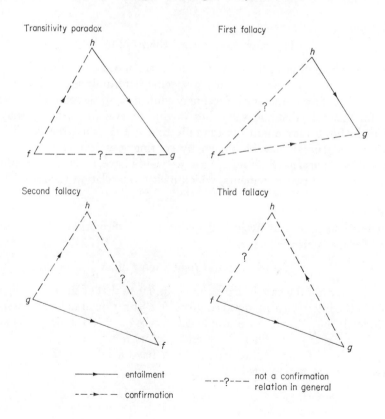

For example, evidence that confirms a generalization may be thought also to confirm a theory from which that generalization is deducible. However, the condition is not generally satisfiable, as Skyrms has shown[1] by considering this counterexample: for any $f$ and $h$, put $g \equiv f \vee h$. Then $f \rightarrow g$, and by the entailment condition $f$ cfms $g$; also $h \rightarrow g$. Hence by converse consequence $f$ cfms $h$. But a non-trivial relation of confirmation cannot subsist between every pair of statements, and so if entailment is retained, converse consequence must be rejected. It should be noticed, however, that the counterexample does not go through for converse entailment, which is a weaker case of converse consequence, nor for double entailment such as would be assumed in the case of theory-generalization-instance in a deductive theoretical system, since entailment is transitive.

Expressing this fallacy in terms of positive relevance confirmation, we have

$$\text{If } p(g/f) > p(g), \text{ and } h \rightarrow g, \text{ then } p(h/f) > p(h)$$

It is intuitively clear that this is not true in general, since both $p(h/f)$ and $p(h)$ are in general less than $p(g/f)$ and $p(g)$ respectively, and $h$ may contain elements not required in the deduction of $g$ which $f$ does not confirm.

### Second fallacy

$$\text{If } g \text{ cfms } h, \text{ and } g \rightarrow f, \text{ then } f \text{ cfms } h$$

This fallacy sometimes arises when $f$ and $g$ are more or less specific determinations of the evidence. The more specific determination $g$, for example that the value of a parameter $c$ is 0·50, may confirm $h$, while something entailed by $g$, for example that the value lies in the interval (0·49, 0·51), may not confirm $h$, for $h$ may entail for example that $c \geqq \frac{1}{2}$. That the condition is fallacious in general may be shown by putting $g \equiv f \& h$. Then $g \rightarrow h$, and hence $g$ cfms $h$; also $g \rightarrow f$. So if $f$ cfms $h$, we would have a relation of confirmation between any two statements, which makes the relation vacuous.

### Third fallacy

The condition produced by interchanging the confirmation relations in the second fallacy is also fallacious.

$$\text{If } g \text{ cfms } h, \text{ and } f \rightarrow g, \text{ then } f \text{ cfms } h$$

This is clearly not the case in an example similar to that of the last paragraph, for if $c$ in the interval (0·49, 0·51) confirms $h$, $c = 0·5$ need not confirm $h$, and may indeed contradict it, for example if $h$ entails $c > \frac{1}{2}$. To show the fallacy in general, put $g \equiv f \vee h$. Then $h \rightarrow g$ and $g$ cfms $h$ by converse entailment; also $f \rightarrow g$. So if $f$ cfms $h$, we would again have a relation of confirmation between any two statements.

[1] *Ibid.*, 238.

## III. Suggested resolutions

It is the transitivity paradox arising from special consequence that has the most serious consequences for confirmation in a scientific deductive system. The paradox was first noticed by Hempel and discussed by Carnap, and has been approached in various ways in the literature. Carnap, and later Hempel, proposed to adopt the positive relevance criterion for confirmation, which entails converse entailment, and hence entails abandoning special consequence, but neither Carnap nor Hempel considered the *prima facie* importance of special consequence in theoretical contexts, and therefore they did not discuss the implications of its abandonment. Other suggestions for resolving the paradox have included: restricting the standard rule of entailment pre-supposed in the application of special consequence; replacing the positive relevance criterion by the $k$-criterion of confirmation; and supplementing confirmation by concepts of 'acceptance' or 'explanation', both of which stand in need of explication at least as badly as does confirmation itself. All these suggestions appear undesirably formal, for the paradox does seem to be a sufficiently fundamental and material consequence of our inductive intuitions to require a type of resolution suggested by concrete cases rather than by formal considerations. In particular, the first suggestion[1] entails abandoning probabilistic confirmation in either the positive relevance or $k$-criterion form, and seems in any case too radical for explication of scientific inference which undoubtedly includes examples of standard entailment in its logical structure. The suggestions regarding acceptance and explanation will receive further attention in a later chapter, but the suggestion to replace positive relevance by the $k$-criterion needs a little more discussion here.[2]

The $k$-criterion for confirmation satisfies special consequence but not converse entailment. For, taking converse entailment in the form

$$\text{If } h \rightarrow g, \text{ then } p(h/g) > k$$

the condition is evidently not always satisfied, for if $h \rightarrow g$, $p(h/g) > k$ only if $p(h)/p(g) > k$, which is not true in general. However, taking special consequence in the form

$$\text{If } p(h/f) > k \text{ and } h \rightarrow g, \text{ then } p(g/f) > k$$

we have $p(h/f) > k$ and $p(g/f) \geq p(h/f)$; hence the condition is satisfied. Moreover, this version of special consequence may seem at first sight closer to our intuitions about transmission of confirmation in a deductive system. It does at least incorporate the idea that a well-confirmed hypothesis must pass this confirmation on to its logical consequences. But a closer look at the

---

[1] Discussed but not adopted by B. A. Brody, 'Confirmation and explanation', *J. Phil.*, **65** (1968), 282, and W. Todd, 'Probability and the theorem of confirmation', *Mind*, **76** (1967), 260.

[2] See H. Smokler, 'Conflicting conceptions of confirmation', *J. Phil.*, **65** (1968), 300.

way in which the condition is satisfied shows that the $k$-criterion does not satisfy it for these intuitive reasons. We are supposed to be given $p(h/f) > k$, but this does not imply that $f$ has raised the probability of $h$ from its initial value; it may indeed have lowered it, for since the $k$-criterion does not satisfy converse entailment, any relation between $p(h/f)$ and $p(h)$ is consistent with $p(h/f) > k$. Thus the idea that a hypothesis $h$ has 'intervened' to pass on confirmation from $f$ to $g$ is misleading, because to know that $p(h/f) > k$, we have implicitly to know already that all entailments of $h$ are given probability greater than $k$ by $f$, and therefore we could have known $p(g/f) > k$ without mentioning $h$. This objection to taking the $k$-criterion as the measure of the confirmation relation may not be conclusive, for we shall see that in certain limiting cases it is little different in effect from the positive relevance criterion. But if the $k$-criterion were adopted in general to avoid the paradox, we should still have the problem not only of explicating examples of transitivity which appear to require both converse entailment and special consequence to be satisfied, but also the problem of how generalizations and theories are confirmed by their own entailments.

In considering possible resolutions of the paradox, it is interesting to notice that there are some limiting circumstances in which transitivity of confirmation in a weaker form than the special consequence condition is valid for the positive relevance criterion, as well as for the $k$-criterion which satisfies the stronger special consequence condition in any case.

Consider the following form of transitivity condition, where entailment relations between hypothesis and evidence are replaced in general by probability relations.

If (i) $p(h/f) > \alpha$, (ii) $p(g/h) > \beta$, and (iii) $f$ is not negatively relevant to $g$ given $h$, that is, $p(g/f\&h) \geq p(g/h)$, then $p(g/f) > \alpha\beta$.

*Proof:* 
$$\begin{aligned} p(g/f) &= p(g/f\&h)p(h/f) + p(g/f\&\sim h)p(\sim h/f) \\ &\geq p(g/f\&h)p(h/f) \\ &\geq p(g/h)p(h/f) \quad \text{by (iii)} \\ &> \alpha\beta \end{aligned}$$

This theorem can be interpreted in terms of the positive relevance criterion as follows.

If  (i) $p(h/f) > \alpha$, where $\alpha$ can be as near to $1$ as we like
   (ii) $p(g/h) > p(g)/\alpha$
   (iii) $p(g/f\&h) \geq p(g/h)$
then   $p(g/f) > p(g)$.

That is to say, for positive relevance of $f$ and $g$, $f$ must highly confirm $h$, $h$ must be more than just positively relevant to $g$ to a certain degree depending on $p(h/f)$, and $f$ must not be negatively relevant to $g$ given $h$.

The principal application of this theorem is to statistical hypotheses whose initial probabilities are non-zero, and which can be indefinitely confirmed by accumulating evidence. Then with sufficient evidence, $\alpha$ can be made sufficiently near to 1 to ensure satisfaction of (ii) whenever $h$ cfms $g$, and if $f$ and $g$ are probabilistically independent given $h$, (iii) is also satisfied. Thus under the same conditions in which convergence of opinion is valid, transitivity in this form is also valid in the limit. However, as we have seen, the conditions under which convergence can be relied upon are not very common in theoretical inference in science, and therefore occasions of this use of limiting transitivity are also likely to be rare.

In the case of an entailment relation between $h$ and $f\&g$, the limiting transitivity theorem reduces to a simple application of (PC4). For then we have (ii) and (iii) satisfied by $p(g/h) = 1 = p(g/h\&f)$, and we need only to satisfy (i). But if $p(h/f) > \alpha$, it already follows by (PC4) that $p(g/f) > \alpha$, since $h \rightarrow g$. Consider first the cases of eliminative or near-eliminative induction discussed in section I(v). Here the probability of all statistical generalizations and all but a finite number of competing universal generalizations is assumed to be zero or vanishingly small, and by the Bacon–Mill methods the probability of the disjunction of true universal generalizations can be increased indefinitely. It follows at once by limiting transitivity that the probability of all consequences of these generalizations are also increased indefinitely.

More generally, where it is not assumed that $\alpha$ becomes indefinitely large, we have positive relevance of $f$ and $g$ if $\alpha = p(f)p(g)$, for if $p(h) > p(f)p(g)$, then, by (PC4),

$$p(f\&g) \geq p(h) > p(f)p(g) \qquad (6.1)$$

It follows by contraposition that, if $f$ and $g$ are *not* positively relevant, then we have $p(f\&g) \leq p(f)p(g)$, and hence, for all $h$ entailing $f\&g$, $p(h) \leq p(f)p(g)$.

Thus a sufficient condition for positive relevance of $f$ and $g$ is 'there exists an $h$ such that $p(h) > p(f)p(g)$'. This is, however, usually no help in *proving* positive relevance, since in order to know that the condition is satisfied we already have to know so much about $p(h)$, $p(f)$ and $p(g)$ as effectively to constitute knowledge of positive relevance. Also, (6.1) is not a *necessary* condition for all $h$ entailing $f\&g$ when $f, g$ are positively relevant, for there are always an indefinite number of hypotheses entailing $f\&g$ whose probabilities are *less* than $p(f)p(g)$, which can be constructed, for example, simply by conjoining further premises with $h$. It is therefore unlikely that, given some $f\&g$, we shall be able to identify just that part $h$ of a general theory consisting of several premises, such that the entailments of $h$ are sufficiently limited in content to satisfy $p(h) > p(f)p(g)$. The general theories we are likely to consider will be of greater content and lower probability. However, the theorem does help to make intelligible, if not justifiable in the light of the transitivity paradox, the intuition that we ought to look for theories with the highest possible probability consistent with entailing $f\&g$, where $g$ is the prediction required, since a theory may have a high probability partly because it contains an $h$ satisfying (6.1). This intuition becomes fully justified, of course, if and only if the conditions of near-eliminative induction are met.

In the next and subsequent chapters, I shall attempt to resolve the transitivity paradox in an illuminating and non-formal way, by examining in particular three types of inductive inference which can be regarded as special cases of the intuitions leading to the proposed special consequence condition. These are: inferences concerned with *enumerative induction,* namely general laws, instance confirmation and the nature of causal relations; theoretical inference from *analogies* and *models*; and the requirement of *simplicity.* I shall argue that these can be interpreted in terms of a unified theory of Bayesian inference which evades the transitivity paradox, and results in a finitist reinterpretation of the function of general laws and theories. These cases will also, incidentally, help to identify some theoretical statements which do have high enough probability on the evidence to ensure satisfaction of the limiting transitivity theorem by satisfying (6.1).

# CHAPTER SEVEN

# Universal Generalizations

## I. Exchangeability and clustering

The special consequence condition seems to be required in cases where it is desired to show that two consequences of a hypothesis are positively relevant to each other, and the transitivity paradox consists in the apparent irrelevance of the hypothesis to this dependence. This immediately suggests that Carnap's notion of *instance confirmation* of general laws, or Mill's *eduction* from particulars to particulars,[1] are paradigm cases of the sort of inference concerned here. Let us therefore first examine the problems surrounding confirmation of general laws.

Take a universal generalization $h$ in the form 'All objects that are $P$ are $Q$', and consider first only the class of individuals $a_1$, $a_2$, ... $a_n$ which are $P$s, thus evading for the moment the 'raven paradox' to be discussed later, in which individuals which are neither $P$ nor $Q$ are considered as evidence for $h$. Call those individuals in the $P$-class which are also $Q$s *positive instances* of $h$, and individuals which are $P$ and $\bar{Q}$ *negative instances*.

A single negative instance falsifies $h$ and sends its posterior probability to zero. It is usually assumed that a positive instance should confirm $h$, and an increasing number of positive instances should increasingly confirm $h$. A clause must of course be added to the effect that there is no other relevant evidence, to avoid confirmation of generalizations like 'I wake up every morning at 7 o'clock' by evidence consisting of sixty years' worth of mornings on which I have woken up at 7 o'clock.

Assume the initial conditions $P(a_1)$, $P(a_2)$, ... $P(a_r)$, ... to be given, and put their conjunction equivalent to $i$. Also put $e_r \equiv Q(a_r)$, and $E_r \equiv e_1 \& e_2 ... e_r$, $(r = 1, 2, ... n)$. Then, provided neither $p(h)$ nor $p(E_r)$ is zero, since $h \& i \rightarrow E_r$, and $i$ is given, by converse entailment $h$ is confirmed by $E_r$. It also follows that if $p(e_{r+1}/E_r) < 1$, all $r$, then $h$ is increasingly confirmed as $r$ increases. This increasing confirmation does not of course have limit 1 unless $h$ is *true*, for if $h$ is false a negative instance will eventually be reached, and will send the posterior probability of $h$ to zero.

---

[1] 'Eduction' is the term used by W. E. Johnson for Mill's induction from particulars to particulars (*Logic*, vol. 3, Cambridge, 1924, 43f). Carnap's instance confirmation is defined below, section III.

So far the intuitive requirements of enumerative induction seem to be satisfied, but several things about this apparent explication must be noticed immediately. First, it presupposes $p(h) > 0$, and as we shall see later this condition may be difficult to satisfy. Second, there are other generalizations besides $h$ which are increasingly confirmed by the same data, and they are generalizations which conflict with $h$ in at least some of their predictions. For example, consider $g \equiv$ 'All objects that are $P$ up to the $n$th to be observed are $Q$, and thereafter they are $\bar{Q}$'. This can be made into a generalization of the same form as $h$ by the following grue-like manoeuvre: define 'quiggle' to mean '$Q$ if applied to all $P$s to be observed up to the $n$th, and $\bar{Q}$ if applied to any other $P$'. Then $g \equiv$ 'All objects that are $P$ are quiggle'. Clearly, evidence up to the $n$th $P$ confirms both $h$ and $g$, but whether it confirms 'The $(n+1)$th $P$ will be $Q$' cannot depend on the confirmation it gives $h$, since it may give equal confirmation to $g$. Even if $h$ is confirmed by an infinite number of positive instances, a conflicting generalization may still be constructed having the same form and equally well confirmed by these instances. Consider the generalization $f \equiv$ 'All $P$ are quod', where 'quod' means '$Q$ if occurring on the even numbered intervals on a line, and $\bar{Q}$ if occurring on the odd intervals'. $f$ and $h$ share an infinite number of positive instances, but also conflict in an infinite number.

The confirmation given by $n$ instances of a generalization to the next instance must depend on the probabilistic relevance between the first $n$ instances and the rest. In dealing with gruified generalizations we have seen how the rest of the theoretical system may be exploited to exclude certain kinds of generalizations whose positive instances are not mutually positively relevant. In this chapter we shall deal with the question of what counts as positive instances and as positive relevance between them, as this has been raised in the raven paradoxes. In both cases restrictions on the merely formal characteristics of a generalization are not sufficient to ensure that just and only those predictions to future instances that ought to be confirmed are confirmed.

The third and most cogent objection to the simple account of enumerative induction just proposed is that it is useless for prediction. Indeed the argument that since $h\&i$ is confirmed by $E_r$, and $h\&i \rightarrow e_{r+1}$, therefore $e_{r+1}$ is confirmed by $E_r$, is a fallacious argument of just the kind that has been exposed in connection with the special consequence condition. However, since $h\&i \rightarrow e_{r+1}$ but not conversely, and therefore $p(e_{r+1}/E_r\&i) > p(h\&i/E_r\&i)$, $p(e_{r+1}/E_r\&i)$ may attain a positive and even a high value, even when $p(h) = 0$.

We have $e_{r+1}$ confirmed by $E_r$ given $i$ if and only if

$$p(e_{r+1}/E_r\&i) > p(\sim e_{r+1}/E_r\&i) \tag{7.1}$$

or alternatively

$$(pe_{r+1}\&E_r\&i) > p(\sim e_{r+1}\&E_r\&i) \tag{7.2}$$

Whether this is the case or not will depend only on the initial probabilities of the $e_r$s and their conjunctions given $i$. Assuming for the moment for simplicity that $p(e_r) = p(\sim e_r)$, all $r$, two further conditions are jointly sufficient for (7.1). I shall call them the *exchangeability* and *clustering* conditions.

The exchangeability condition is the condition of randomness of selection of individuals. Carnap calls it *symmetry* of individuals, and states it thus:

> A [probability] function is called symmetrical if it has the same value for any state-descriptions . . . such that one is constructed from the other by replacing individual constants with others.[1]

De Finetti takes the example of the probability of a particular sequence of coin tosses, where each toss constitutes a trial for heads or tails:

> . . . it is particularly interesting to study the case where the probability does not depend on the order of the trials. In this case every result having the same frequency $r/n$ on $n$ trials has the same probability, . . . if this condition is satisfied, we will say that the events of the class being considered, e.g., the different tosses in the example of tossing coins, are *exchangeable* (in relation to our judgment of probability).[2]

In particular, if tests are exchangeable, the initial probability of a given outcome of a single test of a $P$-object for $Q$-ness or $\bar{Q}$-ness will respectively be the same for each individual $P$-object, that is $p(e_1) = p(e_2) = \ldots = p(e_r)$. The tests are in this sense random over the individuals.

Exchangeability is not sufficient, however, to ensure that a conjunction of $r+1$ positive instances is more probable than a conjunction of $r$ positive instances with one negative instance, as we require to satisfy (7.1). Exchangeability tells us only that, if there is one negative instance, it is equally likely to turn up anywhere in the sequence, not that it is less likely that we obtain a sequence with one negative instance than all positive instances. Exchangeability is concerned only with the order of tests, not with the specific outcomes of those tests. Thus the assumption that the world is more likely to be homogeneous than heterogeneous, or as we shall say, *clustered*, in the sense that $r+1$ positive instances are initially more probable than $r$ positive and one negative, must be explicitly built into the initial distribution, and cannot be derived from the probability axioms and exchangeability alone. Indeed the counter-inductive instruction to predict negative instances after some finite fixed number of positive instances can be represented by Bayesian methods as easily as can enumerative inductive methods, simply by giving sequences of $n$ positive instances and the rest negative, higher initial probability than

---

[1] *Logical Foundations of Probability*, 483.
[2] B. de Finetti, 'Foresight: its logical laws, its subjective sources', *Studies in Subjective Probability*, ed. H. E. Kyburg and H. E. Smokler (New York, 1964), 121.

F

uniform sequences of positive instances.[1] The *clustering postulate* (or rather, the specific application of it in this case) states that: Given $r$ instances of $P$s, it is initially more probable that none or all will be positive instances of 'All $P$ are $Q$', than that there will be any other proportion of $Q$s. In particular the clustering postulate entails (7.2), or equivalently

$$p(PQa_1 \& PQa_2...PQa_{r+1}) > p(PQa_1 \& PQa_2...P\bar{Q}a_{r+1}) \qquad (7.3)$$

The postulate is stated in very limited form at present, and we shall see later that it has more general application than merely to enumerative induction. Meanwhile it should be noticed that the clustering postulate is directly contrary to the result that would be obtained if the $P$s were supposed to consist of a collection of *independent* $Q$s and $\bar{Q}$s equally weighted, for then the most probable composition of $r$ $P$s would be $r/2$ $Q$s and $r/2$ $\bar{Q}$s. In these matters of induction it must be assumed that God is not a coin-tosser.

What events are to be regarded as sufficiently 'of the same kind' to be exchangeable? For De Finetti this is a matter of personal belief.

> Let us consider, to begin with, a class of events (as, for example, the various tosses of a coin). We will say sometimes that they constitute the *trials* of a given phenomenon; this will serve to remind us that we are almost always interested in applying the reasoning that follows to the case where the events considered are events *of the same type*, or which have *analogous* characteristics, without attaching an intrinsic significance or a precise value to these exterior characteristics whose definition is largely arbitrary. Our reasoning will only bring in the events, that is to say, the trials, each taken individually; the analogy of the events does not enter into the chain of reasoning in its own right but only to the degree and in the sense that it can influence in some way the judgment of an individual on the probabilities in question.[2]

De Finetti's characterization needs clarifying in two respects. First, what he calls the relation of 'analogy' between events, or the judgment that they are of the same kind, leads in his example of the tossed coin to the assumption of exchangeability. But this assumption does not entail clustering, as we just saw in connection with counter-induction. In what follows, I shall generally assume that where there is a judgment of analogy or similarity, there is also a clustering postulate. Secondly, De Finetti holds that the judgment of analogy is 'largely arbitrary'. This is clearly not sufficient for an intersubjective confirmation theory, and must be supplemented by a positive account of intersubjective similarity such as that given in chapter 2.

---

[1] Though, as J. Dorling hints in 'Bayesianism and the rationality of scientific inference', *B.J.P.S.*, **23** (1972), 182, Bayesian methods cannot represent *general* counter-inductive instructions, if by this is meant *always* expecting the proportion of future positive instances to be the inverse of that observed, because this would lead in Reichenbach's terminology to a non-convergent and hence incoherent rule. In any case, it is doubtful whether a general counter-inductive rule can be stated without internal inconsistency (see R. S. McGowan, 'Predictive policies', *Aris. Soc. Supp. Vol.* 41 (1967), 57).

[2] *Op. cit.*, 120.

Our picture of enumerative induction is now as follows: a large number $r$ of instances of the same kind reveal themselves as positive instances of a certain universal generalization. In the absence of any negative instance in the data, the probability that the next instance of the same kind as the antecedent of the generalization is also an instance of the consequent, increases with $r$, and the same holds for the prediction that any finite number of next instances likewise satisfy the generalization. This result is independent of whether the generalization itself has zero or finite probability. Enumerative induction is a relation between individual instances in virtue of their similarity relations. From this point of view the universal generalization itself may be seen as that generalization which picks out positive relevance among its instances. *If the* instances are positively relevant, *then* they are suitable to be understood as positive instances of a significant generalization, but not every expression having the form of a universal generalization has positively relevant positive instances, as we saw in connection with grue-like transformations, and so not everything having the form of a generalization is scientifically significant.

We shall later consider what this account of instance confirmation involves for the notion of causality appropriate to general laws. Meanwhile it is convenient to give the account some immediate and more superficial support by considering how it deals with the so-called raven paradoxes which follow from the criteria of confirmation as so far stated.

## II. The raven paradoxes

In their original formulation the raven paradoxes follow from three apparently innocuous assumptions.

(1) The equivalence condition for confirmation.

(2) 'All $P$ are $Q$' is logically equivalent to 'All $\bar{Q}$ are $\bar{P}$', and to 'Everything which is $P$ or $\bar{P}$ is either $\bar{P}$ or $Q$'.

(3) Nicod's criteria: for any $P$ and any $Q$, (i) an object that is $P\&Q$ confirms $h \equiv$ 'All $P$ are $Q$'; and (ii) a $\bar{P}\&\bar{Q}$, and (iii) a $\bar{P}\&Q$, are respectively irrelevant to $h$.

Consider $h \equiv$ 'All ravens are black', or for short 'All $R$ are $B$'. By (3) this is confirmed by a black raven, and a non-black non-raven is irrelevant to it. But by (2) $h$ is equivalent to $h' \equiv$ 'All non-black things are non-ravens', which by (3 i) is confirmed by non-black non-ravens. By (1) anything which confirms $h'$ confirms $h$, hence non-black non-ravens after all confirm $h$, contrary to (3 ii). This constitutes the first paradox. By a similar argument a second paradox follows from (1) and (2), namely that anything that is a black non-raven confirms $h$, contrary to (3 iii).

The second paradox can be disposed of fairly easily. From (1) and (2) and case (CE3) of converse entailment (p. 136), it follows that if anything that is $R$ or $\bar{R}$ (that is, anything at all) is also *either* $\bar{R}$ or $B$, that thing confirms $h$. But it does not follow from this that anything which is *both $\bar{R}$ and B* confirms *h*. To suppose that it does is to commit what I have called the third fallacy of transitivity, namely 'If $f \to g$, and *g* cfms *h*, then *f* cfms *h*'. Putting $f \equiv \bar{R}Ba$ and $g \equiv (\bar{R} \lor B)a$, $f \to g$ and *g* cfms *h* by (1) and (2). But *f* does not necessarily confirm *h*.[1] To know that an object is either not a raven or is black is to have quite different data from the knowledge that it is a black non-raven, and no paradox arises from supposing that something which is only known to be either a non-raven or black confirms that all ravens are black. Indeed if a random sample of such objects were collected, that is *RB*s, $\bar{R}B$s and $\bar{R}\bar{B}$s, with no $R\bar{B}$s, such a sample would surely be regarded as confirming *h*. $\bar{R}\bar{B}$s on their own, however, as we shall see later, may be regarded as positively or negatively relevant or irrelevant to *h*, depending on other assumptions or background information. A similar argument disposes of the apparent paradox that something known to be a non-raven or known to be black confirms *h*, for again it does not follow that such instances confirm *h*, even though each entails *g*, and *g* confirms *h*.

The first paradox, however, cannot be disposed of so easily. Most discussions of it have accepted assumptions (1) and (2), and concentrated on breaking (3). (1) indeed seems intuitively inescapable, and is certainly demanded by a probabilistic confirmation theory because of the equivalence axiom for probability. It has also been generally accepted that (2) is required for a scientific law, whether or not laws are analysed as universal quantifications over material implications, that is, as $(x)(Rx \supset Bx)$. Any stronger form of connection between *R* and *B*, for example one expressing a causal modality, would certainly have to satisfy (2). The upshot of the present discussion for causal relations will be examined later; meanwhile I shall, in common with most recent writers on the paradoxes, accept (2) as well as (1).

To avoid contradiction it is therefore (3) that must be modified, and indeed investigation has tended to show that no one of Nicod's criteria is correct in all circumstances. The type of solution to the paradoxes that has found most favour depends on considering the further background knowledge we have, and in particular the facts that there are more non-black things than black in the universe, and far more non-ravens than ravens.[2] It is assumed that (3) does not even appear plausible unless the universe were thus disproportionate,

---

[1] In Hempel's first statement of the paradoxes ('Studies in the logic of confirmation', *Aspects*, 19), he proposed a solution of this type for both the first and second paradoxes. However, in a later discussion he almost falls into the third fallacy of transitivity himself: ' "All non-black things are non-crows", is confirmed, *inter alia*, by a white shoe, or rather, by the evidence sentence "*s* is not black and *s* is not a crow", where *s* is a name of some white shoe' ('The white shoe: no red herring', *B.J.P.S.*, **18** (1967), 239).

[2] See especially J. Hosiasson-Lindenbaum, 'On confirmation', *J.S.L.*, **5** (1940), 133; D. Pears, 'Hypotheticals', *Analysis*, **10** (1950), 49; H. G. Alexander, 'The paradoxes of confirmation', *B.J.P.S.*, **9** (1958), 227; J. L. Mackie, 'The paradox of confirmation', *B.J.P.S.*, **13** (1963), 265; and the very thorough development of this type of solution in R. G. Swinburne, 'The paradoxes of confirmation: a survey', *Am. Phil. Quart.*, **8** (1971), 318.

for if all these proportions were equal, we should accept non-black non-ravens as confirming instances of 'All ravens are black' by symmetry considerations alone. Supposing then that the proportion of ravens in all observed objects is $x$, the proportion of black objects in all observed objects is $y$, where $x < 1 - y$, both $x$ and $y$ small, and assume further that the initial probability of $a$ being a raven, given this background information $i$, is $p(Ra/i) = x$; the probability of $a$ being black is $p(Ba/i) = y$; and that the probability of $a$ being both a raven and black is $p(RBa/i) = xy$. The simplest assumption then to make about $p(RBa/h\&i)$ is that it is equal to $x$, since, assuming the observed proportion of $R$s to be the proportion in the population, and given that 'All $R$ are $B$', then the proportion of $RB$s is also $x$. This assumption is not, however, entailed by the other probabilities assumed. Now we obtain the following table of values of the relevant probabilities.

| $e$ | $RBa$ | $R\bar{B}a$ | $\bar{R}Ba$ | $\bar{R}\bar{B}a$ |
|---|---|---|---|---|
| $p(e/i)$ | $xy$ | $x(1-y)$ | $y(1-x)$ | $(1-y)(1-x)$ |
| $p(e/h\&i)$ | $x$ | $0$ | $y-x$ | $1-y$ |

It is clear from this table that $p(RBa/h\&i) > p(RBa/i)$, and hence by Bayes' theorem that $p(h/RBa\&i) > p(h/i)$, and also that $p(\bar{R}\bar{B}a/h\&i) > p(\bar{R}\bar{B}a/i)$ and hence $p(h/\bar{R}\bar{B}a\&i) > p(h/i)$. Thus $RBa$ and $\bar{R}\bar{B}a$ both confirm $h$ given $i$, and so far this is consistent with the first paradox. However, it is also apparent that, with the assumed values of $x$ and $y$, $RBa$ confirms $h$ (and of course $h'$) *more* than $\bar{R}\bar{B}a$ does, in the ratio $x/xy : (1-y)/(1-y)(1-x)$. This has been held to be a sufficient resolution of this paradox, and an indication that (3 ii) requires modification. It also follows from the table that $\bar{R}Ba$ *dis*confirms $h$, since $y - x < y(1 - x)$. Thus (3 iii) also seems to need modification.

By generalizing the assumed posterior probabilities in the second row (which is possible because they are not determined by the first row), Swinburne has shown that there are some circumstances of background information under which each of Nicod's criteria are intuitively incorrect. He argues that it may even sometimes be correct to hold that observation of $PQa$ *dis*confirms 'All $P$ are $Q$'; for example, when the hypothesis is 'All grasshoppers are located in parts of the world other than Pitcairn Island', there are assumed initially to be few grasshoppers, and observation of numbers of grasshoppers elsewhere makes it decreasingly plausible that Pitcairn Island, which is otherwise similar to other parts of the world, should be an exception in the case of grasshoppers.[1] Swinburne also shows with respect to the second paradox, concerned with black non-ravens, that with general values of the probabilities in the second row replacing those assumed above, nothing can be concluded

---

[1] Examples similar to this are used to establish the same conclusion by I. J. Good, 'The paradox of confirmation', *B.J.P.S.*, **11** (1960), 145, and **12** (1961), 63; and 'The white shoe is a red herring', *B.J.P.S.*, **17** (1966), 322.

in the absence of other evidence about whether $\bar{R}Ba$ confirms or disconfirms or is irrelevant to $h$.

Swinburne's general solution is an ingenious one to the paradoxes as originally proposed. However, several things should be noticed about any solution of this type (call it the proportionality solution) and about the statement of the paradoxes to which it is a solution. First, Nicod's criteria postulate instances found *by chance* out of the universe. But as Watkins and others have pointed out in this context,[1] in scientific observation we usually deliberately seek instances of a certain kind, that is, we look for *ravens*, rather than for any object at all, to confirm 'All ravens are black'. However, if we do restrict consideration to deliberately selected *positive* instances of a generalization, we come up against a condition which is more difficult to get rid of than Nicod's criteria, but which still generates a paradox. This is the converse entailment condition of confirmation. Let us replace Nicod's criteria (3) by a consequence of that condition.

(3′)                    If $h'\&j \rightarrow e$, then $e$ cfms $h'$ given $j$

Now putting $h' \equiv$ 'All $\bar{B}$ are $\bar{R}$', and $j \equiv \bar{B}a$, we have $e \equiv \bar{R}a$, and $\bar{R}a$ confirms $h'$ given $\bar{B}a$, so that to confirm $h \equiv h'$ it looks as though we can as well search deliberately for non-black objects to discover if they are ravens as search for ravens to discover if they are black. This result certainly still looks paradoxical, and cannot be excluded by reference to the random nature of Nicod's criteria.

If we adopt the proportionality solution, however, the appearance of paradox can be removed as before, and without abandoning (3′). Given that we only examine ravens, we have from the values assumed $p(Ba/Ra\&i) = y$, and $p(Ba/Ra\&h\&i) = 1$. Given that we only examine non-black objects, we have $p(\bar{R}a/\bar{B}a\&i) = 1-x$, and $p(\bar{R}a/\bar{B}a\&h\&i) = 1$. Since $x < 1-y$, black ravens confirm $h$ more than do non-black non-ravens, as before.

The second problem with this type of solution is more fundamental. So far the statement of the paradox and its solutions have all been concerned with the confirmation of universal generalizations, and not with confirmation of next instances. This would be unobjectionable if confirmation of $h$ entailed confirmation of next instances, but as we have seen this is not the case, for confirmation and disconfirmation of a universal generalization on present evidence are both logically compatible with disconfirmation or confirmation of even an infinite number of its next instances, and hence concern merely with the confirmation of generalizations is quite vacuous with respect to prediction. Moreover, assumptions made about proportions of objects in the population, and consequent assignments of probabilities as in the proportionality solution, are quite independent of assumptions required for next

[1] J. W. N. Watkins, 'Between analytic and empirical', *Philosophy*, 32 (1957), 112; J. Agassi, 'Corroboration *versus* induction', *B.J.P.S.*, 9 (1959), 311.

instance confirmation regarding the clustering of individuals with respect to their properties. For example, in all forms of the proportionality solution it is assumed that $p(RBa/i) = xy = p(Ra/i)p(Ba/i)$; that is, that having property $R$ is probabilistically *independent* of having property $B$. This entails that there is no probabilistic relevance between $R$ and $B$ as such, and hence that mere observation of a $R$ has no effect in confirming that it is a $B$. The question has never been raised in connection with this type of solution, as to how the probability of a given object or set of objects having $RB$ affects the probability of other finite sets of objects having $RB$, but it is just this question that has to be answered if we are to give an account of the confirmation of predictions.

In order to consider this question, suppose we accept that the initial probabilities of the first raven and the first black object to be observed are $x$ and $y$ respectively, and that their conjunction in a given object $a$ is initially probabilistically independent, so that $p(RBa) = xy$. Now, what is the probability that the next raven to be observed will be black? The assignment of probabilities so far has told us nothing about the initial probability of $RBa\&RBb$; indeed the proportionality solution of the paradox would be consistent with the properties of $a$ being probabilistically independent of those of $b$, so that the probabilities of both $Bb$ and $\bar{B}b$ given this evidence remain at their initial values $y$ and $1-y$ respectively. What is required is to assign probabilities to represent positive relevance between $RBa$ and $RBb$. Since the assumptions of the proportionality solution are not sufficient to do this they can throw no light on the character of a universal generalization as indicating positive relevance among its instances.

Let us, however, along the lines previously suggested, forget the confirmation of universal generalizations, and suppose that the only finite concern we can have with $h$ in a finite time is to confirm finite sets of predictions of the kind '$a$ is a raven, hence $a$ is black'. Suppose we have observed, as we may well have, many black ravens and no non-black ravens, and, as we certainly have, many non-black non-ravens. How much of this evidence supports the prediction 'Raven $a$ is black'? First, it is unlikely that we regard the test of raven $a$ for blackness as *exchangeable* with colour tests of herrings, polar bears, mail boxes and the like, since there is no reason to suppose that the probability of obtaining a given proportion of black to non-black objects in such a sequence of tests depends only on their number and not on their individual character as ravens, bears, etc. Second, there seems no reason to suppose that this motley collection forms a *cluster* in the sense of the clustering postulate. The intuition that led to the paradox in the first place is in fact just the intuition that particular non-ravens are *not* exchangeable for or clustered with ravens in testing 'All ravens are black'. But now we are stipulating that our assignment of probabilities explicitly depends on such intuitions, and is not dictated by logical equivalences of universal generalizations.

It is of course necessary to show explicitly that such intuitions can be satisfied consistently with the probability axioms. Absence of exchangeability can be interpreted in terms of low analogy of properties in the following sense. Suppose we are given a particular non-black non-raven, for example a white shoe, which has a number of other properties $C$, $D$, $E$, ..., and we are given in addition a raven with properties $\bar{C}$, $\bar{D}$, $\bar{E}$, ..., but we do not know its colour. Is the raven exchangeable with the shoe with respect to the probabilities of properties $R$ and $B$? Clearly not, because we have evidence of extreme disanalogy between these objects in respect of almost all known properties, and before objects can be regarded as exchangeable, all evidence must be taken into account. Not only are shoes and ravens not exchangeable, but it is to say the least questionable whether we should regard ravenness as positively relevant to blackness on the evidence of the white shoe. There seem to be two possible lines of argument here. We may adopt a kind of *negative analogy* principle, according to which the more different two objects are known to be, the more different they are likely to be in respects as yet unknown. This would dispose us to expect that the raven will be *non-white* (not necessarily *black*, since there is no reason on the evidence stated to distinguish among non-white colours). Such a negative analogy principle is equivalent to assuming that on the whole classes of objects do not overlap in respect of single properties, that is to say, if there is known to be a class of non-black non-ravens, then it is more likely than not that the class of ravens which does not overlap this class in respect of $R$ will not overlap it in respect of $B$ either, that is, that all ravens will be black. The negative analogy assumption tends to 'polarize' the world into classes defined by sets of uniformly co-present or uniformly co-absent properties. There does not seem to be much evidence that our intuitive methods do accord with such a negative analogy postulate except, as we shall see below, where there are strong assumptions yielding the methods of eliminative induction. On the other hand, in the absence of other evidence in this case, a strong *clustering* postulate would suggest that there is as much total homogeneity in the world as is consistent with the evidence, and in that case, given the white shoe and the raven, it will be more likely that the raven is as much like the shoe as possible, that is, that it is white.

Secondly, even assuming that sufficiently close analogy is present between the ravens and non-ravens to permit exchangeability (suppose the non-ravens are white crows, for example), it still does not follow from the probability axioms that we must adopt postulates which yield confirmation of black ravens by non-black non-ravens. Suppose we are given evidence $E$ consisting of $r$ objects that are $RB$, and another object $a$ that is $R$, and we assume as before that $Ra$, $Ba$ are positively relevant given $E$, that is

$$p(Ba/Ra\&E) > p(\bar{B}a/Ra\&E)$$

or equivalently

$$p(RBa\&E) > p(R\bar{B}a\&E) \tag{7.3}$$

It does *not* follow from this that $\bar{B}a$ is positively relevant to $\bar{R}a$ given $E$. We have

$$p(\bar{R}a/\bar{B}a\&E) = p(\bar{R}\bar{B}a\&E)/p(\bar{B}a\&E)$$
$$p(Ra/\bar{B}a\&E) = p(R\bar{B}a\&E)/p(\bar{B}a\&E)$$

and we may well wish to adopt a clustering postulate such that $r$ black ravens and one non-black raven is a more probable, because more homogeneous, state of the world than $r$ black ravens and one non-black non-raven. With that assumption

$$p(R\bar{B}a\&E) > p(\bar{R}\bar{B}a\&E) \tag{7.4}$$

from which it follows that $\bar{B}a$, $\bar{R}a$ are *negatively* relevant given $E$.

(7.4) would surely be the appropriate assumption if the $\bar{R}$s are crows, for given that something is as similar to a raven as a crow, and that so far all ravens have been black, in the absence of other evidence we are likely to predict that the crow will be black.

Conversely, if we are given evidence $E'$ consisting of $r$ non-black non-ravens, it does not follow that $Ra$, $Ba$ are positively relevant given $E'$, even if we postulate that $\bar{R}a$, $\bar{B}a$ are positively relevant given $E'$. For then we have

$$p(\bar{R}\bar{B}a\&E') > p(R\bar{B}a\&E') \tag{7.5}$$
$$p(Ba/Ra\&E') = p(RBa\&E')/p(Ra\&E')$$
$$p(\bar{B}a/Ra\&E') = p(R\bar{B}a\&E')/p(Ra\&E')$$

and consistently with (7.5) we may very well adopt a clustering postulate such that

$$p(R\bar{B}a\&E') > p(RBa\&E') \tag{7.6}$$

since $R\bar{B}a$ is more similar to the $\bar{R}\bar{B}$s in $E'$ that is $RBa$.

These results hold even in the special case of symmetry of predicates and their values, that is, on the assumption that all probability values are invariant respectively to interchange of $R$, $\bar{R}$, of $B$, $\bar{B}$, and of $R$, $B$. For then we still have that the relevance of $R$, $B$ given $E$ or $E'$, is independent of that of $\bar{R}$, $\bar{B}$. In this case, however, *if* the clustering postulates (7.3) and (7.4) are adopted, then (7.5) and (7.6) are entailed by symmetry of predicates. But freedom to choose or not choose (7.3) and (7.4) in the case of symmetry contrasts sharply with the restricted choice available in the proportionality solution for universal generalizations. For in that solution, symmetry is equivalent to the initially expected *equal* proportions of ravens and non-ravens, and black and non-black objects, that is, $x = y = \frac{1}{2}$, and not only is 'All $R$ are $B$' confirmed by $\bar{R}\bar{B}$s, but also the *amount* of confirmation by $RB$s and by $\bar{R}\bar{B}$s turns out to be the same. Such equal confirmation has generally been taken to be non-paradoxical in this case. In terms of next instance confirmation, however, it is quite plausible, and consistent with probability, to suppose that a universe consisting of $r+1$ black ravens is initially *more* probable than that consisting of a motley collection plus one black raven, even where expected frequencies of ravens and non-ravens, black and non-black objects are equal.

To summarize these results on the raven paradox, we have considered three different types of problem.

(1) Confirmation of the generalization $h \equiv$ 'All $P$ are $Q$' by instances of $\bar{P}\bar{Q}$ as well as of $PQ$ appears paradoxical. But by adopting the proportionality solution the appearance of paradox is removed by showing that with small proportions of $P$s and of $Q$s, though $\bar{P}\bar{Q}$ does confirm $h$ it does so much less than does $PQ$. Confirmation of $h$ has, however, in itself no implications for confirmation of next instances of either type.

(2) In general, next instances of $PQ$s are confirmed by evidence consisting of $PQ$s if and only if exchangeability and clustering postulates are adopted. $\bar{P}\bar{Q}$s may not even be regarded as exchangeable with $PQ$s, in which case no decisions about the probability of the next $\bar{Q}$ being a $\bar{P}$ or a $P$ are called for in the absence of further information about the other properties of the objects concerned.

(3) If $\bar{P}\bar{Q}$s are regarded as exchangeable with $PQ$s, however, they still do not confirm $h$ unless a specific clustering of $\bar{P}\bar{Q}$s with $PQ$s is assumed. No such assumption is required by the probability axioms, therefore no paradox arises with respect to confirmation of next instances. This result applies even when the proportion of $P$s, $\bar{P}$s, $Q$s and $\bar{Q}$s are assumed equal, so that $P$, $\bar{P}$ and $Q$, $\bar{Q}$ are symmetrical with respect to pro-bability values. In this case the proportionality solution demands that $\bar{P}\bar{Q}$s and $PQ$s confirm 'All $P$ are $Q$' equally; the next instance solution on the other hand does not demand that $\bar{P}\bar{Q}$s confirm that the next $\bar{Q}$ is $\bar{P}$ to any degree, unless a specific clustering postulate is adopted to entail this.

There is a further type of case, which has not been discussed in this section, in which confirmation of 'All $P$ are $Q$' by instances of type $\bar{P}\bar{Q}$ is not only non-paradoxical but actually welcomed. This is the case of the method of difference or co-absence in eliminative induction, where, as we have seen, very strong assumptions about the initial probabilities of uniform and statistical generalizations permit all but one universal generalization to be effectively eliminated by evidence. The difference between this kind of case and generalizations about ravens concerns so-called 'causal' assumptions, and hence raises wider issues with regard to Bayesian methods than do the raven paradoxes themselves. These issues will be discussed later, after we have considered the application of the clustering postulate to analogy arguments. But first it is convenient to compare the Bayesian methods already discussed with Carnap's explication of the same types of inference in his probabilistic confirmation theory.

## III. Clustering in Carnap's confirmation theory

Carnap's confirmation theory was developed almost entirely in terms of a first-order language containing a finite set of non-logical monadic predicate constants. In his first publications on the theory, the predicates are two-valued; later he extended the theory to many-valued predicate families in some special cases. Further extensions of the theory to relational and continuously variable predicates soon led into extreme mathematical difficulties, however, and although some of his attempts to deal with these are available in unpublished form, he was never able to complete anything like a general theory for unrestricted types of language.[1] The theory is developed far enough, however, to accommodate universal and statistical generalizations of the type we have been considering, and I shall describe its principles in sufficient detail to show how Carnap incorporates something corresponding to the clustering postulate for inductive inference.

Carnap's elementary language system consists of:

(i) As underlying logic, the first-order functional logic with identity and with only individual variables.

(ii) A finite or enumerable set of individual constants $\{...a_k...\}$.

(iii) A finite set of $\pi$ monadic two-valued non-logical predicate constants, the *P-predicates* $\{...P_j...\}$, whose values will be denoted by $P_j$, $\bar{P}_j$. The distinction between the name of the determinable predicate $P$ and its positive value will always be clear in context. There are $k = 2^\pi$ conjunctions of P-predicates which may apply to any individual. Each of these conjunctions will be called a *Q-predicate*.

Now we define a *state description* to be an assignment of a specific Q-predicate to each individual of the language, and a *structure description* to be a disjunction of all those state descriptions which are obtained from each other by mere permutation of the individual constants. In a language of $n$ individuals and $K$ Q-predicates there are $K^n$ state descriptions and $\tau = (n+K-1)!/n!(K-1)!$ structure descriptions. $K^n$ and $\tau$ are finite or enumerable according as $n$ is finite or enumerable.

Let the set of structure descriptions be the mutually exclusive and exhaustive set $\{h_1, h_2, ..., h_\tau\}$. Each structure description $h_s$ is specifiable by an ordered set of integers

$$\{q_{s0}, q_{s1}, ..., q_{s,K-1}\}$$

[1] See *Logical Foundations of Probability*, second edition (London, 1962), with a new preface on predicate families, and *Continuum of Inductive Methods*. In *L.F.P.*, 124, Carnap comments on the unknown mathematical properties of the combinatorial algebra of relations that would be needed to extend his theory to many-adic predicates. *Cf.* Suppes' remarks on the 'combinatorial jungle' resulting from the attempt to apply Bayesian methods strictly ('Concept formation and Bayesian decisions', *Aspects of Inductive Logic*, ed, J. Hintikka and P. Suppes, (Amsterdam, 1966), 41).

where $\sum_t q_{st} = n$, and $q_{st}$ is the number of individuals belonging to the $Q$-predicate $Q_t$ according to $h_s$. The $q$s will be called *q-numbers*.

Carnap introduces four important symmetry assumptions for the initial probability distribution over sentences of the language, of which the first three or their equivalents are almost universal among confirmation theorists. The fourth assumption generates Carnap's own distinctive $c^*$-distributions, and was later abandoned in order to develop alternative initial distributions and hence alternative inductive methods, by means of a parameter $\lambda$ varying from zero over all positive real numbers. The four symmetry assumptions are as follows.

### (1) *Symmetry of individuals*

This is the assumption that all probability values are invariant to mere permutation of the names of individuals, that is, of the individual constants. Thus in particular all state descriptions in a given structure description must be assigned the same probability value, and if there are $m$ states in the structure $h_s$ and the probability of the structure is $p_s$, then the probability of each state in $h_s$ is $p_s/m$. Symmetry of individuals is equivalent in De Finetti's language to *exchangeability*.

### (2) *Symmetry of P-predicates*

This is the assumption that probability values are invariant to interchange of $P_j$ and $P_k$, and of $P_j$ and $\bar{P}_j$, for all $j$, $k$. This implies that each value of $P_j$ is given initially equal weight $p(P_j a) = p(\bar{P}_j a) = p(P_k a) = \frac{1}{2}$, and every assignment of a $Q$-predicate to an individual has the same initial probability $1/K$.

The assumption may be criticized as importing an undesirable element of indifference into the specification of initial probabilities. How in practice, it may be asked, would we identify the basic stock of predicates to which equal weights are to be given? But if, as in the present account, we are attempting only a working model of a probability system sufficient to explicate common inductive inferences, it may be good policy to start with such a simplifying assumption, leaving the question of identification of the predicates in applications to a later stage of development of the theory. This at least seems to have been Carnap's view, for he regards all questions of application as part of the *methodology* of inductive inference, and as separable from the *logic* of inductive inference in which only the syntax of probability functions is discussed. In any case the idealized character of the symmetry assumption is somewhat mitigated by noticing that if we have reason to think that some predicates, or their values, should be weighted differently from others and we know what the weights should be, we can define a new set of fictitious predicates in terms of the old, and give these equal weights. This procedure might require use of predicates having more than two values, but Carnap has shown at least in principle how to provide such an extension of the theory.

## (3) *Symmetry of Q-predicates*

This is the assumption that probability values are invariant to permutations of $Q$-predicates, or, to put it another way, that the $q$-numbers are not ordered.

This assumption is much more controversial than the first two types of symmetry, for it implies an indifference between situations that are *syntactically* different, and not different only in semantic interpretation of what counts as an individual or a property. For consider two structures $h_r$, $h_s$ in a language of two predicates $P_1$, $P_2$, where $Q_1 \equiv P_1 P_2$, $Q_2 \equiv P_1 \bar{P}_2$, $Q_3 \equiv \bar{P}_1 P_2$, $Q_4 \equiv \bar{P}_1 \bar{P}_2$, and $h_r$, $h_s$ are specified respectively by the ordered $q$-numbers $\{q_{r1}, q_{r2}, 0, 0\}$, $\{q_{s1}, 0, 0, q_{s4}\}$ (all $q_r$s, $q_s$s non-zero). $h_r$ is a structure in which all individuals are similar in $P_1$ and different in $P_2$; $h_s$ is a structure in which all individuals are different in both $P_1$ and $P_2$. In other words, $h_r$ represents a more highly clustered universe than $h_s$. To adopt a symmetry assumption under which $q$-numbers are unordered is to obliterate precisely the distinctions we have found important in considering inductive inference, for all information about similarity relations between individuals other than possession of an *identical* set of properties (identical $Q$-predicates) has been lost. We shall see in the next section how this symmetry assumption affects the possibility of confirming analogical inferences.

## (4) *Symmetry of structure descriptions*

From symmetry of $Q$-predicates it follows that all structures having the same unordered $q$-numbers have the same initial probability, but it does not yet follow that *all* structures have the same initial probability. This is the extra symmetry assumption Carnap makes to generate his $c^*$-probability distributions over sentences of the language. He first discusses an indifference assumption which is even more simple, namely that all *state* descriptions have the same initial probability (his $c_0^+$-function). It is clear at once, however, that this assumption yields no inductive inference at all, for since *all* states of the world are equally probable, it follows that whatever evidence has so far been collected, the posterior probability over all states consistent with this evidence is still uniform, that is to say, any prediction from the evidence to a specific future observation is just as probable as any conflicting prediction about that observation. A world for which $c_0^+$ was believed to be the initial probability distribution would be a world in which all events were regarded as 'loose and separate', that is, a Humean world, or the world of Wittgenstein's *Tractatus*. As Carnap puts it, it is a world in which there is no 'learning from experience'.[1] But we do believe we learn from experience; hence $c^+$ is an inadequate explication of inductive inference.

The $c^*$-function derived from symmetry of structure descriptions fares better. It enables us to learn from experience at least in the particular cases of universal or statistical generalizations in finite domains, where the evidence

---

[1] *Logical Foundations of Probability*, appendix, 562, 565.

consists of a number of individuals having identical sets of predicates. To illustrate the difference between the $c^+$- and $c^*$-functions, consider the simple case of two individuals $a$, $b$, and one predicate $P$. The table shows the initial $c_0^+$ and $c_0^*$ values for each state of this world.

| States | Structures | $c_0^+$ | $c_0^*$ |
|---|---|---|---|
| 1. $Pa\&Pb$ | $\{2,\ 0\}$ | $\frac{1}{4}$ | $\frac{1}{3}$ |
| 2. $Pa\&\bar{P}b$ | $\{1,\ 1\}$ | $\frac{1}{4}$ | $\frac{1}{6}$ |
| 3. $\bar{P}a\&Pb$ | $\{1,\ 1\}$ | $\frac{1}{4}$ | $\frac{1}{6}$ |
| 4. $\bar{P}a\&\bar{P}b$ | $\{0,\ 2\}$ | $\frac{1}{4}$ | $\frac{1}{3}$ |

Suppose the evidence is $Pa$. What is the probability according to these two functions that the next individual is $P$ or $\bar{P}$? We have

$$c^+(Pb/Pa) = c_0^+(Pa\&Pb)/c_0^+(Pa)$$
$$= \tfrac{1}{2}$$
$$= c^+(\bar{P}b/Pa)$$

and hence there is no learning from experience. But

$$c^*(Pb/Pa) = c_0^*(Pa\&Pb)/c_0^*(Pa) = \tfrac{1}{3}/(\tfrac{1}{3}+\tfrac{1}{6})$$
$$= \tfrac{2}{3}$$
$$> c^*(\bar{P}b/Pa)$$

and hence there is learning from experience.

The possibility of learning is represented in the $c^*$-function by the equivalent of a clustering postulate which gives greater initial probability to states in which all individuals have identical $Q$-predicates, that is, to states 1 and 4 in the example. Where the $Q$-predicates of different individuals are different, $c^*$ will always assign lower initial probability to the corresponding state, so that where there are identical $Q$-predicates in the evidence, predictions that next instances will be as before will always be confirmed. Carnap's generalized $\lambda$-functions for instance confirmation differ from $c^*$ in the amount of clustering they presuppose in the initial distributions: $\lambda = 0$ corresponds to the assumption of maximum clustering and is equivalent in elementary cases to Reichenbach's straight rule; $\lambda = K$ corresponds to $c^*$; and $\lambda$ infinite corresponds to the unclustered or Wittgensteinian world represented by $c^+$.

Suppose that instead of calculating the probability of the next instance, as in the above example, we ask for the probability of a generalization of the type $h \equiv$ 'All $P_1$ are $P_2$' on the evidence $e$ of $s$ positive and no negative instances. Carnap shows that if the number of individuals $n$ is finite, $c^*(h/e)$ is finite and increases with $s$. But if $n$ is enumerably infinite, $c_0^*(h)$ is zero and hence $c^*(h/e)$ is zero for any $e$ whatever. This result holds in general for the

$c_0^*$ value of any universally quantified hypothesis in an infinite domain, and has been considered the death-blow to any confirmation theory of Carnap's type, since it is generally assumed that universality in infinite domains is an essential characteristic of scientific laws and theories, and must be capable of confirmation. It is easy to see that the same result must follow for *any* method of calculating initial probabilities that depends on indifference among structure descriptions, since the number $\tau$ of structure descriptions is infinite for infinite $n$.

Carnap's own response to the zero confirmation of universal laws is to argue that the application of inductive logic never involves more than finite sets of instances, and so he is content to allow non-zero confirmation values only to what he calls *instance confirmation* (the probability that the next individual will be a positive instance of a law), and to *qualified instance confirmation* (the probability that the next instance satisfying the antecedent of the law will satisfy its consequent).[1] Both these probabilities are calculable and independent of $n$, and hence finite for all $n$. Carnap's proposal has not found much favour, but it is in line with the 'next instance' approach adopted here, for we have already found reasons to doubt the utility of confirmation of universal laws and theories compared with that of next instances. Further arguments in support of this general position will be discussed in the next chapter.

## IV. Extension to analogical argument

So far, the clustering postulate has been applied to generalizations involving only two predicates and their negations. This is, of course, an idealization, since positive instances of such generalizations will certainly always differ in *some other* properties, however trivial. The judgments that instances are exchangeable and clustered are judgments that the respects in which the instances differ are irrelevant to confirmation relations between $P_1$ and $P_2$ and can be neglected. There are, however, other types of case in which objects are known to be different in some properties as well as being similar in some others, and in which nevertheless some *analogical* argument from the properties of one object to the yet unknown properties of the other would be regarded as justified. Such analogical arguments appear at all levels of scientific theory: it is assumed that all solid bodies however otherwise different

---

[1] Carnap speaks of the qualified instance confirmation of the *hypothesis h* as being given by the q.i.c. of its next positive instance, but this is an oversight, since such a function of the hypothesis does not satisfy the probability axioms. The q.i.c. must be the probability of at most finite conjunctions of next instances, not of a universal hypothesis in an infinite domain. Popper comments that the q.i.c. as Carnap understands it is 'squarely hit by the so-called "paradox of confirmation"' (*Conjectures and Refutations*, 282n), just as is the confirmation of universal generalizations, if this is finite. Popper fails to notice, however, that if q.i.c. is interpreted as the probability of the next positive instance, it is *not* hit by the paradox, as we saw in the previous section.

gravitate; that analogous chemical elements produce compounds with analogous properties; that micro-objects and events may be reconstructed from macroscopic models; and many other such inferences.

Goodman has discussed inferences of this type under another name in his theory of 'overhypotheses'.[1] A positive overhypothesis is of the form 'Every bagful of marbles in stack $S$ is uniform in colour', which is supported by every instance of a bagful in $S$ that is found uniform in colour, and in turn increases the support which 'All marbles in bagful $B$ in $S$ are red' derives from its own positive instances. For such increase of support, Goodman argues, it is necessary that the predicates in the overhypothesis be *projectible* (in the sense that 'green' is projectible, and 'grue' is not). This is an example of transitivity of confirmation through a higher-level hypothesis, which yields a justified inference, in fact an analogical inference. There are, however, two shortcomings of Goodman's proposed explication of this case.

(1) He understands projectibility wholly in terms of entrenchment in the language. But we have seen in the case of the entrenchment of 'green' as compared to 'grue' that this is not sufficient to account for projectibility unless entrenchment coheres with the rest of the theoretical network.

(2) His proposal is too weak, since it does not explain why hypotheses containing non-entrenched predicates can sometimes be seen to act as positive overhypotheses even in the absence of a general theory. In a discussion and modification of Goodman's account, Swinburne suggests that it applies to the following case: 'that all specimens of various inert gases behave in a certain way increases the probability that all specimens of another inert gas will also behave in that way'.[2] But 'inert gas' has not been very long in the language, neither have there been other well-entrenched predicates which are co-extensive with it, and yet this would have been a good analogical inference even before this predicate was entrenched at all. A relation of analogy between inert gases, not entrenchment, is what makes this inference justifiable.

Analogical inferences will be interpreted here in terms of clustering of objects whose mutual similarities outweigh their differences for objects in the same cluster, and whose differences outweigh their similarities for objects in different clusters. What is now required is to extend the notion of clustering from simple generalizations to deal with diverse objects predicated by a large number of different properties. Unfortunately, even the simplest possible extension of the next instance method leads at once into intractable mathematical complexity, and it proves impossible to give a general combinatorial theory of clustering for all possible states of the universe. To see why the problem escalates in this way, let us consider how a theory of Carnap's type would have to be extended to deal in general with analogical inference.

[1] *Fact, Fiction, and Forecast*, 110f.
[2] R. Swinburne, *An Introduction to Confirmation Theory* (London, 1973), 179.

In defining his $c^*$-function, as we have seen, Carnap makes the postulate of symmetry of $Q$-predicates. This means that every structure description has the same probability, and if one structure contains fewer state descriptions than another, because some of its individuals have identical predicates, then single states of that structure have higher probability than single states of the other structure. In order to consider a rich enough language for the representation of analogical inference, take four two-valued monadic predicates $A$, $B$, $C$, $D$, where we assume all predicates and both values are symmetrical. Then Carnap's $c^*$-distribution entails that, for two objects $a$, $b$, the state $ABCDa\&ABCDb$, which is the only state in its structure, has higher probability than any state where $a$ and $b$ differ in one or more predicates. The same is true for any state in which both objects have identical predicates. Moreover, the states in which $a$ and $b$ differ all have the same probability, since there are just two of them in each corresponding structure. This distribution satisfies a clustering postulate for simple generalizations, since it satisfies

$$p(ABCDa\&ABCDb) > p(ABCDa\&ABC\bar{D}b)$$

But suppose $a$ and $b$ are known to differ in just one of the properties, and to be similar in two others. A simple analogical argument would be one in which it is regarded as justified to assume *in the absence of any other information*, that since similarities outweigh differences, $a$ and $b$ are to be assumed similar in the remaining one of the four properties. That assumption corresponds to a probability distribution such that

$$p(ABCDa\&AB\bar{C}Db) > p(ABCDa\&AB\bar{C}\bar{D}b)$$

However, it follows from the symmetry of $Q$-predicates that these two expressions are equal, and similarly for any distribution of predicates in which at least one $P$ is known to apply to $a$ and its negation to $b$. A highly counterintuitive consequence follows, namely that however greatly similar or greatly different two objects may be, so long as they differ *at all*, however little, analogical argument from one to the other has the same justification in terms of confirmation, that is, no justification at all. It would certainly be more intuitively plausible to have a probability distribution for which confirmation of the next similarity *increases* with known similarity, and perhaps decreases with known difference. For example, writing now $S_1ab$ for one $P$-predicate similar between $a$ and $b$, and so on for two, three, four predicates similar, we might require

$$p(S_4ab) > p(S_3ab) > p(S_2ab) > p(S_1ab) > p(S_0ab)$$

No theory which accepts symmetry of $Q$-predicates can satisfy this condition or anything corresponding to it. Carnap has indeed remarked on this inadequacy of his $c^*$ and $\lambda$-theories, and has suggested (with Stegmüller) a theory (the $\eta$-theory) which permits analogical inference in special cases.[1] But unfortunately practically nothing is known about the mathematical properties of the combinatorial algebra involved in this type of extension of Carnap's system; therefore it is not profitable, at least at present, to pursue such an approach to analogical argument. This means that we cannot hope to find any definite probability distribution of Carnap's type which is anything like adequate to represent scientific inference, and any detailed explication of induction along these lines must await development of new methods.

## V. Causal laws and eliminative induction

One way in which this 'combinatorial jungle' is made manageable both in thought and in logic, is the assumption that under certain circumstances uniform causal laws are more probable than statistical generalizations, and therefore that eliminative or near-eliminative methods of induction are applicable. We have already seen in the previous chapter that the assumptions of eliminative induction justify methods of variety and difference which appear to be quite distinct from those employed in the enumerative inferences discussed in the present chapter. It has frequently been concluded that enumerative and eliminative induction are quite different types of inference, applicable to different kinds of subject matter. I believe this conclusion is mistaken, and that if the conditions of eliminative induction are further scrutinized, they will also be found to be explicatable by Bayesian methods in terms of the clustering postulate.

Eliminative induction requires that the probability that there is at least one pair of properties which are uniformly 'causally' related is unity or nearly unity, that is, that the probability that all relations between properties are statistical is very small. Two questions arise about this assumption. First, does the reference to a uniform *causal* relation imply that some special relation is presupposed in eliminative induction over and above a Humean uniform correlation such as we have assumed in discussing enumerative induction? And second, if there is no such extra feature, can the assumptions of eliminative induction be justified by enumerative methods?

With regard to the first question, I have suggested in the earlier discussion of universals that Humean analyses of universals and of causes are closely related. We need not recognize 'natural kinds' distinct from relatively stable classifications based on similarities and differences, and neither need we assume some special category of causal relations distinct from correlations of presence and absence. Just as I have argued that classifications related in

---

[1] R. Carnap and W. Stegmüller, *Induktive Logik und Wahrscheinlichkeit* (Vienna, 1959), appendix B. The extension of Carnap's theory to accommodate arguments from analogy is discussed in P. Achinstein, 'Variety and analogy in confirmation theory', *Phil. Sci.*, 30 (1963), 207; and M. Hesse, 'Analogy and confirmation theory', *Phil. Sci.*, 31 (1964), 319.

theoretical networks are adequate for the understanding of universal terms, so I shall argue in detail in the next chapter that co-presences and co-absences related to a theoretical network are adequate for the understanding of cause. What *counts* as causal relation is uniform correlation of properties found to be connected in large numbers by different kinds of objects and systems, and related to other such correlations by an acceptable theory.

The reason why it is pragmatically necessary to resort to speaking of causal laws is not that causality brings in any extra element over and above clustering plus systematic structure, but rather that with the background information we generally have, and the number and complexity of the objects involved, it is impossible to hold all correlations in mind separately. The 'causal laws' are nothing but those generalizations which most conveniently summarize this data for the purposes we have in mind. This explains why the examples of clustering and analogy that have so far been considered in this chapter have been so simple as to misrepresent almost all realistic situations. But I hope to show now that they are to be understood as elementary cases of the same kind as those out of which all causal systems in science, however complex, are constructed.

The assumption that certain properties are good candidates for uniform causal relations can be justified on past evidence by Bayesian methods and clustering. Let us consider a classic example of analogy between two systems where the inference expected from a clustering postulate appears to be contradicted by assumptions about uniform causality. This example is the strongest sort of test of the assertion that, when all evidence is taken into account, clustering is a sufficient postulate of induction. The example is that appealed to by all nineteenth-century writers on analogy, namely the putative inference from existence of life on earth to life on Mars, given a strong analogy in some respects between the earth and Mars. Both bodies are planets of the solar system, spherical, move in elliptic orbits, and so on. But their surface temperatures are different, the earth has atmosphere and fluid water, Mars as far as we know does not, and so on. There is a temptation here to argue purely causally as follows: since atmosphere and the rest are causally necessary to life, absence of this property alone would be sufficient to give a negative answer to life on Mars, even if all other properties of the two bodies were similar. Hence any analogy argument from a preponderance of similarities would be fallacious.

This way of putting it, however, neglects to take account of all the evidence. Evidence from *other* systems for the claimed causal law must be included, and this yields uniform correlations of presence of oxygen and life in a great variety of circumstances, supported by a theoretical system which entails that this is a uniform causal law. We shall see in the next chapter that even the phrase 'supported by a theoretical system which entails that this is a uniform causal law' is at best an approximate way of putting the matter, since as it stands it commits a fallacy of transitivity. The support of the theoretical system should rather be expressed by saying that this causal law is confirmed both by deductive connections with more general laws of chemistry, and also, and more importantly, by analogies between this law and other low-level laws relating presence of oxygen with various kinds of oxidation process. Schematically, then, we have a very high probability that the next occurrence of life will be accompanied by presence of oxygen, under all statable circumstances, and hence that absence of oxygen in any instance highly confirms the absence of life in that instance. This confirmation would also be increased by a close analogy in other respects between the earth and Mars, since the conditions under which oxygen and life have always been found to be co-present and co-absent on the earth would then more nearly be approached on Mars. The argument from total clustering here of many already observed systems outweighs the direct analogical argument from planet to planet, which suggests that since Mars is predominantly similar to the earth it should contain life, but which neglects all the rest of the evidence.[1]

These considerations throw some light on the suggestion of a 'negative analogy' postulate mentioned in connection with the raven paradoxes. If two objects *a*, *b* are known to be much more different than similar, is it to be assumed that they will continue to be different in as yet unknown properties? If clustering is the only postulate appealed to, *and there is no other evidence*, the answer is that they will more probably be as *similar* as possible in yet unknown properties. But this consequence seems to be contradicted by some cases in which, from evidence of many $PQ$s in a variety of circumstances, we would predict that the next $\bar{Q}$ will be $\bar{P}$, as in the case of absence of oxygen yielding a prediction of absence of life. However, the two sorts of cases are different, and the appearance of contradiction is illusory. For in the first place, where co-absence is predicted on the evidence of many instances of co-presence, a great deal of evidence of objects *other* than *a* and *b* is taken into account, and this evidence is such as to increase both the probability of $Qa$ given $Pa$, and of $\bar{P}a$ given $\bar{Q}a$. Secondly, a negative analogy postulate would in general have different consequences from the clustering postulate in cases like this. For negative analogy would demand that the more *different* *a* and *b* are known to be, the greater the confirmation of difference in as yet unknown properties, whereas, as we have just seen, if clustering yields

$$p(\bar{P}a/\bar{Q}a\&PQb\&E) > p(Pa/\bar{Q}a\&PQb\&E)$$

[1] The probability relations of such systems of properties, under the assumption of high probability of some uniform correlations between properties, are worked out by J. M. Keynes in terms of property 'groups' (or natural kinds), in his 'Principle of limited independent variety'. (*A Treatise on Probability* (London, 1921), 251f. Cf. the analysis of his method by S. F. Barker, *Induction and Hypothesis* (Ithaca, New York, 1957), 55f.) Keynes also recognizes a 'clustering postulate' in the sense that 'some element of analogy must . . . lie at the base of every inductive argument' (*Treatise*, 222).

where $E$ is the rest of the total evidence, then the more *similar* $a$ and $b$ are known to be in respects other than $P$, $Q$, the greater the confirmation of $\bar{P}a$. This consequence of the clustering postulate is a better account of inductive practice than is negative analogy. It therefore appears that the suggested negative analogy postulate is both unnecessary and undesirable.

Let us summarize this discussion by sharpening up the relation between eliminative and enumerative induction. Taking $h$ to be the hypothesis 'All $P$ are $Q$', we have seen that $h$ can be made increasingly probable by favourable evidence $E_r$ under the conditions of eliminative induction, provided of course that $p(h)$ is initially finite. Then it follows by limiting transitivity that both $p(Pa/Qa\&E_r)$ and $p(\bar{Q}a/\bar{P}a\&E_r)$ can be made as large as we like by increasing favourable evidence $E_r$. It should be noticed that this result can be expressed in weaker form by supposing that the hypothesis is not quantified over an infinite domain, but only over a finite domain which may be specified to be very large. That is, we may replace $h$ by $h' \equiv$ 'All the next $n$ $P$s to be observed are $Q$', where $n$ is a definite number as large as we like, and $a$ is understood to be one of these $n$ instances. This device is adopted here to eliminate difficulties in satisfying the condition $p(h) > 0$, and will be discussed and justified in the next chapter. Now let us consider how the same situation would be formulated as an enumerative induction. Here we assume that the initial probability distribution is such that $p(Pa/Qa\&E_r)$ and $p(\bar{Q}a/\bar{P}a\&E_r)$ have the same values as in the eliminative formulation, the consistency of such a distribution being guaranteed by the existence of the eliminative formulation. This distribution over instances expresses a bias towards uniform correlation of $P$ and $Q$ by asserting that, if all $P$s have been $Q$s in the past under all circumstances, then the next $P$ will most probably maintain this correlation and hence be a $Q$, and the next $\bar{Q}$ will maintain this correlation and hence be a $\bar{P}$. Once the strictly finite $h'$ is adopted, the eliminative and enumerative formulations are entirely equivalent, and it becomes clear that the presupposition of eliminative induction, namely that $h'$ should be increasingly confirmable by $E_r$, rests essentially upon a clustering postulate for the instances of $h'$, rather than upon the dubious notion of an uncompletable elimination of all rival hypotheses. It follows that enumerative induction is more fundamental than eliminative.

# CHAPTER EIGHT

# Finiteness, Laws and Causality

## I. The distribution of initial probabilities

The non-transitivity of confirmation has led to the conclusion that if we are concerned with confirmation of predictions from hypotheses, and if we are not in a position to use limiting transitivity, then the probability of hypotheses is irrelevant to confirmation of predictions, and might be zero without affecting that confirmation. This conclusion may seem to tend towards instrumentalism, and to have disastrous consequences for any realist interpretation of theories or for any strong view of the causality expressed by scientific laws and theories. I shall argue in this and the next chapters that it does not have such consequences. But first it is necessary to consider how far hypotheses can be given *non*-zero confirmation, because apart from confirmation of predictions, there certainly are some types of inference, such as eliminative induction, where finite probabilities of hypotheses seem to be presupposed. I shall first examine some attempts within Carnapian confirmation theory and elsewhere to assign non-zero probability to universal generalizations, and then consider the relation between zero probability and the causality or necessity of laws.

If the basic language of the confirmation theory is in any way infinite, whether with regard to individuals or predicate-values, an infinite number of hypotheses are in principle specifiable, and clearly not all of them can have finite initial probability without violating the axiom that the probability of their total disjunction is 1. Indeed only a finite subset of such hypotheses can be assigned probability greater than some fixed number, however small, and the rest of the infinite set must have initial probability zero. In particular, for a language such as Carnap's, with a finite number of predicates, if the number of individuals is denumerably infinite, any simple kind of indifference assumption over logically possible states of the universe leads to zero initial probabilities for all hypotheses describing particular states. In Carnap's theory this result is not confined to universal generalizations of the type 'All $P$ are $Q$' (call this type (U)), but applies also to statistical generalizations of the type 'A proportion $p$ of $P$s are $Q$s', for given $p$, since in a denumerably infinite universe $p$ may take all rational values between 0 and 1.

That zero probability has to be assigned to universal generalizations in Carnap's confirmation theory, and in all obvious modifications of it, has been thought to be a strong objection to the whole programme of developing a probabilistic inductive logic. It does of course appear to be extremely counter-intuitive, because it means that there are never any better grounds in this theory for asserting 'All $P$ are $Q$' on the evidence of a large sample of $P$s which are $Q$ than on no evidence at all, or, worse, on the evidence of a large sample of $P$s which are not $Q$, since the confirmation of a false assertion is also zero. The same objections count almost as strongly against the zero initial probability of statistical generalizations. Unlike universal generalizations, statistical generalizations cannot be shown conclusively to be false by finite evidence, but surely some evidence should provide some confirmation of them, and some kinds of evidence should provide more confirmation than others. However, replacement of statistical generalizations with point values of $p$ by 'interval' generalizations is common in statistics, and results in finite confirmation values for generalizations of the form 'A proportion $p$ of $P$s are $Q$s', where $1 \geqq p \geqq 1 - \eta$, $\eta$ small, of which there is a finite number. This suggests a classification of the possible ways in which a finite subset of generalizations might be given finite probability, when the domain of individuals is denumerably infinite.

Within Carnap's language there are broadly three such ways.

(*a*) Assign non-zero probability to finitely many hypotheses and zero probability to denumerably many.

(*b*) Assign denumerably many hypotheses non-zero probability, in which case there is no lower limit to these probabilities; in other words we cannot satisfy the condition $p(h) \geqq \varepsilon > 0$, $\varepsilon$ fixed, for all $h$.

(*c*) Assign finite probability to infinite disjunctions of hypotheses, in which case each component of such a disjunction may have zero or vanishing probability, or, what is equivalent, assign finite probability only to hypotheses ranging over *finite* domains of individuals.

Suppose, first, we pick out *a priori* some finite subset of hypotheses for preferential treatment. The most obvious way to do this in conformity with our intuitions about causal uniformities is to give initial preference to universal as opposed to point statistical hypotheses. For finite $\pi$ there is a finite number of hypotheses of type (U), and the initial probability of each may be put equal to $\rho$ ($\rho$ constant for each hypothesis to satisfy symmetry of $P$-predicates). $\rho$ may then be used to determine the initial probability of any state description with one or more $q_i$ equal to zero, that is to say, of any state description in which at least one universal generalization is true. Suppose the sum of the initial probabilities of such state descriptions is $\sigma$, where $\sigma < 1$. The disjunction of all state descriptions in which no universal generalization is true may then be given probability $1 - \sigma$, so that some statistical hypotheses may have finite probability, but denumerably many will not. This is essentially the method adopted by Keynes, Nicod and von Wright,[1] although the two last consider only confirmation of universal generalizations by the instances entailed by them, and hence do not specifically discuss confirmation of statistical generalizations.

The most detailed attempt to develop a confirmation theory along these lines is due to Hintikka.[2] Adopting the Carnapian formulation in terms of structure descriptions and $Q$-predicates, he defines a *constituent* (corresponding to a disjunction of possible worlds) as a sentence asserting of each $Q$-predicate that there is or is not an individual satisfying it. If there is in a given constituent a $Q$-predicate not satisfied by any individual, then at least one universal generalization is true in that constituent. In a language with $K$ $Q$-predicates there are $2^K$ constituents. Hintikka first proposes to assign equal initial probability $1/2^K$ to each constituent. Universal generalizations then have finite initial probability, and he shows that their posterior probabilities behave well qualitatively, but are somewhat over-optimistic quantitatively, leading to unduly high posterior values on little evidence. To rectify this fault, Hintikka next proposes to give higher initial probability to those constituents which have *more* $Q$-predicates satisfied by individuals, that is, to to those which include *fewer* universal generalizations.

In Hintikka's system, as in Carnap's, the symmetry of $Q$-predicates is assumed. We have seen that any system including this postulate is not capable of explicating analogy arguments of a kind fundamental for science. Hintikka's system must therefore be rejected as it stands. It would be interesting to investigate the consequences in this system of dropping this symmetry postulate, but at first sight the proposal to do so does not look promising. Explication of analogy arguments requires that higher initial probability be assigned to relatively clustered universes, that is, in Hintikka's terms, to universes in which relatively *few* $Q$-predicates are instantiated and *many* universal generalizations are true. But this would lead in Hintikka's system to even greater over-optimism in regard to the confirmation of universal generalizations than in his original assignment of equal probability to constituents. Thus there is certainly no simple way of adapting the system to the requirements of analogy arguments.

A quite different type of suggestion is made by Shimony.[3] He effectively

---

[1] J. M. Keynes, *A Treatise on Probability* (London, 1921); J. Nicod, 'The logical problem of induction', *Geometry and Induction* (English trans., London, 1930), 157; G. H. von Wright, *The Logical Problem of Induction* (Oxford, 1957).

[2] J. Hintikka, 'Towards a theory of inductive generalization', *Logic, Methodology, and Philosophy of Science*, ed. Y. Bar-Hillel (Amsterdam, 1965), 274; 'On a combined system of inductive logic', *Acta Phil. Fennica*, **18** (1965), 21; 'A two-dimensional continuum of inductive methods', *Aspects of Inductive Logic*, ed. J. Hintikka and P. Suppes, 113; 'Induction by enumeration and induction by elimination', *The Problem of Inductive Logic*, ed. I. Lakatos (Amsterdam, 1968), 191.

[3] A. Shimony, 'Scientific inference', *The Nature and Function of Scientific Theories*, ed. R. G. Colodny (Pittsburgh, 1970), 79.

rejects the assumption of most Carnap-type confirmation theories that the starting point for assigning an initial probability distribution must be some *a priori* indifference assumptions over possible states of the universe. Instead he accepts the fact that science never finds itself in this *tabula rasa* condition, but that at any given point of time there is some essentially finite set of *seriously proposed hypotheses* which do not exhaust all logical possibilities. To these he proposes to give finite probability; the rest can be regarded as having zero probability. A confirmation theory is therefore relative to the given state of science, at least in its broad features, and can provide an explication of inductive inference as carried out within that framework.[1] Shimony calls his proposal *tempered personalism*: 'personalism' because it takes probability distributions over states of the world to reflect beliefs; 'tempered' because it is recognized that not all possible belief-distributions are open, that only seriously proposed hypotheses are worthy of finite belief, and that there may be other constraints upon the sequence of distributions, for example that transformations should be Bayesian.

As it stands Shimony's suggestion does little to indicate how the seriously proposed hypotheses should be identified at any given stage of science. Some of them, Shimony considers, will be hypotheses with desirable features such as simplicity, which will be preferred to other hypotheses lacking these features. Thus inverse-square hypotheses will be preferred to inverse 2.0001 hypotheses, conic sections to general polynomials, universal to statistical generalizations. Shimony would reject any solution along the lines of method (*c*) and for assignment of initial probabilities which abandons the attempt to confirm any precise universal hypotheses, since it is generally accepted that these have often been particularly significant in science.

In order to spell out conditions for selecting seriously proposed hypotheses with some generality, we should first have to distinguish the adoption of a particular predicate language from that of a particular set of theories within that language. On the question of the basic predicates of the language of the confirmation theory, Shimony is on strong ground, for all confirmation theorists would agree that some set of predicates must be assumed before the conditions of confirmation can be expressed, and few would object to taking these predicates to be the ones 'seriously proposed' in current language. If someone wishes to add to or delete from or otherwise modify this set, his belief distribution will be changed in ways not expressible within his initial confirmation theory, and any grounds he has for this modification must be expressed in terms exterior to that theory. He might, for example, have become aware of a quite new colour, inexpressible in the old language; or a

---

[1] In his later writings Carnap proposed a similar method by allowing for the introduction of material 'meaning postulates' in his system, thus reducing the number of logically possible hypotheses in the language to a smaller set satisfying some non-logical conditions. *Cf. Philosophical Foundations of Physics* (New York, 1966), chaps. 27 and 28.

particular theory, say the phlogiston theory, may have become so improbable in his current theory that he wishes to take its probability henceforth to be zero, and drop the predicate 'phlogiston' from his scientific vocabulary. It may be thought that every interesting theoretical development in science involves this kind of conceptual and therefore linguistic change, and that it is profitless ever to assume a language stable enough to get confirmation going. However, I have already given some arguments to show that this conclusion is mistaken, and that the sort of radical universal and continuous conceptual change sometimes envisaged by philosophers of science is comparatively rare in the history of science.

Suppose, then, for the development of a *local* Bayesian theory, that we assume a basic predicate-set given. Shimony now proposes that external constraints be put upon the probability distribution over states of the world expressed in this language in such a way as to give finite probability to universal hypotheses of certain familiar and significant types in science as we actually find it. The most influential method of doing this so far proposed is that of Harold Jeffreys,[1] which is in fact a method of giving non-zero probability to denumerably many hypotheses, and is therefore an example of what I have called method (*b*).

Jeffreys makes use of a theorem of set theory to the effect that

> If a set of positive numbers is such that the sum of any finite subset of the numbers is less than or equal to 1, then the set is finite or denumerable.

Since the sum of the initial probabilities of any subset of $\{h_s\}$ cannot exceed 1, it follows that at most denumerably many hypotheses must have initial probabilities less than any fixed finite $\eta$, and for these hypotheses at least there will be no guarantee of convergence with increasing evidence unless Bayesian transformations are assumed.

This proposal leaves two problems to be investigated: first, how to reduce the potentially non-denumerable set of possible hypotheses in any language more complex than Carnap's to a denumerable set; and second, how to order them in respect of probability values to give a convergent series of such values summing to 1. Jeffreys' suggestion in regard to the first problem is to restrict the set of hypotheses to differential equations of rational order and degree, in which the different possible values of numerical coefficients are either themselves rational, or, if continuous, are not regarded as generating distinct hypotheses for every value in the continuum. His solution to the second problem is an ordering by simplicity, calculated by some arithmetic combination of the order, degree and number of parameters of the equations, the simplest having the highest initial probability.

These suggestions do provide a method of picking out hypotheses for

---

[1] H. Jeffreys, *Scientific Inference*, second edition (Cambridge, 1957), chaps. 1 and 2; *Theory of Probability*, third edition (Oxford, 1961), chap. 1.

preferential treatment, but there are severe objections to them in the form developed by Jeffreys. First, the restriction to differential equations is inadequate even for physics in its present theoretical state, let alone for the other sciences. Second, it has proved very difficult to define a simplicity ordering which satisfies all our intuitions regarding simplicity. Moreover, it is not clear what justification could be given for the assumption that greater simplicity is correlated with higher initial probability of hypotheses.[1] Some of these points will be investigated in chapter 11, but meanwhile it must be concluded that this example of method (b) at least has not yet been shown to be adequate for a confirmation theory.

Method (c) involves abandoning the attempt to confirm generalizations in infinite domains (let us call these *strictly universal* generalizations). Structure descriptions would now be specified, not by a set of precise $q$-numbers, which would give zero probabilities, but by a set of intervals of $q$-numbers, for example 'A proportion $p$ of $P$s are $Q$s', where $\theta_1 \leq p < \theta_2$, and $(\theta_1, \theta_2)$ is a fixed finite interval in the range 0 to 1.[2] This hypothesis may then have non-zero probability, since it is a member of a finite set of similarly defined hypotheses, and the confirmation of any finite prediction may be made as precise as we like by taking the intervals sufficiently small. Alternatively, the method may be interpreted as in Carnap's instance confirmation, as a way of assigning finite probability only to states of finite subsets of all the individuals, since these states are equivalent to disjunctions of denumerable states involving the rest of the denumerably many individuals. Since there is a finite number of states of a finite number of individuals, each can be given a finite probability.

It has often been argued that, since science necessarily involves strictly universal generalization, a confirmation theory must assign finite probabilities to such generalization, and that therefore method (c) is unacceptable. That this consideration is not relevant to the confirmation of finite predictions has already been argued in connection with the transitivity paradox. But in addition to this argument, I shall now try to show more positively that even though strictly universal generalizations may have an essential role in science, they do not need to be *confirmable*. It will turn out that the zero probability of strictly universal generalizations is not after all counter-intuitive when carefully scrutinized, and that whenever a generalization in an infinite domain appears to be required by scientific inference, it can be replaced

---

[1] Jeffreys explicitly adopts the view proposed here, that confirmation theory is the explication of scientists' intuitions. His justification for the simplicity postulate goes as follows: 'The actual behaviour of physicists in always choosing the simplest law that fits the observations therefore corresponds exactly to what would be expected if they regarded the probability of making correct inferences as the chief determining factor in selecting a definite law out of an infinite number that would satisfy the observations equally well or better, and if they considered the simplest law as having the greatest prior probability.' (*Scientific Inference*, 60.)

[2] Proposed by J. Darlington, 'On the confirmation of laws', *Phil. Sci.*, 26 (1959), 14.

without loss by a generalization in a finite domain, whose probability is non-zero. That is to say, I shall argue that nothing is lost by adopting method (*c*).

## II. The probability of laws

Let us begin by asking what would be implied by ascribing to (U) a non-zero probability. To ask the question this way instead of presupposing that universal generalizations must have finite initial probability is to draw attention to how strong the assumption of finite initial probability really is. For the assumption implies that it is reasonable to believe that (U) has *some chance*, however small, of remaining part of the corpus of acceptable scientific knowledge indefinitely and under all empirically possible circumstances, that is to say, under all circumstances which can in fact obtain in nature. It is not obvious that it is reasonable to believe this. This is not so much because such a belief covers an infinite number of instances, but rather because it may not be reasonable to believe that (U) states a law *accurately* even in one instance. Qualifications may have to be introduced, for example, *P* may in all observed and in almost all as yet unobserved instances co-occur with *P'*, a predicate which we have not included in our language, and perhaps have not even yet noticed. In that case we may, if we pursue science long enough, be forced to modify (U) to 'All *PP'* are *Q*', because we have found instances which are *P* but neither *P'* nor *Q*. Or, we may find that some *P*s are not *Q*, but are predicated by a closely similar *Q'* which was not previously distinguished from *Q*, as when a law containing metric predicates is found to hold not strictly but only approximately. *

Few of the laws which have been cited as typical by philosophers of science have in fact remained unmodified in some such ways. Consider for example Galileo's law of falling bodies, the law that planets move in ellipses, Boyle's law, Newton's law of gravitation, the chemical law of constant proportions and Mendel's laws of inheritance. All of these can be expressed in the form 'For all bodies, if a body is *P* it is also *Q*', and in this form all have either been falsified, or shown to be true only for a finite number of instances, in which form they may of course have non-zero probability. Even an apparently hard case such as 'All molecules of water consist of two hydrogen atoms and one oxygen atom' has already had to be modified to 'All molecules of water (other then heavy water) consist of two (non-heavy) hydrogen atoms and one oxygen atom'. This means that refusal to ascribe non-zero probability to 'All water molecules consist of two hydrogen atoms and one oxygen atom' in more than a finite number of instances could not have eliminated any useful inferences in the past, since the law has in fact turned out to be true only in limited domains. It will not do to reply to this example that the *modified* law about non-heavy water and hydrogen has to be asserted in an infinite domain,[1]

---

[1] *Cf.* M. Spector, review in *Metaphilosophy*, **2** (1971), 251.

because exactly the same argument applies to it, namely that it is inconceivable that this law too will not be modified in some future science, since, among other possible theoretical reasons, there will be somewhere in the universe some entities which we would *now* call non-heavy water which will turn out not to consist of what we would *now* call two atoms of non-heavy hydrogen and one of oxygen. It is not reasonable to suppose that *any* lawlike generalization of the form (U) in current or in any future science will remain forever unqualifiedly true in every instance. Since this is not reasonable, we are committed to the belief that no assertion such as (U) has any chance of being strictly true, and therefore that its probability is always properly zero.

At this point, however, the difficulty seems to arise again at a higher level. Are we not committed by the argument just given to the assertion that at least one universal law must have finite credibility, namely the law that 'All strictly universal laws proposed in science are false'? But this is not an objection if the total number of scientific laws ever proposed or seriously considered in science is finite, for then it is perfectly possible to have a finite degree of belief even amounting to certainty in the falsity of all of them. That the domain of a quasi-sociological law referring to all *actually proposed* laws is finite is likely to meet less intuitive resistance than the proposal that the domains of scientific laws themselves are required to be finite.

Returning to consideration of the scientific laws themselves, some of the examples just given may seem to involve not so much infinite domains of discrete individuals as continua of individuals, for example, space and time points. And even on the very weak assumption that in every interval of such individuals, there is only a countable infinity of points (the rational divisions which we could in principle measure with laboratory instruments), then every generalization over such an interval is an infinite generalization in the sense considered here, and has zero probability. Can we do without so modest a requirement in our domain of individuals? Consider the generalization $(s)(Es \supset Fs)$, where $s_0 \leqq s \leqq s_1$, and $s_0$ to $s_1$ is an interval of at least rational numbers. Let us call this generalization (V). An example would be 'For all spatial points, the electric force $E$ implies a mechanical force $F$', where, to ensure that this is an empirical law, $E$ is defined in terms of charges and distances and not in terms of mechanical forces. To say that this has probability zero is to say that it is unreasonable to believe that it has any chance of being true at all rational points in the interval $s_0$ to $s_1$. As in the case of generalizations over infinite domains of discrete objects, let us first consider what it would mean to believe that it *has* some chance of being true at all such points. This would mean that however closely we peer at the interval $s_0$ to $s_1$, taking smaller and smaller intervals within it, still we shall find that $Es$ materially implies $Fs$. Now it is of course not possible *operationally* to investigate this assertion for more than a finite number of subintervals, but it is dangerous to assume that what cannot operationally be demonstrated to be true is either eliminable from science or false. I certainly do not want to make any such assumption. We can peer more and more closely at diminishing spatio-temporal intervals by means of theoretical as well as operational instruments.

Let us rather proceed as we did in the case of infinite domains of objects, and investigate the truth of (V) not for *all* points, but for *any given* point, say $s_i$. As we consider smaller and smaller intervals converging on $s_i$ it is surely unreasonable to believe that the theory in terms of which the predicates $E$ and $F$ are understood will remain accurately true. Small-scale heterogeneities may arise, as when a macroscopically continuous fluid is found to break up in the small scale into discrete molecules; or the theory may be found altogether inadequate in the small scale, as when quantum and nuclear theories superseded classical physics. If we believe that this process has in principle no end, then we shall believe that putative laws such as (V) have indeed zero probability, and that the laws which we in fact make use of in inferences involve only quantification over finite sets of intervals, where these can be made as small as is required. Thus, I conclude, generalizations of type (V) can be made amenable to confirmation theory for the same reasons and under the same conditions as those of type (U) previously discussed.

Let us pause here and try to bring out the significance of these arguments. The emphasis has not been laid on the impossibility of knowing of an infinite number of instances of $P$s whether all are $Q$s, but rather on the impossibility of knowing in *any single instance* that all relevant predicates have been taken into account, because of the near-certainty that scientific development will not leave the present list of predicates unchanged. This should not, however, be taken to imply that laws do not of their nature have universal and, where appropriate, infinite reference. It is not the potential reference of true laws to infinite domains that is here in question, but the possibility of knowing, even of a single instance, of what true law it is an instance. Neither is this point the same as Jonathan Bennett has made in relation to a Humean versus a Kantian account of laws.[1] He has suggested that the difference is essentially that a Humean will not, and a Kantian will, reject the idea of a natural law in an infinite domain with isolated exceptions. The point being made here is indifferent to such Humean or Kantian accounts, because it implies that on both these views infinite generalizations will have zero probability, but that on either of them generalizations in finite domains may have finite probability. Nothing is here implied about what may really be the case about laws for someone who knows what the true laws are; what is implied is concerned only with what can reasonably be believed about laws by us, whose knowledge is limited. For the same reason, it is not an objection to the present account that it implies that the probability of certain existential statements is unity. In an

---

[1] J. Bennett, 'The status of determinism', *B.J.P.S.*, **14** (1963), 106.

infinite universe, if the zero probability of the generalization (U) corresponds to our intuitions, then the existential 'There is a $P$ which is not $Q$' has the highest possible probability. But it surely cannot be held that such an empirical existential statement is *certain*. This, however, is not the correct interpretation of a probability value of unity. Probability values are concerned with our *beliefs*, not directly with what actually is the case in the world; thus a probability value of unity implies only that we have the best of all possible reasons for believing that there is, somewhere in the infinite universe, a $P$ which is not a $Q$. The arguments for this belief are the same as those for belief in the falsity of (U).

A possible objection to this account has been made by Kneale.[1] He points out that the fact that we generally prefer when necessary to modify our description of a single instance, rather than abandon the search for general laws, does indicate that such laws play a specially important role in scientific inference. That is to say, if we find a $P$ that is not $Q$, we may prefer an un-refuted general law 'All $P_1$ and $P_2$ are $Q$' to the unrefuted statistical generalization '$p$ per cent of $P$s are $Q$s'. This point, however, is not relevant to confirmation theory as such, because the theory cannot dictate or explicate the preference. If confirmation theory is to be relevant we can assign initial probabilities such that both these generalizations have finite probability, and that then the confirmation of the first on the given evidence is greater than that of the second. Whether this is satisfied or not will depend on the nature of the total evidence and the assignment of initial probabilities. Nothing that has so far been said entails that this condition cannot be satisfied, and hence the preference for general laws as opposed to statistical generalizations explicated within a confirmation theory in finite domains.

A related objection is raised by Shimony in connection with his proposal to give non-zero probability to precise universal hypotheses of particularly significant types such as inverse-square laws. It certainly does seem to be a travesty of Newton's account of the universal law of gravitation, or of Dalton's law of simple multiple proportions, to suppose that these scientists mentally restricted their belief in these hypotheses to an interval of values only approximately corresponding with the simple mathematical expressions in the laws, however small the interval be taken. There are several aspects to this objection, however, which should be carefully distinguished. Firstly, to hold that confirmation theory is a theory of *belief* is not to claim that it is a theory of Newton's or Dalton's *psychology*. It is rather to claim that a reconstruction of coherent and rational beliefs can be given only if certain conditions are fulfilled, and if these conditions turn out to be fulfilled most conveniently by setting the probability of universal generalizations equal to zero, then so long as this does not violate essential features of scientific inference, the fact that it

[1] W. C. Kneale, *Probability and Induction* (Oxford, 1949), 73.

may violate scientists' unconsidered and possibly incoherent beliefs (even Newton's and Dalton's) is irrelevant.

Secondly, it is not even clear that reflective scientists, if pressed, would regard such mathematically elegant laws and theories as having unique claim upon their beliefs. Those who adopt some metaphysical attitude towards the 'simplicity of nature', as did the Pythagoreans (ancient and modern), Kepler, Galileo and Einstein, would of course make such a claim. (Incidentally, all their specific theories, with the possible exception of some of Einstein's, have already proved inaccurate.) But others, for example Newton, Dalton, Maxwell, may equally well be interpreted as developing the simplest mathematical relationships approximately consistent with the data and adopting simple laws rather as a pragmatic convenience than as a tribute to Pythagoras. Even Newton never *claimed* more than inductive support for the law of gravitation, and he was certainly aware of the imprecision of inductive data.

There is, however, another and more important consideration which may effectively reconcile Shimony's argument with the zero probability of point hypotheses. This is that the significance of precise mathematical hypotheses in physics almost always stems from *qualitative* rather than quantitative considerations, and hypotheses ascribing qualitative predicates may always be precise. For example, suppose an early scientist were interested in a law relating the number of rotations of a perfectly cylindrical log of given radius which would take it down a non-slippery plane hillside of given length. His theory will involve a precise numerical constant $\pi$, but this has emerged from the qualitative specification of the cross-section of the log as circular and it is calculable from that specification; it is not adopted inductively as a particular point hypothesis. Similarly, if the theory of gravitation includes a qualitative model of isotropic flux from a point source diverging into space, the inverse-square law is entailed precisely, as it is also if planetary orbits are asserted, qualitatively, to be elliptic. It should be noticed that this argument is independent of the possibility of knowing accurately that a physical object *is* circular—this kind of inaccuracy will introduce approximations into the application of any theory—the point is rather that *if* the data are taken as adequately described by certain non-metrical predicates such as 'circular', *then* precise numerical hypotheses are entailed. In cases like this it is the use of ideal geometric models in physical theory that makes some precise mathematical laws significant, and incidentally gives them non-zero probability, because although a continuum of values of the ratio of measures across and round an irregularly shaped object are logically possible, only a finite number of ideal geometric shapes may figure in the qualitative language. This is once more to throw the onus of restricting possible hypotheses back on to the choice of language, and out of the jurisdiction of particular probability assignments in a particular language. It concedes Shimony's point with regard to dependence on the existing language, but supplements it by indicating how the 'seriously

G

proposed hypotheses' arise from the language rather than from particular initial distributions of belief which restrict the possible hypotheses in a given language.

There are, however, some apparently precise numerical hypotheses for which the numerical constants do not appear to be calculable from qualitative considerations and have to be derived inductively, but which nevertheless are regarded as point rather than interval hypotheses. Consider, for example, 'All electrons have mass $9 \cdot 1 \times 10^{-28}$ g', or 'The velocity of light in vacuum is $3 \times 10^{10}$ cm s$^{-1}$'. It should be noticed first that some physicists, notably Eddington, have regarded such hypotheses as comparable with hypotheses containing the constant $\pi$, in that they claim that certain universal constants should be calculable, up to assignment of conventional units, from the non-numerical mathematical properties of an adequate physical theory. Meanwhile, however, the numbers are derived experimentally, and are therefore subject to experimental error. The hypothesis that is in fact believed and should be given non-zero probability is after all an *interval* hypothesis of the kind 'All electrons have mass $9 \cdot 1 \times 10^{-28}$ g $\pm \theta$g'. But this consideration does not remove a certain discomfort about the analysis of such hypotheses in terms of inductive confirmation. For it is not the number of electrons examined that tends to increase confirmation of the hypothesis within narrower and narrower intervals, but rather increasing accuracy of the experiment, even if it were physically possible to carry this experiment out on a single reidentifiable electron.

This discomfort ought, however, to be traced to beliefs about a different hypothesis, which is non-numerical, namely 'All electrons (unlike roses, crows, samples of natural water, etc.) have identical physical properties'. It is indeed difficult to claim, in line with the present argument, that we are certain that *this* hypothesis is strictly speaking false, and similarly with the hypothesis 'Light travels with constant velocity in a vacuum'. Both these hypotheses might indeed be held to have a quasi-analytic status in current scientific language, for both are necessary presuppositions of all currently viable particular theories. It must be granted, of course, that the language in which current science is expressed is unlikely to be adequate for future science, much less that it is going to be acceptable for all time. But is it not reasonable to believe that as long as science uses a particular basic language, *some* universal generalizations will remain acceptable, and that these generalizations will only be overthrown or modified if new predicates are introduced or old ones discarded? For example, the generalization 'All $P_1$ are $Q$' may not be eternal, but if $P_1$ and $Q$ are primitive predicates of language $L_1$ in which it is proposed, whereas $P_2$ or $Q'$ do not occur in that language, it may seem strange to hold that 'All $P_1$ are $Q$' is false in $L_1$ simply because another generalization 'All $P_1$ & $P_2$ are $Q$' or 'All $P_1$ are $Q \vee Q'$' replaces it in a richer language $L_2$.

This objection, however, presupposes that truth is concerned with the best a given language can do, and not with a state of the world which is independent of language. If the situation is as just described, then it may objectively be the case that 'All $P_1$ are $Q$' *is false* both in $L_1$ and in every other language containing the predicates $P_1$ and $Q$. It is false because there may be $P_1$s which are not $Q$s, either because there are $P_1$s which are not $P_2$, or because there are $P_1$s which are $Q'$ and not $Q$. For example, what we now refer to as 'electrons' will almost certainly be found to include some objects whose physical properties are *not* identical with those of other electrons, just as some hydrogen atoms turned out to be 'heavy'. Our current terms have a certain generally understood reference, but the specification of what exactly should count as part of that reference is necessarily incomplete and subject to modification with changing theories. This uncertainty applies as much to our current fundamental postulates about 'electrons' and 'light' as to all other generalizations.

This is not to say that nothing true can be said with the predicates of a given language, for since the infinite set of generalizations exhausts the possibilities of truth as that language carves up the world, at least some of them must be true of the world. But it is to say that if the domain of application is infinite, then it is maximally probable ('almost certain') that we shall never in a finite time light upon an exact universal or statistical generalization that is in fact true of the world described in that language.

Another objection related to the linguistic basis of a confirmation theory has been made by Nagel in comments on Carnap's theory of induction.[1] Nagel remarks that universal laws may be indispensable even in recognizing the *evidence* for a law. For example, confirmation of the law 'All swans are white', even for a finite domain, requires acceptance of some statement '*a* is a swan', and this in turn presupposes universal laws implicit in the universal term 'swan'. Similar points have been made by Popper,[2] who would seem to hold that no evidential statement of this kind is in principle a singular statement, nor can be known conclusively to be true, because the observation report '*a* is a swan' can always be challenged and further tested, and such tests may result in withdrawal of the statement—the light was bad; being a logician I was obsessed with the concept of 'swan'; what I saw flying across the lake was actually a goose. If there is in principle no end to this process of challenge and test, then indeed a statement such as '*a* is a swan' is not singular, but has consequences which are properly tested by events other than the observation which prompted its utterance. It is not clear, however, that these consequences are such that '*a* is a swan' must be taken to be an elliptic expression of one or more universal generalizations in infinite domains. Indeed, on Popper's view at least, it would be highly paradoxical if this were so, because if we can never know such a universal generalization to be true or even probable,

---

[1] 'Carnap's theory of induction', *The Philosophy of Rudolph Carnap*, ed. P. A. Schilpp (La Salle, Ill., 1963), 785.
[2] E.g. in *The Logic of Scientific Discovery*, 94.

and if every statement of evidence both of positive and negative instances of a generalization itself presupposes such knowledge, it is difficult to see how any generalization can even be known to be *false*. Why, except for entirely pragmatic reasons, should we accept, even provisionally, one general statement rather than any other?

This problem is not so pressing on Nagel's view, for presumably Nagel does believe that we can know or be reasonably confident in the truth of a statement such as '*a* is a swan'. In that case it may be suggested that he is over-ambitious in requiring confirmation theory to explicate the conditions of applicability of its own basic language. To adopt a language containing certain extra-logical predicates in terms of which to explicate the mutual inductive relations of its statements is indeed to presuppose that the predicates are universals, and we cannot give finite sets of necessary and sufficient conditions for the applicability of universal terms. But it does not follow that the process of applying universal terms is like that of coming to accept universal laws in depending on probabilities in infinite domains. Indeed the difficulties involved in any such construal of universals may well suggest that here as well as in our account of induction, we ought to eschew infinities altogether.[1] At worst, Nagel's objection means that confirmation theory does not give an account of the logic of the learning of its own basic language, but it is not at all clear why it should be expected to do so. To require this would be something like requiring a formalization of arithmetic to define uniquely the natural numbers, a task which, though originally worth attempting, we now know to be impossible, but we do not thereby regard the formalization of arithmetic as an impossible or worthless enterprise.

Nagel has a related objection, based on the supposition that experimental tests always themselves depend on acceptance of the universal laws governing the test apparatus and surrounding test conditions. But here, surely, infinite domains need not be involved, only high probability over finite domains which are defined to be as large as is necessary. Generalizations concerned with instruments do not even need to be universally *true*—the laws of classical optics, for example, may provide a sufficient *approximation* to the behaviour of a telescope, even when it is testing predictions derived from a theory which contradicts classical optics. It follows that any presupposition involved in accepting the evidence as highly probable can always be stated without loss in terms of high confirmation of generalizations in finite domains.

Another objection can be dealt with equally briefly. This is that if all infinite generalizations have zero probability, then a generalization which has not yet been refuted is no better off than one which is known to be false. And yet we must consider the status of the former as different from the latter. My reply to this would be that we do consider it as different, but only in virtue

---

[1] We have already seen in chapters 1 and 2 that fixed (and presumably also universally applicable) natural kinds are not required for a viable account of universal terms.

of the confirmation of the generalizations over limited domains which are its consequences. That is to say, what we actually use in theoretical inference are statements of the form 'All the next $n$ $P$s in a limited region of space and time are $Q$', where $n$ and the limited region may be as large as we like to specify. Such statements have non-zero probability in any well-behaved confirmation theory. The value of that probability will, however, be very different according to whether the evidence contains instances of $P$ which are not $Q$. In other words, our confidence in predicting of the next million $P$s that they will be $Q$ will be much diminished if the strictly universal generalization 'All $P$ are $Q$' has already been refuted.

## III. The necessity of laws

So far, we have considered only objections to the zero probability of laws arising from what might be called a Humean account. There is, however, an influential construal of laws in which it is claimed that they are asserted not only for a potentially infinite number of actual instances, but also for a potentially infinite number of 'conditional' or 'counterfactual' instances. That is, not only are all (perhaps a finite number) of actual ravens asserted to be black, but it is also asserted that if anything (out of an infinite number of things) *were* a raven (and some are and some aren't), then it *would be* black. There is much dispute over the question whether such counterfactual inferences are in fact required in science. I am one of those who believe that they are,[1] thus making the confirmation problem as difficult for myself as possible. However, I think it can be argued that this view does not after all make the problem more difficult than it is already, because once the controversial introduction of counterfactual inferences has been performed, no new problems arise about their confirmation.

The difficulty can be put like this. If counterfactual inferences from laws are required, then we require finite confirmation, not only of singular assertions of the form 'This raven is black', but also of 'If this thing were a raven (though it isn't), it would be black'. Leaving aside the question of how we would supplement the logic of material implication to allow this latter assertion to be non-vacuous, it usually seems to be assumed that it would be well confirmed if it was entailed by the universal law 'All ravens are black', which was itself well confirmed. Those who assert that such counterfactual inferences are required in science also assert that they are entailed by universal laws, but to assume that therefore any confirmation of the universal law is passed on to its entailment is to commit the transitivity fallacy.[2] Moreover,

[1] *Cf.* P. Alexander and M. Hesse, 'Subjunctive conditionals', *Aris. Soc. Supp. Vol.* 26 (1962), 185 and 201.
[2] I committed this fallacy myself in the first publication of the present section in 'Confirmation of laws', *Philosophy, Science, and Method*, ed. S. Morgenbesser, P. Suppes and M. White (New York, 1969), 84.

the second part of the requirement cannot be met either, if the universal law cannot have confirmation other than zero in an infinite domain. Does the assertion of the counterfactual presuppose an infinite domain? It is not clear that it does. On the view being defended here, the assertion which does have finite probability is of the form 'All ravens in the next million observed in this fairly large space–time region are black' (call this (W)). Given that the reason this is confirmed by the evidence of black ravens is not that it is not lawlike, but that it does not presuppose that we know the whole truth about ravens everywhere and everywhen, it seems reasonable to maintain that *if* (and no doubt *only if*) 'All ravens are black' entails 'If this were a raven this would be black', *then* 'If this object in this space–time region were a raven it would be black' is confirmed by black ravens for the same reason that (W) is, namely that the instances are or would be clustered. Whatever may be understood by counterfactual instances, they do not imply the existence of infinite domains any more than factual instances do.

There is, however, a deeper objection to the whole idea of explicating confirmation in terms of probability, which has been expressed most forcefully by Kneale in *Probability and Induction*. Kneale's interpretation of scientific laws is that they are principles of physical necessitation, which he distinguishes on the one hand from logical necessity, and on the other from universal co-occurrence, or accidental generality. This notion of physical necessity is itself far from clear, but at least the consequences which Kneale claims that it has for a probabilistic theory of confirmation can perhaps be evaluated without raising the much more difficult issue of necessity itself. It should be noticed first that Kneale himself does not claim that the issue of necessity versus accidental generality is co-extensive with the difference between strictly universal and statistical generalizations. The latter may also be necessary or they may be accidental. They may be necessary either because they are consequences of necessary and strictly universal laws (as in classical statistical mechanics), or because they are concerned with objective and irreducible probabilistic propensities (as we believe to be the case with the fundamental particles). Thus this notion of necessity must not be confused with the quite different notion of strictly universal determinism, and whenever Kneale uses the expression '$P$-ness necessitates $Q$-ness' he is to be understood as subsuming expressions of the form '$P$-ness necessitates $Q$-ness in $p$ per cent of the cases, and not-$Q$-ness in the rest'.

Kneale's argument is briefly as follows.[1] If the generalization 'All $P$ are $Q$' is a law of nature, it is properly construed as '$P$-ness necessitates $Q$-ness'. However, in a probabilistic theory of confirmation in which the probability of a law on given evidence is taken to be some function of the number of

---

[1] *Probability and Induction*, 211f.

worlds in which law and evidence are both true, and of the total number of worlds in which the evidence is true, we have to contemplate 'All *P* are *Q*' as true in some out of a number of possible worlds. But if *P*-ness *necessitates* *Q*-ness, there cannot *possibly* be other worlds in which it is not true. In the same way the truth of a mathematical theorem cannot be just one of the possibilities, because either it is true, and the alternatives are impossible, or it is false, and then not possibly true.

This argument undoubtedly has some initial appeal, because the conception of physical necessity sits uneasily with that of random sampling on which probabilistic methods are based in statistics. But if we draw a distinction between the necessity of a law as it is in nature, and the apparent randomness of the ways in which we may come to know it, the argument can, I think, be shown to depend on a pun on the word 'possible'. The situation to which confirmation theory is relevant is that of contemplation of a potential law 'All *P* are *Q*' when we do not know on the basis of the evidence available whether this generalization is true or not. If it is a law, then on Kneale's view *P* necessitates *Q*, and then in one sense it is impossible that *P* should be not-*Q*, but it does not follow, since by definition we do not *know* that '*P* necessitates *Q*' is a law, that it is impossible that '*P* necessitates *Q*' should be *false*, and hence that there are other possible lawlike statements, some of which may be true. And Kneale has not noticed that if his objection were conclusive it would also count against the use of probability even for accidental universal generalizations, for if it is a cosmic *fact* that all *P*s are as a matter of fact *Q*s, it is likewise not 'possible' that a *P* should not be a *Q*. But this rests more obviously on a pun on 'possible', and Kneale is not deceived by it, for he accepts the ordinary scientific use of probability for chance occurrences. The case of logical necessity to which he appeals is not conclusive for his argument either, for when we do not know whether a potential mathematical theorem is true or false (as with Goldbach's conjecture, for example), there is one sense of 'possible' in which it is possibly true and possibly false. But in this case, since mathematical theorems, unlike scientific laws, are matters of *proof*, it is not likely that our degree of belief in Goldbach's conjecture is happily explicated by probability functions. Whether confirmation of laws is so explicable is a question of whether our degree of reasonable belief in them behaves like a probability function in the light of confirming evidence, and cannot be decided *a priori* by arguments about physical necessity, however that may be understood.

There is implicit in Kneale's discussion another serious question about laws *vis-à-vis* accidental generalizations. Since confirmation theory depends merely on the syntactic relation between evidence statements and law statements, it cannot distinguish laws from accidental generalizations merely in virtue of evidence of their instances, and in particular it cannot reflect our greater willingness to allow confirmation to a potential law than to an accidental generalization on that evidence.[1] For example, considered merely as a relation between instances and generalization, a sample of viruses all of which have caused Asian flu in persons whose birthdays lie in June, is no better evidence for the generalization 'All these viruses cause Asian flu' than it is for 'All these viruses attack only people whose birthdays are in June'. And yet we should be much more inclined to regard the former as well-confirmed than the latter.

The first remark to be made about this difficulty is that it is not reasonable to expect a confirmation theory to decide on the basis of generalizations and instances alone how we should distinguish laws from accidental generalizations. It is almost universally admitted that the necessity of laws, however that is to be construed, at least does not reveal itself by our peering more closely at the events which may be instances of the laws. Neither is the problem removed by building into a confirmation theory a logical connective equivalent to 'necessitates' as well as the usual connective of material implication, for in such a theory the problem would be that unless prior distinctions were made about what kinds of statements are potential laws, all accidental generalizations not yet contradicted by the evidence would be as highly confirmed as potential laws. Instead of all laws looking like accidental generalizations, all accidental generalizations would look like laws.

One widely accepted way of making the distinction is the view that the generalizations we are prepared to accept as potential laws are those that follow from the conjunction of appropriate boundary conditions with more or less systematic theories, which are in turn supported by a variety of other laws. The generalizations that we regard as accidental, on the other hand, are those which are not so connected with other generalizations by theories. This view demands of a confirmation theory that when laws are consequences of a good theory they should be given confirmation higher than the confirmation they receive from their instances alone, for they share the latter possibility of confirmation with accidental generalizations, but not the former.

The view is, however, open to the general objection that some potential laws can be distinguished from accidental generalizations even when no systematic theory is present; for example, we may be quite ignorant of any explanatory theory of *how* the virus causes Asian flu, and yet be prepared to accept this correlation as lawlike. This is one reason why it is not satisfactory to adopt this view as it stands as a basis for the confirmation of laws. There is also a more fundamental reason why it indeed *cannot* be accepted as it stands as defining the conditions of adequacy of a confirmation theory. This is a further consequence of the non-transitivity of confirmation, namely that in a probabilistic theory using positive relevance as the criterion of confirmation, the confirmation of a law $l$ cannot be increased above its value on the evidence of its own instances merely in virtue of its deducibility from a theory $t$, unless

---

[1] This was the original context of Goodman's discussion of the grue paradox (*Fact, Fiction, and Forecast*, chaps. 1, 2).

the confirmation is also increased by all the evidence for the theory independently of explicit formulation of the theory itself. If probabilistic confirmation theory is to explicate the relation between explanatory theories and laws, theories must satisfy stronger conditions than those required on the deducibility account, because this account is not sufficient to show that theories are relevant to the confirmation of laws. I shall suggest in the next chapter that the further conditions to be imposed on $t$ must be derived from relations of *analogy* between the objects specified in the total evidence $E$, so that $t$ is not so much one among many possible postulate systems having $E$ among its consequences, as the explicit statement of the analogies which exist among a number of empirical systems including $E$, the best theory being the one which embodies the strongest and most extensive analogy. A theory would therefore be established when the number of analogous systems which are instances of it is large, and this would incidentally have the effect that, given appropriate differentiating boundary conditions, the number of its empirical consequences is also large, that is, that a good theory has high 'content', as is also required in the deductive account.

On such a view it is at once possible to understand why we accept the correlation of viruses with Asian flu as lawlike, and the correlation of viruses with June birthdays as accidental. The former correlation does, and the latter does not, have analogies with many other empirically established correlations, and this can be known even before more sophisticated theories of the disease are available. In order to ensure that the probability $p(l/E)$ is greater than $p(l)$ it is necessary only to impose upon the probability assignment the requirement that evidence comprising data about a large number of analogous objects or systems of objects yields in general higher probability values for $l$ than evidence comprising isolated objects or systems which are instances of $l$ only.

The view that at least some types of explanatory theory are best regarded as explicit statements of analogies among instances is a very natural extension of the view of laws which has been maintained here, for exactly the same considerations apply to the deducibility of instances from general laws as to the deducibility of laws from theories. Moreover, in this account it may be said that the concept of causality is restored to both laws and theories in the following manner: firstly by recognizing a primitive relation of similarity or analogy between their instances, and secondly by embedding laws in a systematic way in a set of other laws which are themselves seen as analogous in the light of a theory. Both these features will be developed in the next chapter. Meanwhile let us summarize the reasons which have so far led to the apparently counter-intuitive proposal to interpret theories for purposes of confirmation as quantified only over strictly finite domains.

(1) It has been argued that it is reasonable to have zero degree of belief in any universal generalization in an infinite domain, as this is actually expressed at any given stage of science.

(2) Inductive inference (and also use of universal terms) can be analysed without loss into inference from particulars to particulars, or at most to generalizations in finite, though large, domains.

(3) This proposal does not detract in any way from the special features that have been ascribed to general laws, and specifically not in the following respects.

(*a*) It does not imply that statistical generalizations should be preferred to universal laws in finite domains.

(*b*) It does not imply that confirmation theory cannot assign finite confirmations to counterfactual inferences.

(*c*) It does not imply that laws are not in some sense causal or necessary.

(*d*) It does not imply that laws cannot be distinguished from accidental generalizations.

(*e*) It does not empty the notion of law in an infinite domain of all significance; indeed it is even necessary that a vanishingly small proportion of such laws in a sufficiently adequate finite language must be true. The proposal merely denies that we can ever have any reasonable grounds for believing such a law to have any *probability* of being true. But this does not, of course, preclude use of a law-expression in strictly universal form from being compendious and convenient, and this fact is enough to explain the use of such expressions throughout the natural sciences. Where such expressions are used in examples in the following chapters, they are always to be understood as shorthand for equivalent expressions in finite domains, where the domains are left unstated but assumed large enough for whatever scientific purpose is currently in question.

(4) The proposal does not devalue theories. It rather reinterprets theories as expressions of the analogies between their instances, in virtue of which analogical inferences can be made to other finite sets of instances. Thus the proposal is not instrumentalist, for the analogy relations are *real* and sometimes recognizable relations between things without which such analogical inference would not be justified.

(5) The proposal reinterprets the notion of causality in a form stronger than Hume's, firstly by appealing to primitive similarity or clustering relations which are assumed both to exist objectively among objects and systems, and in some cases to be recognizable without inference, though not necessarily infallibly. Secondly, as we shall see in more detail later, this notion of causality depends on the systematic character of theories understood as expressing analogies between laws, whose formal properties of simplicity, etc., can be used to correct mistaken ascriptions of similarity.

Finally, we shall see in the next chapter that a strictly finite view of confirmable theories does not preclude the possibility of introduction of novel theoretical concepts, although it will be argued that these must generally be introduced by analogues or models from existing concepts. Thus, it will be shown that although the *universality* commonly regarded as necessary for theories has been discarded in this account, at least three other characteristics commonly required for a realist account of theories can be retained as essential: namely a reinterpreted *causality*, the introduction of *novel theoretical concepts*, and a *realist interpretation* of these concepts as assertions of causally important common properties or relations between explananda covered by the theory. This account of theories and laws is not instrumentalist.

# CHAPTER NINE

# Theory as Analogy

## I. Some false moves: 'acceptance' and 'explanation'

The problem of non-transitivity of confirmation through theories to their predictions has led to a variety of suggestions for explicating the related notions of support, acceptability and explanation, as understood in theoretical contexts. The problem was implicitly pointed out in a striking way by Putnam, in comments on the adequacy of Carnap's confirmation theory.[1] Putnam takes the example of the predictions that were made about the character of the first atomic bomb explosion, which were subsequently tested with success. Here a body of evidence drawn from physics and chemistry supported nuclear theory, and from this theory were derived in great detail, and hopefully with great confidence, predictions about what would occur in a type of high-energy situation so far wholly unrealized in any previous experience, namely the slamming together of two subcritical masses of uranium to produce an explosion of calculable magnitude and effects. Clearly a 'pure theoretician's' attitude of inferring from a theory to interesting (and 'severe') predictions which might or might not succeed is quite inadequate to explicate the nature of the inferences involved in this example. Unless something corresponding to high confirmation or degree of belief had been justified by the existing evidence, the disutility of failure, even under the historical conditions then prevailing, would surely have prevented the test being carried out at all.

Putnam's own comment on his example is insufficiently radical. He claims that in a confirmation theory of Carnap's type, posterior probabilities depend only on data, and data are expressed in an observational vocabulary. But in nuclear theory many theoretical terms also occur, and must be taken account of in any explication of confidence in the prediction. Putnam is right in pointing to the importance of theoretical terms, however they are understood, but he has overlooked the fact that even if *no* theoretical terms enter, as in the simple generalizations considered in chapter 7, transfer of confirmation to predictions is illicit. A confirmation theory adequate to take account of such examples is clearly going to involve some radical reinterpretation of what is to to be understood by theoretical inference.

[1] H. Putman, ' "Degree of confirmation" and inductive logic', *The Philosophy of Rudolph Carnap*, ed. P. A. Schilpp, 779.

Some proposed reinterpretations have required abandoning either the entailment or converse entailment conditions of confirmation, or the criterion of positive relevance. I have already given arguments for doubting the efficacy of such moves. There are, however, less radical possibilities, which involve *adding* to the basic conditions of confirmation one or more concepts which restore some kind of transitivity of support for theories. In particular, the notions of *acceptance* and *explanation* have been invoked for this purpose. Let us consider these in turn.

It is clear that *if* there are circumstances in which we are justified in accepting, or *treating as true*, a given hypothesis, then we must, in any useful sense of these terms, also accept and treat as true its logical consequences. The notion of acceptance here must be put strongly enough to demand treatment as true, to avoid falling again into the transitivity fallacy. It is not sufficient to talk in terms of 'tending to acceptance', as for example Mackie does in a discussion of non-transitivity, for although he recognizes the existence of the fallacy of transitivity in the case of positive relevance, and dissociates acceptance from confirmation by converse entailment, he also holds that

> ... if we accept a proposition we must accept its deductive consequences, and therefore whatever tends to justify our accepting it tends also to justify our accepting its consequences.[1]

This implication is certainly false if 'tends to justify acceptance' is explicated as 'gives high probability', because then it is a version of the transitivity fallacy. And since on any account in which 'tendency to justify acceptance of $h$' is a matter of degree, and is little affected by tacking on to $h$ some other $h'$ of low content such that the evidence still tends to acceptance of $h\&h'$, any such account will still make Mackie's implication fallacious.

There are indeed cogent reasons for doubting whether any useful concept of acceptance can be defined in a self-consistent manner at all. We never *know* that a theory is true, but do we ever have enough reason to *treat it as* true? This question has been found to raise unexpected difficulties. The obvious first suggestion to make is that if the posterior probability of a theory on its evidence is high enough, we are entitled to accept it. Suppose we set a level $k$, near to 1, for acceptance, that is, adopt as criterion

(A1) Accept $h$ if and only if $p(h/e) > k$

We also intuitively require acceptance, as 'treatment as true', to satisfy two more conditions.

(A2) Acceptance of $h$ entails acceptance of its logical consequences

(A3) Acceptance of a set of hypotheses entails acceptance of their conjunction

---

[1] J. L. Mackie, 'The relevance criterion of confirmation', *B.J.P.S.*, **20** (1969), 38.

These conditions immediately lead to the so-called 'lottery paradox'.[1] Consider the random drawing of one lottery ticket out of 200, so that the chance of drawing a particular ticket is 0·005, and suppose we have set $k = 0·990$. Then according to the conditions we must accept the hypothesis $h_1 \equiv$ 'Ticket 1 will not be drawn', since $p(h_1/e) = 0·995 > 0·990$. But the same applies to the hypothesis that any given ticket will not be drawn, so accepting one such hypothesis entails by (A3) accepting that no ticket will be drawn. But this contradicts the conditions of the problem.

Many suggestions have been made for avoiding this paradox, but none appears adequate for a concept of acceptance which would also solve the transitivity problem. For example, in a detailed discussion of possible criteria of acceptability, Risto Hilpenen concludes with the following criterion.

> A singular hypothesis '$A(a_i)$' is acceptable on the basis of $e$ if and only if '$(x)(Ax)$' is (on the basis of $e$) an acceptable generalization.

And a generalization $g$ is acceptable on evidence $e$ if and only if (i) $p(g/e) > 1 - \varepsilon$, and (ii) $n > n_0$, where $n$ is the number of observations and $n_0$ a suitable 'threshold' of acceptability.[2] This suggestion evades the lottery paradox, since $\varepsilon$ can be set so low, or $n_0$ so high, that at most one generalization is accepted. But since acceptability of singular predictions is made to depend on the probability of a generalization, the rule clearly begs the transitivity questions at issue here.

In another discussion of acceptance by Keith Lehrer, acceptance is permitted only of the 'weakest' of a number of competing hypotheses, that is, that unique hypothesis whose probability on evidence is greatest.[3] This proposal, however, would never permit acceptance of a hypothesis in the circumstances we are considering. For consider a theory which is confirmed within a certain domain of experiments (say, for example, at low energies). We want to extrapolate to make predictions in a different domain (say high energies). This extrapolation would always be stronger than the hypothesis restricted to the original domain, and hence could not be accepted on Lehrer's criterion. Any acceptability criterion which permitted such an analogical inference would have to allow that certain kinds of extension of the weakest hypothesis entailing the evidence may also be accepted. But to specify what kinds of extension these are would be to raise exactly the same problems as arise in specifying a probability distribution which favours relevant as opposed to irrelevant predictions.

A second type of suggestion regarding confirmation of predictions involves

[1] H. E. Kyburg, 'Probability, rationality, and a rule of detachment', *Logic, Methodology and Philosophy of Science*, ed. Y. Bar-Hillel (Amsterdam, 1965), 305.
[2] R. Hilpenen, 'Rules of acceptance and inductive logic', *Acta Phil. Fennica*, **22** (Amsterdam, 1968), 74, 64.
[3] K. Lehrer, 'Induction, reason, and consistency', *B.J.P.S.*, **21** (1970), 103.

restricting the hypotheses involved to those which are *explanations* of the evidence and predictions. Brody, for example, proposes to restore something like the converse and special consequence conditions by replacing converse consequence by 'If $f$ cfms $g$, and $h$ *explains* $g$, then $f$ cfms $h$', and special consequence by 'If $f$ cfms $g$, and $g$ *explains* $h$, then $f$ is evidence for $h$'. The first suggestion, he claims, 'allows for all the inferences for which the converse consequence condition was used and was needed, but does not allow those inferences which gave rise to Hempel's and Skyrms' paradoxes'.[1] For example, evidence $f$ which confirms some observational generalization $g$ will also, according to the new condition, confirm some theoretical hypothesis $h$ only if $h$ explains the generalization. But if the account of explanation is such that $h \rightarrow g \& g'$ does not *explain* $g \& g'$, though it entails it, then $f$ does not in general confirm $g \& g'$ on the new condition; much less does it confirm $g'$ which may be any statement whatever. Thus Hempel's paradox is blocked by the modification of converse consequence, even if special consequence is retained in its original form. Skyrms's paradox is also blocked, since if $f$ confirms $f \lor g$ by entailment, and explanation is such that $g$ does not explain $f \lor g$, although it entails it, then $f$ does not in general confirm $g$ which may be any statement whatever.

Brody's proposal of course depends crucially on what is to be understood by 'explanation'. But he gives us no help here, and it must be noticed that the concept of explanation appealed to cannot be merely the deductive-nomological explanation of the Hempel–Oppenheim model,[2] because this model involves conditions which make the proposed explication either useless or circular. First, the deductive model requires only deductive relations between hypothesis and evidence, together with the proviso that no part of the hypothesis be redundant for deduction of the evidence. This certainly would ensure that, for example, $g \& g'$ does not explain $g$ merely in virtue of entailing it. But the condition of non-redundancy, as it stands, is impossible to satisfy for theories strong enough to yield any interesting predictions. For, as we have seen in Putnam's example, a theory is often required to be strong enough to yield predictions in new domains not mentioned in the evidence, and such a theory will certainly involve component hypotheses that are not necessary for deduction of the evidence. A confirmation theory that failed to confirm such a theory would be useless. Secondly, for the concept of deductive explanation to be at all applicable, the explanatory theory has either to be known to be true, a condition which is never satisfied, or, as Hempel later wrote, it has to be 'strongly supported or confirmed'.[3] This leads to patent circularity of explication of confirmation and explanation; moreover, it is still inadequate, since a postulate involving merely high confirmation of the

[1] B. A. Brody, 'Confirmation and explanation', *J. Phil.*, **65** (1968), 297.
[2] 'Studies in the logic of explanation', *Aspects of Scientific Explanation*, 245.
[3] *Aspects*, 338.

hypothesis says nothing about its consequences, and therefore cannot by itself provide the required replacement for special consequence.

It must be concluded that neither the concept of acceptance nor that of explanation has yet been shown to provide solutions of the transitivity paradox. We shall now consider some alternative suggestions.

## II. Deduction from phenomena

A classic and influential approach to theoretical inference, which might also be called a false move in the attempt to resolve the transitivity paradox, is the programme which Newton called 'deduction from the phenomena'.[1] As is well known, Newton was unwilling to admit that his law of universal gravitation was a *hypothesis* in the sense that gravitational force was a postulated concept not found in the phenomena. Whatever may have been the case in some of his speculations concerning the aether, in *Principia Mathematica* he wished gravitation to be taken as a mere description of observable motions, and the law of gravitation to be interpreted as a deduction from the mechanical laws of motion (containing only observables) and the phenomenal laws of planetary motion.

The notion of deduction of general laws from particular phenomena must appear logically bizarre, and clearly no such deduction can be possible, or has indeed ever been claimed, without the intervention of general premises of some kind. The interest and importance of such claims to deduction will therefore depend on the character and cogency of the general principles appealed to. But if some acceptable general principles can be found, which, together with statements of phenomenal evidence, do actually entail certain general laws, then the transitivity paradox is evaded. For since the general principles and the evidence are taken as true, all their consequences, including the entailed general laws and *their* consequences, will also be taken as true.

---

[1] The terms 'deduce', 'induce', 'derive' and 'infer' were used very imprecisely in the seventeenth century, although some of the modern distinctions can be found in the applications to which they refer, as we shall see below. Some of Newton's expressions which imply a process of reasoning *from* phenomena *to* generalizations are the following: 'In the third [Book] I derive from the celestial phenomena the forces of gravity with which bodies tend to the sun and the several planets. Then from these forces, by other propositions which are also mathematical, I deduce the motions of the planets, the moon, and the sea.' (*Mathematical Principles of Natural Philosophy*, English trans., ed. F. Cajori (Cambridge, 1934), preface to first edition, xviii.) 'I frame no hypotheses; for whatever is not deduced from the phenomena . . . have no place in experimental philosophy. In this philosophy particular propositions are inferred from the phenomena, and afterwards rendered general by induction.' (*Ibid.*, General Scholium to second edition, 547.) '[The method of] Analysis consists in making experiments and observations, and in drawing general conclusions from them by induction. . . . And although the arguing from experiments and observations by induction be no demonstration of general conclusions, yet it is the best way of arguing which the nature of things admits of, and may be looked upon as so much the stronger, by how much the induction is more general.' (*Optics* (fourth edition, reprinted London, 1931), Query 31 to second edition, 404.)

In the case of the law of gravitation, for example, it then follows that evidence which together with the principles is sufficient to entail the law, will also be sufficient to entail, and *a fortiori* confirm, further consequences of the law. No appeal to either the converse or the special consequence conditions is required.

It is not difficult to show that Newton does exhibit a kind of deducibility from phenomena in his derivation of the law of gravitation, in Book III of the *Principia*, provided only the following principles be accepted.

(1) The laws of motion and Kepler's laws, which form the premises of the deduction, are derived from particular phenomenal motions, 'generalized by induction', and are provisionally taken to be true. Then the law of gravitation for the particular bodies mentioned in Kepler's laws can be shown to be just a deductive consequence of these phenomenal laws or, rather, to be an alternative mathematical expression of the accelerations involved, where acceleration is proportional to force by the second law of motion.

(2) The process of universal generalization by induction takes place according to a general principle that Newton calls Rule III of his Rules of Philosophizing: 'The qualities of bodies . . . which are found to belong to all bodies within the reach of our experiments, are to be esteemed the universal qualities of all bodies whatsover'.[1] This rule permits the extension of the law of gravitation for particular pairs of bodies to all bodies in the universe, in virtue of the 'analogy of nature, which is wont to be simple, and always consonant to itself'. Newton explains the application of the rule as follows.

> . . . if it universally appears, by experiments and astronomical observations, that all bodies about the earth gravitate towards the earth, and that in proportion to the quantity of matter which they severally contain; that the moon likewise, according to the quantity of its matter, gravitates towards the earth; that, on the other hand, our sea gravitates towards the moon; and all the planets mutually one towards another; and the comets in like manner towards the sun; we must, in consequence of this rule, universally allow that all bodies whatsoever are endowed with a principle of mutual gravitation.

(3) The purely deductive character of the argument given these two principles must be supplemented by the remark that the converse deductive

---

[1] *Mathematical Principles*, Book III, Rules of Philosophizing. In this quotation I have omitted the significant and difficult phrase 'which admit neither intensification nor remission of degrees', for which see J. E. Maguire, 'The origin of Newton's doctrine of essential qualities', *Centaurus*, **12** (1968), 233. This paper and the same author's 'Force, active principles, and Newton's invisible realm', *Ambix*, **15** (1968), 154, contain detailed analyses of Newton's Rules of Philosophizing. For the logic of 'deduction from phenomena', see the important series of papers by J. Dorling: 'Maxwell's attempts to arrive at non-speculative foundations for the kinetic theory', *Studies in Hist. Phil. Sci.*, **1** (1970), 229; 'Einstein's introduction of photons: argument by analogy or deduction from the phenomena?', *B.J.P.S.*, **22** (1971), 1; 'Henry Cavendish's deduction of the electrostatic inverse square law from the result of a single experiment', forthcoming; and 'Demonstrative induction', forthcoming.

process, from the law of gravitation to the phenomenal laws, is found to *modify* these laws, thus introducing a circular self-correcting element into the argument. This self-correction is possible because the phenomenal laws are initially stated in a simplified form, sometimes deliberately, as when the orbit of the moon is said to be approximately circular, and sometimes as a result of unconscious experimental approximation, as when Kepler's third law *l* is stated in the form $a^3 \alpha T^2$, where *a* is the semi-major axis of a planet's elliptic orbit, and *T* its periodic time in that orbit. From *l* together with the laws of motion, the law of gravitation *g* is deduced. But when the converse deduction of the phenomena from *g* and the laws of motion is performed, initial conditions are introduced which were neglected in the previous deduction of *g*. For example, it is found that Kepler's third law follows only if it is assumed that the sun's mass is infinitely large compared to that of the planet. The sun's mass is, however, known *not* to be infinitely large, and this correction yields a relation between major axis and periodic time that differs somewhat from the third law, though still within the range of observational approximation. And similarly *g* is found to be consistent with a slightly elliptic as opposed to a truly circular orbit of the moon.

The possibility of self-correction implied in this principle has been thought by both Duhem and Popper[1] to nullify Newton's claim to have inferred the law of gravitation by any type of reasoning different from the ordinary hypothetico-deductive method, and *a fortiori* to refute his claim to have arrived at the law deductively from phenomena. But this is too strong a reaction, for in spite of the restrictions on its purely deductive character, the form of inference used by Newton is certainly not hypothetical in the sense of introducing postulates or concepts wholly different from those of the phenomena. It proceeds rather, as I have suggested all generalizations and scientific laws proceed, by induction from particular to particular, firstly from data-point to data-point to yield an elliptic orbit, and secondly from planet to planet, and hence to all pairs of bodies, to yield the general applicability of the inverse-square law of gravitational force, which was already implicit in the mutual accelerations of each particular observed pair of bodies. I have already indicated in chapter 5 how self-corrections such as are involved in this process can be accommodated in a probabilistic, though not in a strictly deductive, model of theory-structure.

Newton was fortunate in not having to go outside the phenomena of his explananda, or beyond a simple rule of analogy, for the 'deduction' of his general law. At least two further important attempts at deduction from phenomena in nineteenth-century physics were not so fortunate. One of these was Maxwell's attempt to 'deduce' the displacement current of electromagnetic theory from experiments, which will be considered in detail in a

---

[1] P. Duhem, *The Aim and Structure of Physical Theory*, second edition, trans. P. Wiener (Princeton, 1954), 193; K. R. Popper, 'The aim of science', *Ratio*, **1** (1957), 24.

later chapter. The other was Ampère's attempt to emulate Newton's 'deduction' of the law of force between masses by a similar deduction of the law of force between moving charges.[1] Ampère starts with some highly idealized 'phenomena' derived from experiments with electric current-carrying circuits fitted on a torsion balance. These phenomena are

(1) Equal and opposite parallel currents placed close together cancel each other's electromagnetic effects.

(2) The effects of a current-carrying circuit bent in small zigzags are the same as those of a straight current between the same points.

(3) No closed current can move a conductor along the conductor's length.

(4) If three circuits in size ratios $n : 1 : 1/n$, placed in line at distances $n : 1/n$ from the middle-sized circuit, carry the same current, then the middle circuit is in equilibrium.

With the aid of a general principle of isotropy of space, Ampère uses some ingenious symmetry arguments to conclude that, for charges moving in small conducting circuits, the law of force between pairs of charges is, as for Newton, an inverse-square law. No hypothesis underived from the phenomena is involved, except the very general principle of isotropy, and no 'theoretical concept' is introduced. But the weakness of Ampère's 'deduction' lies in the non-obvious character of these 'phenomena', with their sophisticated idealizations of experimental concepts and apparatus, which already presuppose a good deal of theoretical interpretation of the data.

If the purely deductive character of this method could be maintained, it would solve the problem of transitivity. But three features of the method count against this simple resolution: first, the phenomena cannot in general be held to be conclusively given just as stated; second, the general principles upon which the deduction depends are themselves hypothetical, or at best normative principles of simplicity or symmetry not necessarily true of the world; and third, the method leaves no opening for introduction of theoretical concepts not present in the descriptions of the phenomena. It has been my contention in chapter 1 that all theoretical concepts are present in descriptions of *some* phenomena, but the method of 'deduction from phenomena' precludes the possibility of looking for these concepts anywhere else but in the explanandum itself. More frequently, however, they come by analogy from *other* domains of phenomena. For all these reasons the method of deduction from phenomena cannot be regarded as an adequate general solution of the transitivity paradox.

[1] A. M. Ampère, 'Théorie mathématique des phénomènes électro-dynamiques', *Mem. de l'Institute*, 6 (1823), 175. J. C. Maxwell (*A Treatise on Electricity and Magnetism*, second edition (Oxford, 1873), section 528) describes Ampère's theory thus: 'The whole, theory and experiment, seems as if it had leaped, full grown and full armed, from the brain of the "Newton of electricity"'.

## III. Whewell's consilience of inductions

Various suggestions have been made about a concept of 'explanation', richer than the hypothetico-deductive one, which might indicate what kind of hypotheses are fit to intervene between data and predictions, and so validly transmit confirmation. We have seen that the notion of 'unity' of logical form has proved to be a tricky one in the case of universal generalizations, and is likely to be even more elusive in the case of highly developed theories. One of the best attempts ever made to capture this notion is to be found in William Whewell, in his conception of 'consilient inductions'. It is worth examining this conception in some detail.

Whewell's view of induction is one of *colligation* or 'tying together' of facts by fundamental *ideas* and their particular applications as *conceptions*. Induction is not mere juxtaposition of facts, as 'Mercury describes an elliptical path, so does Venus, so does the Earth . . .' etc., but is the 'superinducing' upon the facts thus reported of the new conception of 'ellipse' which is initially not in the facts themselves. (Or perhaps better, not *seen to be* in the facts, for Whewell rejects a radical distinction between fact and theory, and allows that while facts are what we are conscious of, 'all facts involve ideas *unconsciously*'.[1]) Thus he suggests as the Inductive Formula: 'These particulars, and all known particulars of the same kind, are exactly expressed by adopting the conceptions and statement of the following Proposition: Mercury, Venus, Earth, Mars, Jupiter, Saturn, describe elliptic paths', making clear that the induction involves adoption of the conception as well as the facts.

Some conceptions constitute good inductions and theories, some do not. When the inductions as it were 'jump together' to produce a unified, coherent, simple theoretical system, Whewell speaks of *consilience of inductions*,[2] and regards the conceptions responsible for consilience as thereby self-validating. It is impossible, he thinks, to doubt the truth of a hypothesis in which conceptions tie together facts in a way that produces good fit, especially when the chances of a good fit being produced by accident are very small. Whewell appeals to the classic analogy of deciphering an unknown language: it is inconceivable that after many texts have been made sense of by a given decoding, there could be any other code that gives a different answer. The unique sense found must be the true sense. Similarly in the history of science, Newton's theory clearly 'consiliates' and simplifies a variety of natural motions in a way that the Ptolemaic theory does not, and similarly Lavoisier's chemistry produces a simplification absent from the last stages of the phlogiston theory. In general, false theories require a new *ad hoc* supposition for

---

[1] W. Whewell, *The Philosophy of the Inductive Sciences*, second edition (London, 1847), vol. 2, 88, 443; *cf.* 85, 469. For Whewell's view that facts are 'theory-laden', and in general his necessitarian view of 'ideas', see G. Buchdahl, 'Inductivist *versus* deductivist approaches', *Monist*, **55** (1971), 343.

[2] Whewell, *op. cit.*, 65.

every new phenomenon and become more and more complex as attempts are made to explain more data; true theories become more simple relative to their comprehensiveness. Unity and simplicity are the signs that the true and clear conceptions of a given science have been arrived at: *force* and *matter* in mechanics; the *medium* in sound, light and heat; *polarity* in electricity and magnetism; *affinity* and *substance* in chemistry; *symmetry* in crystallography; *likeness* in the classificatory sciences; and so on.[1] Such successful conceptions can even be said to be 'necessary' in their respective domains.

The tests for a good hypothesis according to Whewell can be summarized as follows.[2]

(1) It explains two or more already known classes of facts or laws.

(2) It successfully predicts 'cases of a *kind different* from those which were contemplated in the formation of our hypothesis'.[3]

(3) It successfully predicts or explains phenomena which, on the basis of previous background knowledge, are surprising, that is, would not have been expected to occur.

(4) It produces unity, coherence, simplification, in the total theory.

(5) It shows that facts previously thought to be of different kinds are after all in essence of the same kind.

Whewell's analysis can be adapted to the probabilistic account as developed here, with some crucial and illuminating differences. First, the two accounts agree in requiring that good theories indicate a 'tying together' of the facts in a stronger relation than the mere entailment of their conjunction by the theory. And Whewell's notion of colligation in terms of the 'right' conception for a given domain of phenomenon is comparable with the notion of clustering or analogy. Indeed Whewell's historical cases give useful examples of how the analogical relations derive from common properties and relations such as 'force', 'matter', 'polarity', specified in the conceptions. However, Whewell's account is largely descriptive, and he attempts little explication of the logical relations between its various elements. His repeated assertions of the validity and even necessity of the conceptions constituting consilient theories are not backed up by cogent argument. Let us see how far his criteria for good theories (which are in themselves unexceptionable) can be explicated in probabilistic confirmation theory, and where they fall short of a complete account.

That a good theory accounts for or includes the known facts, can be represented in terms of converse entailment, and that it *successfully predicts* facts of a different kind, and surprising facts, has also been shown to increase

---

[1] See Whewell's account of the development of these sciences in terms of their respective colligating conceptions in *ibid.*, vol. 1.

[2] Similar summaries of Whewell's tests have been given in R. Butts, *William Whewell's Theory of Scientific Method* (Pittsburgh, 1968), introduction, and L. Laudan, 'William Whewell on the consilience of inductions', *Monist*, **55** (1971), especially 371.

[3] *The Philosophy of the Inductive Sciences*, vol. 2, 65.

its confirmation, the increase being greater the greater the initial *im-probability* of the facts predicted. However, there is a crucial ambiguity in Whewell's account here, since he does not distinguish between the confirming effect on the theory of *past* successfully tested predictions, and the confirmation that as yet *un*tested predictions of a good theory may acquire from the evidence. Indeed Whewell hardly discusses the latter case, although without it the emphasis on *prediction* is somewhat empty, since apart from the psychological effect of a surprisingly successful prediction, that a fact was predicted before it was observed should not in itself affect the final judgment on a theory for which it is evidence. Let us distinguish the two problems.

(i) *Whewell's problem:* Theory $t$ is suggested to explain evidence $e_1$, and is supported by $e_1$; how much is our confidence in $t$ increased if it also predicts $e_2$, and $e_2$ is *subsequently* directly confirmed?

(ii) *The transitivity problem:* $t$ is suggested to explain $e_1$, and is supported by $e_1$; $t$ also entails $e_3$; how much is our confidence that $e_3$ *will* turn out a correct prediction increased by the fact that $t$ entails it?

The transitivity problem is, as we have seen, independent of the solution to (i). However, it is inconceivable that Whewell would have regarded the conditions of a good consilient hypothesis as complete unless they also provided a solution to (ii). His points (4) and (5) above provide the clue as to how this might be done. First, unity is to be effected by *simplicity* of theory. Simplicity is a criterion which raises questions to be dealt with in the next chapter. More immediately relevant is the second criterion, namely that a good theory exhibits apparently diverse domains as having similar causally essential characteristics, showing the domains after all to be of the same kind. This point is admittedly something of an interpretation of Whewell's actual words. His clearest expression is to be found in the summary aphorisms at the end of *Philosophy of the Inductive Sciences* where he says (his italics): '*The several Facts* are *exactly expressed as one Fact, if, and only if, we adopt the Conceptions and the Assertion* of the Proposition'.[1] There is, however, still an ambiguity here, in that Whewell sometimes explicitly denies that the unifying conception can be *seen in* the facts antecedently, as has been assumed in the present account of clustering and analogy, and claims that it only comes to be known, and sometimes seen, as a result of being first suggested by a clear (theoretical) idea of the facts.[2] Whewell does not seem to exclude the view that the correct conception is *really* in the facts, but his account of how it comes to be *known* differs from that adopted here just in the crucial point that would enable him to resolve the transitivity problem.

---

[1] *Ibid.*, 470. *Cf.* 70: 'all the suppositions resolved themselves into the single one, of the universal gravitation of all matter'.

[2] Especially in his dispute with Mill in 'Mr Mill's logic', *The Philosophy of Discovery* (London, 1860), 259f.

To see this, consider his critique of Newton's Rules of Philosophizing.[1] Of the second rule, that to effects of the same kind, causes of the same kind are generally to be ascribed, Whewell objects that this rule is no help in discovering conceptions, since until we have clear ideas and correct facts (which presuppose conceptions 'superinduced upon the facts'), we do not know what constitute 'same effects'. For example, are planets of the same kind as bodies in free space? Newton says yes, but Descartes had said no. Again, Whewell considers that Rule III, regarding the analogy of nature, cannot be applied until we clearly know the relations of ideas that constitute the true analogy of nature. If this were a necessary condition of application of Rule III, however, we would never be in a position to predict that a conception will be true of a new domain of phenomena until we had tested it to determine that it was both a clear expression of the domain, and also consilient with domains which share that conception. In other words, we would never be in a position to predict with high probability that a conception is true of a new domain in virtue of its (perhaps initially unclearly) seen analogy with domains to which the conception is already known to apply. What I have called Whewell's problem would be solved, but the transitivity problem would not. Moreover, presumably only the unique, true conception will help us to consiliate the facts; different analogies yielding the same predictions in a given domain will not be mutually reinforcing. And yet analogical inferences from distinct models may assist prediction where they overlap, even though most or none of them turn out ultimately to be the 'true' expression of causally essential similarities. Consider the successes of a fluid model of heat, a corpuscular model of light, and both particle and wave models of quantum physics. Whewell's account collapses as soon as it is recognized that we *never* arrive at the perfectly clear, true conception.

## IV. The analogical character of theories

Relations of analogy in a theoretical system may be seen most clearly in the elementary case of a classification of objects in virtue of monadic properties ascribed to them. Suppose objects fall into discrete and well-defined classes, classes of classes, and so on, in virtue of clusters of properties which constitute the necessary and sufficient definition of each class. An Aristotelian taxonomy of species defined by the 'essential properties' of their individuals, and of genera including species distinguished by differentiae, is the classic example of such a system. Two aspects of the system can be seen to be equivalent.

(i) There is a possibility of arguing by analogy from the essential properties of one object to those of another recognized to be in the same species, and

---

[1] *The Philosophy of the Inductive Sciences*, vol. 2, 287.

from the properties of objects of one species to those of another in the same genus. For example, the fossil remains of an extinct animal, and sufficient knowledge of the skeleton structure of animals of related species, can yield a probable reconstruction of the whole fossil skeleton.[1] Such analogical argument is explicated by adopting a clustering postulate for the initial probability distribution over objects and properties.

(ii) The classification system itself constitutes the theory which 'picks out' the analogies between objects in virtue of which they form an economical taxonomy, and the theory categorizes these analogies by naming species, genera and differentiae.

An Aristotelian-type classification is of course far too simple to represent realistic classifications even in traditional taxonomy, but the same principles can be extended to more sophisticated classification methods. With a general FR-type classification in which classes are defined not by necessary and sufficient clusters of properties, but by some criteria of high average similarity among members of a class and low average similarity between members and non-members of that class, there will be equivalence between the possibility of probabilistic analogical argument from object to object, and the existence of the taxonomic categories themselves.[2]

Whewell's question about the priority of recognition of analogy or theory can be raised sharply in relation to such taxonomic examples. Do we first, as Aristotle thought, recognize individuals as instances of a universal, or as Whewell might say, categorize them in terms of a prior conception brought to bear upon them from outside, and subsequently take them to be analogous in virtue of that universal or conception? Or do we first recognize the pairwise analogies, and then build these up into a theoretical taxonomy? Put thus, the question does not have a sharp answer. The development of advanced science with highly comprehensive and integrated theoretical structures puts emphasis on the first aspect, and modern classification methods put emphasis on the second. The priority of one aspect or the other depends on circumstances, and ought not to be seen as a profound epistemological question. Reverting to the example of universal gravitation, Whewell may at least be credited with insight into one aspect of the historical situation. It is indeed the case that stones and planets were not confidently 'seen' as significantly analogous, or even

---

[1] Extensive use of such homologies and analogies was made in nineteenth-century morphology. It was, however, then generally assumed that analogical inference is justified only if co-occurrences of properties indicate some stronger relation between properties which would constitute a causal law linking individuals in a species, or species in a genus. Both common evolutionary ancestry and functional correlation of parts required for viability in similar environments were appealed to, to provide such causal relationships. Causality has been defined here essentially as a matter of more comprehensive correlations over the total evidence, but whatever be the interpretation of causality, the common assumption remains that analogical argument from morphological similarities is possible. For further discussion, see my *Models and Analogies in Science* (Notre Dame, Ind., 1966), 61, 81–5.

[2] See the references in note 1, p. 47 above.

analogous at all, before the analogy was made the strikingly successful basis of Newton's theory. Perhaps the historical recognition of a new analogy may be accounted for in the following stages: First an analogy relation or principle of clustering is imagined and tested in apparently diverse domains which do *not prima facie* seem analogous, as in the case of the planetary orbits, the moon's orbit and falling bodies near the earth. At this stage there is no justified confidence in before-test predictions, but the theory is increasingly supported by each successful test. Then the new analogy 'catches on': it becomes possible to 'see' that those respects in which planets, the moon and falling stones are now shown by the theory to be analogous, are also respects in which there are now recognizable analogies between all these bodies and the water of the oceans, comets and their tails, and subsequently and unquestionably, earthy material projected into the solar system by rocketry.[1] At this stage, in virtue of the antecedent analogies now drawn attention to by the theory, predictions of high before-test probability can be made in new domains. Then, even a law previously accepted, for example Galileo's law of falling bodies, may be corrected by Newton's theory, and we have more justified confidence in the corrected law than in the previous experimental approximation to it, although the correction may still be experimentally undemonstrable.

Let us now consider in more detail the probability relations involved here, taking as example Newton's theory of gravitation $t$, Kepler's laws $e_1$ as initial data, and a prediction $e_2$ that a body will fall in the neighbourhood of the earth with a certain (non-constant) acceleration as derived from $t$. The prediction will be justified if the relation between $e_1$ and $e_2$ satisfies the positive relevance condition

$$p(e_2/e_1) > p(e_2). \tag{9.1}$$

Let us express Kepler's laws and the predicted acceleration relation very schematically as

$$e_1: \ (x)(F(x)\&G(x) \supset P(x)\&Q(x))$$
$$e_2: \ (x)(G(x)\&H(x) \supset Q(x)\&R(x)) \tag{9.2}$$

Here a relation of 'analogy' has been assumed between $e_1$ and $e_2$ in the following sense: the predicates $F$, $G$ represent properties of the planets asserted by Kepler's laws to have certain motions which we denote by the conjunction of predicates $P$, $Q$. (It would of course be necessary to use small finite intervals of metric predicates if this were a realistic reconstruction, but we simplify drastically for purposes of exposition by considering only monadic predicates.) $e_2$ is expressed in similar fashion, for the bodies referred to in $e_2$ share some properties with those referred to in $e_1$ but not all. All these bodies are solid, massive, opaque, and so on; but bodies near the earth differ from planets in

---

[1] Compare Goodman's example of 'educated recognition' in the case of the forged Vermeers quoted on page 50 above.

size, shape, chemical composition, and so on. In the same way, Kepler's laws can be thought of as describing motions which are in some respects the same as and in some respects different from the motions described in $e_2$: all the orbits are ideally conic sections, but they are traversed at different speeds, and about different foci.

The interpretation of inference to be adopted here is that the confidence we have in the prediction of $e_2$ is due to the relation of analogy between $e_1$ and $e_2$ which is constituted by the repetition of predicates $G$ and $Q$ in the expressions of $e_1$ and $e_2$. We regard $e_2$ as confirmed by $e_1$ because the bodies described by $e_2$ are sufficiently similar in some respects to those described by $e_1$ to justify the inference that their behaviour will also be similar. Explication of this inference therefore requires probability distributions which will yield the inequality (9.1) in particular when $e_1$ and $e_2$ are as specified in (9.2), and in comparable cases.

It has already been remarked, however, that it is impossible in practice to specify a suitable probability distribution for this relation of analogy within a language even of monadic predicates. Such an attempt not only meets unsolved mathematical difficulties, but is also liable to strain intuition too far, for we do not have completely clear intuitions about the analogical inferences which would be justified in complex cases. The best tactics seem therefore to be to take some simple cases, such as the one just discussed, in which it is clear that the inference would generally be acceptable, and find what conditions these cases would impose upon a confirmation theory, and what their consequences would be.

Assuming, then, that the judgment of analogy in this case is represented by satisfaction of (9.1), the question that immediately arises concerns the place of the theory $t$ in this explication. We have not needed to mention $t$ either in the expressions $e_1$ or $e_2$, nor in the statement of inequality of the probability-functions. Is $t$ then wholly redundant? Further inspection of (9.2) reveals that this is not the case, for it is implied in (9.2) that there is a theory, indeed more than one theory, which has the traditional relation to the data and prediction of entailing their conjunction. In particular, if $t$ is

$$(x)(F(x) \supset P(x)) \& (G(x) \supset Q(x)) \& (H(x) \supset R(x)) \qquad (9.3)$$

then $t \rightarrow e_1 \& e_2$. Furthermore, $t$ has the desirable characteristic of 'saying more than' $e_1 \& e_2$, since it is not the case that $e_1 \& e_2 \rightarrow t$. What $t$ does in effect is to pick out from $e_1$ and $e_2$ the predicates $G$, $Q$ which are in common between them, and to assert that the essential correlation in both cases is that bodies which are $G$ are also $Q$, and that the properties of the two domains of phenomena which are different are due to two other laws, one of which (relating $F$ and $P$) applies only to the $e_1$-domain, and the other (relating $H$ and $R$) only to the $e_2$-domain.

It might be noted at this point that when we speak of Newton's theory as 'explaining' Kepler's laws and the law of falling bodies, we do not as a rule claim that Newton's theory includes a deductive explanation of *all* the differences between planets and falling bodies, that is, we do not include in the explanatory theory *laws* $(x)(F(x) \supset P(x))$ and $(x)(H(x) \supset R(x))$ which mention all the properties the bodies do *not* share. Newton's theory contains laws explaining why some features of the motions of planets are different from those of falling bodies, but not all such features are mentioned in the theory; for example, their different chemical compositions do not appear in the antecedent of any law of Newton's theory, nor do their different initial velocities appear in the consequent of any such law. Kepler's laws, on the other hand, if they are considered as data to be explained, do not imply the distinctions between properties which are in the later light of Newton's theory regarded as 'relevant' or 'irrelevant' to the search for explanation. Kepler's own understanding of planets in the assertion 'All planets move in ellipses' certainly included for example the assumption that planets have magnetic properties, which he considered specially relevant to explanation of their motions. This aspect of the explanandum is, however, not mentioned in Newton's 'explanation' even in the initial conditions for Kepler's laws.[1] It follows that the expression (9.3) for $t$, which was used above in deference to the requirement that the explanandum be *deducible* from the explanans plus initial conditions, is too strong to reproduce the real situation, in which an 'explanation' is *not* required to entail the explanandum *as that was originally formulated*, but is already the result of assumptions of relevance and irrelevance which are rarely made explicit in deductivist accounts of theories. Before the deductive account can be made to work at all, irrelevant features must in fact be dropped from the explanandum as unexplainable by that theory (although they may of course be explained by another theory). The analogical account of theories which has just been suggested has the merit of making these assumptions of irrelevance explicit from the beginning. According to this account, we should regard the theory $t$, not as in expression (9.3), but rather as $(x)(G(x) \supset Q(x))$, together with the statements of initial conditions which differentiate the $e_1$-domain as an application of $t$ from the $e_2$-domain.

The pattern of theoretical inference we have been studying now takes on a different aspect. We are no longer concerned with a dubious inductive inference from $e_1$ up to $t$ and down to $e_2$, but with a direct *analogical* inference from $e_1$ to $e_2$. And $t$ does not provide the upper level of a deductive structure, but rather extracts the essence from $e_1$ and $e_2$, that is to say it reveals in these laws the relevant analogies in virtue of which we pass from one to the other inductively.

---

[1] Another example of the same kind is the dropping from the explanation of the Michelson–Morley experiment of the information that it was carried out *on the earth*. This information was crucial to classical physics, where an absolute standard of rest was assumed, but irrelevant in Einstein's relativity theory.

## V. The function of models

These confirmation conditions for theoretical inference can now be extended to give an account of the function of analogical models in science, for models of theories also provide examples of inference in which we have stronger logical relations between $e_1$ and $e_2$ than can be included in the usual formal deductivist schema.[1]

Consider an expression of a theoretical system in a domain of entities $S$ in which all the constant theoretical terms $T_1$, $T_2$, ... whose 'meanings' are problematic, have been replaced by the variables $\tau_1$, $\tau_2$, ....

$$(x)(y)...[(x, y, ... \ \varepsilon S)\phi(\tau_1, \tau_2, ... O_1, O_2, ...)] \qquad (9.4)$$

This is a representation of the theoretical calculus, together with a set of observation predicates $O_1$, $O_2$, ..., so that at this stage it is only a *partially* interpreted system, the $\tau$s remaining uninterpreted. Now there will be a model of this system (let us call it the Q-model) which is represented by replacing the $\tau$s again by the problematic theoretical terms $T_1$, $T_2$, ..., although it is rather difficult to see that this is in the ordinary sense an *interpretation*, because the problem of the meaning of theoretical terms arises precisely from the fact that we do not know what constant predicates the $T$s are, and so do not know what domain of entities and predicates satisfies this model. However, *if* we knew this, then in the usual logician's sense the Q-model would be a model of the system (9.4). In what follows 'model' will be used of *linguistic* entities (systems of laws, theories, etc.), *not* of the sets of entities and predicates which satisfy these systems.

It seems that what a *physicist* normally means by a model for a theory is not the Q-model, but rather a system of laws satisfied by a set of entities and predicates different from the set of entities and predicates which were to be explained when he set up his theory. When he refers to a set of Newtonian particles as a model for gas theory, this is a set of entities different from the gases, whose behaviour is to give the causal explanation of the gas laws. When a crystallographer builds a structure of coloured balls and steel rods on the laboratory bench, this is a set of entities different from the organic molecules he is attempting to construct a theory for. So it is necessary to talk in terms of *two* domains of entities: $S$ will now be, not a universal domain, but what I shall call the *domain of entities of the Q-model*, and I shall denote by $S^*$ the *domain of entities of the P-model*, where the P-model is a model in the physicist's sense just indicated.

Let us elaborate expression (9.4) in order to take account of the relation of the Q-model to the P-model. We shall suppose that there are two sets of

---

[1] This section was originally written as a reply to the formal-deductivist construal of theories developed in G. Maxwell, 'Structural realism and the meaning of theoretical terms', *Minnesota Studies*, vol. 4, ed. M. Radner and S. Winokur, 181.

observation predicates, $O_1$, $O_2$, ..., and $O_1'$, $O_2'$, ..., where the first set enter into laws *known to be true* in *S and S\**, and the second into laws known to be true only in *S\**, and not yet examined in *S*. We shall suppose the Q and P-models to have an *analogical relation* in the following sense: they share the $O$ and $O'$-predicates, and the Q-model involves also predicates $M_1$, ... not applicable to *S\**, and the P-model involves predicates $N_1$, ... not applicable to *S* (the 'negative analogy' of the two models). We assume that the laws known to be true of *S\** and *S* are respectively

$$e_1^*: (x)(y)...[(x, y, ...\varepsilon S^*)\psi(N_1, O_1, ...)] \qquad (9.5)$$

$$e_1: (x)(y)...[(x, y, ...\varepsilon S)\psi(M_1, O_1, ...)] \qquad (9.6)$$

and the laws known to be true of *S\** and unexamined in *S* are respectively

$$e_2^*: (x)(y)...[(x, y, ...\varepsilon S^*)\psi'(N_1, O_1', ...)] \qquad (9.7)$$

$$e_2: (x)(y)...[(x, y, ...\varepsilon S)\psi'(M_1, O_1', ...)] \qquad (9.8)$$

The P and Q-models can be expressed as

$$t^*: (x)(y)...[(x, y, ...\varepsilon S^*)\phi(P_1, N_1, O_1, O_1', ...)] \qquad (9.9)$$

$$t: (x)(y)...[(x, y, ...\varepsilon S)\phi(T_1, M_1, O_1, O_1', ...)] \qquad (9.10)$$

where the P-model is a true interpretation of the partially interpreted expression corresponding to (9.4), with the $P$s as observable in *S\**, and we have for the P and Q-models respectively

$$(x)(y)...[\phi(P_1, N_1, O_1, O_1', ...) \rightarrow \psi(N_1, O_1, ...)\&\psi'(N_1, O_1', ...)] \quad (9.11)$$
$$(x)(y)...[\phi(T_1, M_1, O_1, O_1', ...) \rightarrow \psi(M_1, O_1, ...)\&\psi'(M_1, O_1', ...)]$$

All predicates in these expressions are constants.

The problem is to show that there is a justified analogical inference, in the type of confirmation theory we have discussed, from $e_1^*\&e_1\&e_2^*$ to $e_2$. But before considering this, there is an obscurity about the notion of observability which ought to be cleared up at this point. There is not only the very important distinction between observable *predicates* and observable *entities* but there is also a distinction concerning observability of predicates *in different domains*. It may very well be the case that in the P-model we have predicates (the $P$s), such as 'mass', 'velocity', 'radius', which are observable in the domain of macroscopic physical objects, but not in that of microscopic objects. Where the P-model is used as a model for the theory about gases, the corresponding predicates in the Q-model (the $T$s) are not observable in any domain, and are only given as it were courtesy-titles when we refer to them as 'mass', 'velocity', etc. If we are to use these adjectives at all in relation to the Q-model, we must at least recognize that they name properties unobservable in the domain *S*, though observable in *S\**. The domains of both Q-model and P-model, however, contain the $O$-predicates, and these are observable in both domains. (The $O'$-predicates may not be observable in $S^*$, as we shall see presently.) In $S^*$ the $O$s will include the average pressure of a cloud of macroscopic particles hitting a surface, such as hailstones striking a wall horizontally. In $S$ they will include the pressure of the gas measured by manometers. 'Pressure' is the same predicate, observable in both domains.

The difference between the deductivist's construal and my own emerges when we consider the $T$-predicates. He generally wishes to say that these are entirely undetermined except by the Q-model, whose status as an interpretation is, as we have seen, highly problematic. Although there may be another model, the P-model, of the same calculus, there is for the deductivist no relation between the Q and P-models other than that they are models of the same calculus, and share the same $O$-predicates. Therefore, in this view, if we do use the words 'mass', 'velocity', etc., in relation to the $T$s, this is an equivocal use when compared with their use in relation to the $P$s. We cannot know whether for God the $T$s are the same predicates as the $P$s or not. Talk of the $T$s therefore seems at best to define a *class* of models of the partially interpreted system (9.4), and it is not clear that replacing the variable $\tau$s by constant $T$s has added anything to the content of (9.4), because we do not know what these putative constants are and have in principle no means of finding out (unless of course the $T$s later become *observable*, but in the deductivist's view this could not be a *general* solution to the problem of the interpretation of theoretical terms, because not all such terms will become observable—if they did *his* problem would dissolve).

If we consider the deductivist construal in the light of the considerations about theoretical inference in the preceding sections, it is not clear that it has any resources for explicating this kind of inference. Even if we waive the difficulties about interpretation for the moment, and suppose that the Q-model is an interpreted theory in the usual sense, we have shown above that inferences from one subset of observable consequences of this model to another subset are not in general inductively justifiable. In particular there would be no justifiable prediction from a set of experimental laws about gases, say Boyle's and Charles's laws, via the kinetic theory to other laws about gases if the kinetic theory is understood as a Q-model, that is, if there is no more than an equivocal sense in which we can speak of it being 'about' masses and velocities of molecules. Suppose, however, we now bring in the P-model. In the gas example this is a model of Newtonian particles whose laws of motion are known to be true, or at least accepted for purposes of exploitation in the theory of gases. We suppose the models expressed as in (9.9) and (9.10), and that $\phi$ entails the known experimental laws shared by both models as in (9.11). It should incidentally be noticed that since $\psi'(N_1, O_1', \ldots)$ is entailed by $\phi$, it is known to be true in $S^*$ even if it has not been directly examined, or even if it is for all practical purposes unobservable in $S^*$. For example, it is unlikely that the analogue of Boyle's law in Newtonian particle mechanics has ever been *observed* to be true; nevertheless it is believed because Newton's laws are believed in that domain.

In a confirmation theory of the type described in section IV we may have a relation of analogy between $e_1{}^*$ and $e_1$ as expressed in (9.5) and (9.6). (Compare the $e_1$, $e_2$ of (9.2), where $N_1$, ... stand for $F$, $P$; $M_1$, ... stand for $H$, $R$; and $O_1$, ... stand for $G$, $Q$.) It is very important to be clear at this point that the relation of analogy here spoken of is *not only* the formal analogy in virtue of the fact that both models are models of the same calculus, but includes what I have elsewhere[1] called a *material analogy* in virtue of the sharing of the $O$-predicates. Similarly, we may have material and formal analogies between $e_2{}^*$ and $e_2$ expressed in (9.7) and (9.8), in virtue of the sharing of $O'$-predicates and their relations expressed in $\psi'$. Since $e_2{}^*$ is true, and if the negative analogy between $S$ and $S^*$ is not too strong, there will then be a justifiable analogical inference to $e_2$, which is strengthened by the truth of both $e_1{}^*$ and $e_1$. Moreover, the same argument yields a justifiable inference to

$$(x)(y)...[(x, y, ...\varepsilon S)\phi(P_1, M_1, O_1, O_1', ...)] \qquad (9.12)$$

where the $P$-predicates are observable in $S^*$ but unobservable in $S$.

Comparing (9.12) with (9.10), we see that the analogical inference leads to the suggestion that the $T$-predicates should be *identified* with the $P$-predicates, rather than being regarded as problematic constants whose reference is in principle unknown. Although their referents in $S$ are unobservable, their 'meaning' is derived from the observables of $S^*$, that is, they mean the same as they do in the P-model, and satisfy the same laws. With any construal of the $T$s not involving some identification of this kind, it is not clear that there can be justifiable analogical inference to $T$-statements in $S$. (We shall return to this point in the next section.) With the identification, however, the inference to predictions can be made even stronger, for as was remarked above, it is not necessary that the $O'$-predicates should be observable in $S^*$, only that the $P$s should be. In such a case it is clear that the inference to $e_2$ depends essentially upon knowing the truth of $t^*$ empirically, and making an analogical inference to the probable truth of $t$ and hence $e_2$, and that this depends upon the identification of the $P$ and $T$-predicates in $S$.

The diagrams opposite help to elucidate the structure of these inferences and to relate them to what has been said in section IV about the status of the theory in predictive inference. In the diagrams arrows on the lines indicate alleged justifiable inferences in the deductive and analogical construals respectively, and dotted lines indicate relations of analogy. In the case of the deductive construal, we have seen that there is no justifiable inference from $e_1$ to $e_2$ unless there is an analogical relation between these laws independent of $t$. If no such relation is apparent, however, we may be able to make an

---

[1] *Models and Analogies in Science*, chap. 3.

inference to $e_2$ if we can find a P-model as represented in the right-hand diagram. Here the inference depends on analogy between the laws $e_1$, $e_1{}^*$, and assertion of the truth of $t^*$. Then, since $t^*$ states the causes of $e_1{}^*$, there may be a justifiable analogical inference to $t$ as stating the causes of $e_1$, and hence to $e_2$ as a consequence of $t$. Truth is, as it were, fed into the *theory* of S from the P-model, and so passed on to $e_2$, whereas in the deductive construal there is no justifiable inference via the theory of $S$ because this theory acquires no probable truth from any source other than its own entailments. Theories cannot be pulled up by their own bootstraps, but only by support from external models.

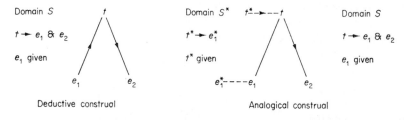

Deductive construal     Analogical construal

It will immediately be objected to this account that there are many examples for which analogical inferences of the kind described would not only be un-justified in the light of further evidence, but would never be regarded as justifiable even before further evidence is collected. Analogical arguments are notoriously weak and liable to failure and must generally be treated with extreme caution. This of course is true, but it must be borne in mind that the above examples have presupposed the principle of total evidence. *If* all the evidence we have is summarized as in $t^*$ and $e_1$, then the inferences may be intuitively reasonable. But if we have other evidence to the effect, for example, that $S$ and $S^*$ differ from each other in many further characteristics, or if we know of other domains in which the inference to $\psi'$ breaks down, then such information may well weaken the inferences to the point of disconfirma-tion. I suspect that when apparent counterexamples to these inferences are produced, they will be found to involve one or other of these types of additional evidence. In principle an adequate confirmation theory must be capable of dealing with such complexities, and must explicate the weakness as well as the strength of analogical arguments.

## VI. Identification of theoretical predicates

In the absence of a comprehensive confirmation theory it is not possible to be precise in reply to objections of the kind just mentioned. But it is perhaps permissible to speculate a little further upon the characteristics which any confirmation theory would exhibit. In particular, it may be possible to suggest some compromise between the position outlined here and the deductive

H

account, though still within the framework of a confirmation theory. There are two ways in which my position has been opposed too sharply to deductivism and should now be modified.

First, it may be doubted whether we wish to *identify* a theoretical predicate such as 'mass' of a molecule or electron, with 'mass' of a macroscopic particle. On the other hand I think it has been correct to say that 'mass' cannot be simply *equivocal* without destroying the possibility of theoretical inference. What we need to reconstruct is a notion of analogical meaning of the word 'mass' in the two domains, where 'analogical' is used as a middle term between 'univocal' and 'equivocal', as in some Thomist philosophy. 'Mass' is not always used 'in the same sense' when predicated of different systems, but it is not on the other hand a *pun* when it is used of positrons, neutrons, quasars and the like.[1] The problem of determining how far the meaning of 'mass' can be extended analogically, and the problem of deciding what analogical inferences are justifiable, are closely related problems. Predicates can be stretched just as far as analogical argument remains justifiable, and conversely. How far this is would have to be decided by looking at the whole complex of evidence in all domains in which the predicate is applied. If, for example, the difference in domain is only one of scale, we shall probably be quite satisfied simply to identify the $P$ and $T$-predicates, but when the difference of scale is accompanied by other differences as radical as those between, say, the macroscopic and the nuclear domains, we may become increasingly unwilling to allow any analogical extension of meaning from one to the other.

The second respect in which my account might be modified in the direction of deductivism is in regard to the attributes of the $S$ and $S^*$ domains which are allowed to weigh in analogical inference. It is convenient here to refer to a discussion by Sellars,[2] in which he argues that to identify the theoretical predicates with the $P$-predicates from an antecedent observation language is to fall into the 'myth of the given', and to misrepresent the *novelty* which may be introduced by using $P$-models in connection with theories. Sellars objects against the construal of $T$-predicates favoured by Nagel in *Structure of Science* that it makes the $T$s new but not meaningful, and against my construal in *Models and Analogies in Science* that it makes them meaningful but not new.

If I have understood Sellars correctly, his proposal for the resolution of this dilemma is as follows. Not all analogy is analogy of *particulars* in virtue of their sharing identical attributes, as has been assumed in the discussion of analogy in section IV above. Attributes themselves may be similar or analo-

---

[1] I have discussed analogy, context-meaning, and related questions in 'Aristotle's logic of analogy', *Phil. Quart.*, 15 (1965), 328; and 'The explanatory function of metaphor', *Logic, Methodology, and Philosophy of Science*, ed. Y. Bar-Hillel (Amsterdam, 1965), 249; both reprinted in *Models and Analogies in Science*. *Cf.* also chaps. 1, 2 above.

[2] W. Sellars, 'Scientific realism or irenic instrumentalism', *Boston Studies in the Philosophy of Science*, ed. R. S. Cohen and M. W. Wartofsky (New York, 1965), 171.

gous, that is, first-order predicates may themselves be predicated by second-order predicates, and may be analogous in virtue of sharing such second or even higher order predicates. Sellars gives as examples the second-order predicate 'perceptible', which applies to first-order predicates, and the second-order predicate 'transitive', which applies to first-order relations such as 'before', 'to the left of'. Such second-order attributes can either be *mentioned*, as in ' "transitivity" is true of *before*', or they can be *shown*, as by exhibiting the transitivity postulate satisfied by 'before'. The function of a P-model is to introduce second and higher order predicates which are shared with the Q-model, and thus to convey some interpretation to the *T*-predicates.

> Thus, as a first approximation, it can be said that models are used in theory construction to specify new attributes as *the attributes which* share certain higher order attributes with attributes belonging to the model, fail to share certain others (the negative analogy)—and which satisfy, in addition, the conditions laid down by the relevant correspondence rules.[1]

Thus, Sellars claims, both meaning and novelty are allowed for in the relation of Q and P-models, and the P-model remains heuristically useful at least so long as its higher order attributes remain implicit. When they are themselves formalized, presumably the P-model can be abandoned, and postulates representing higher order attributes can be explicitly added to the Q-model and corresponding calculus.

I confess that many features of this suggestion remain obscure to me. In the first place, if the higher order attributes can be referred to by intensional expressions such as 'transitive', 'perceptible', it is not clear that we have escaped the 'myth of the given'. These expressions are already in the descriptive language, and a more sophisticated analysis of 'analogy', involving a type-logic, could presumably take account of analogies depending on the sharing of these predicates, as it can of first-order predicates. If novelty depends on introduction of new predicates, there is no novelty here. If, on the other hand, the introduction of novelty depends essentially on the P and Q-models merely *exhibiting* analogy of higher order attributes, it is not clear that anything other than *formal* analogy has been introduced. *Each* model of a calculus 'exhibits' such analogy with every other model of the same calculus; indeed once the calculus has been fully expressed, mention of models other than the Q-model is wholly redundant to this kind of analogy, since the higher order attributes of the Q-model are already shown by the calculus itself. And in this case the construal of the *T*s as 'the attributes which share certain higher order attributes with attributes belonging to the [P-]model' does not seem to differ in principle from Nagel's account, which Sellars rejects, nor from the deductivist's which we have seen reason to reject above. For in both these accounts it follows from the status of the Q and P-models that there are *some* second

---

[1] *Ibid.*, 181.

or higher order attributes they share, namely the relations exhibited in the calculus of which they are models.

It is possible, however, that Sellars has in mind a situation which is somewhere between the two extremes just mentioned, namely a P-model which tacitly introduces higher order attributes in virtue of which we vaguely accept an 'analogy' between it and the explanandum, but which have so far been unanalysed, and for which we may not have names in the language. For example, we may recognize an analogy between a loud noise and a bright flash, and may exploit it in a P-model for light drawn from the phenomena of sound, without necessarily having in our language a concept 'intensity' which applies to both noise and flash. However, the question now arises whether the P-model is introduced here because of antecedent recognition of an analogy, even though this was inexpressible in the existing language, or whether the adoption of this P-model itself *introduces* a new higher order attribute which sound and light phenomena share. The answer to this question is, surely, 'six of one and half-a-dozen of the other'. But such liberality must not be taken to the point of admitting *any* model as a candidate for the P-model. Unless *some* analogy of predicates, whether first or higher order, is recognized, which is not merely the relation of isomorphism between two models of some same calculus, use of a P-model in theoretical inference becomes vacuous, as I shall now try to show.

Lying behind Sellars' attempt to reconcile the meaningfulness of theoretical concepts with the possibility of novelty, there is some obscurity about the function of models in *inference*. Sellars has not given sufficient weight to the fact that my plea in *Models and Analogies in Science* for recognition of the logical role of models depends essentially upon taking *predictive power* as a necessary condition for theories. From this point of view, acceptance of his shared higher order predicates as ingredients in the role of models in relation to theories will depend upon whether or not we regard such shared predicates as justifying analogical inference. Take the example of transitivity, and assume that the P-model contains the first-order relation 'before', and that the Q-model contains a relation $R$ which is either said or shown to be transitive. According to Sellars we need know nothing about $R$ except that it satisfies the postulates of the theory and is transitive. Waiving now the question of what relation $R$ *is*, we must nevertheless ask how the P-model helps us to make analogical inferences in the theory. The answer is, surely, not at all. For suppose we risk an analogical inference from statements of the P-model involving 'before' to statements of the Q-model involving $R$. There is a large class of relations, all members of which are consistent with what is known about $R$, but clearly the analogical inference will not be valid for all of them. Suppose $R$ is, in God's private eye, 'larger than'; then an analogical inference involving $R$ would be equally as justifiable or unjustifiable as a similar inference involving 'smaller than', but the conclusion cannot be true in both cases.

Any analogical inference may of course lead to false conclusions for *empirical* reasons, but in this case one or other conclusion must be false for *logical* reasons. On the other hand, if $R$ is *known to be* a particular transitive relation having other affinities with 'before', such as 'to the left of', there might very well be a justifiable inference from the P-model to the Q-model if the theory were concerned, for example, with a geometry of space–time. In other words, whether higher order predicates can function to justify analogical inference is a question only to be decided by examining particular predicates, and the possibility of doing this presupposes either that the predicates are already in the language, or that they can be coined as required to name particular attributes in virtue of which an analogy is suspected. It cannot be the case that every shared higher order predicate is sufficient to generate a justifiable analogical inference, for this would certainly lead to inconsistency, as in the 'transitivity' example above, and would even be vacuous, if it could be shown that any system shares *some* higher order attributes with every other system, a proposition which should not be too difficult to prove if our ontology is generous enough.

Finally, a remark about novelty and the myth of the given. Sellars' objection to identifying the predicates of the Q and P-models seems to follow from his rejection[1] of the assumption (which he ascribes to Nagel) that there is a one-to-one correspondence between predicates and the extralinguistic attributes to which they apply. If this assumption were true, then indeed it would be difficult to see how analogical inference from observable predicates of the P-model would leave room for any theoretical novelty. Sellars instead wants to allow for enrichment and revision of the observational vocabulary by its interaction with theory. But use of models as analogues neither presupposes such a one-to-one correspondence nor rules out the kind of interaction of observation and theory that Sellars requires. Indeed, if we understand predicates as *analogical* in their applications to different situations rather than as either univocal or equivocal (as suggested above and discussed in the references there given), the possibility of novelty is safeguarded by the indefinite variety of analogical extensions of existing predicates. It is, after all, rare for new descriptive predicates to be *coined*, and much more common for new situations to be described by complex combinations of old predicates. I do not wish to deny, however, in anything I have said here, that totally new concepts may sometimes be required in theoretical science. It only seems to me to follow from the foregoing arguments that *if* totally new concepts are introduced, there is no possibility of theoretical inference of the kind discussed here, and consequently I suspect that such occasions of total novelty are rare.

In the two previous chapters the concept of a universal generalization over an infinite domain of individuals has been replaced by the notion of analogy

[1] *Ibid.*, 184.

or clustering between finite sets of individuals in respect of particular properties, and it has been shown how this strictly finite conception can be
represented in a probabilistic confirmation theory so as to explicate inductive
predictions. In this chapter the same conception has been generalized to cover
theories and their explananda. Firstly I have argued that theories should be
reinterpreted in a manner suggested by Whewell, as statements of unifying
analogies between domains of phenomena which had been taken prior to the
theory to be of different kinds, and secondly I have interpreted the function
of physical models for theories as yielding analogical argument from given
empirical systems to empirical predictions, where the relata of the analogical
argument are stated to be analogous by the theory. There then arises the
further question whether the unifying theory or the 'perception' of the analogy
comes first. This is of course partly a historical question; however, with
respect to the logic of theories, I have argued that where the theory is regarded
as justifying inductive inference from the evidence to further predictions,
then the theory must be taken to be an assertion, for the empirical instances
it covers, of an *antecedently* perceived analogy between those instances.

CHAPTER TEN

# *Simplicity*

## I. Subjective and notational simplicity

The notion of simplicity of theories is frequently invoked by both scientists and philosophers as a major criterion for theory choice. It is a criterion appealed to explicitly by Einstein in developing his special and general theories of relativity. Quine makes it a decisive factor in choosing between possible theory–observation and analytic–synthetic networks.[1] It is the presupposition of the statement and of any solution of Goodman's 'grue' paradox. Yet formal studies of the concept of simplicity are few, and intuitive appeals to it seem peculiarly resistant to application.[2]

Even superficial investigation of simplicity soon reveals that there is not one but many types of concept involved. The first distinction that needs to be made is between *subjective* and *objective* concepts. Subjective simplicity may be taken to refer to pragmatic or psychological preferences for certain kinds of theories because they are imaginatively or mathematically easy to work with, or because they can be conveniently computed. Conditions of these sorts may of course be correlated with more objective senses of simplicity, but by themselves they are difficult to make precise and are not even philosophically very interesting. Convenience is a characteristic that depends greatly on the psychological make-up and circumstances of different people, on the logical and material tools available to them, on the purposes they have in view, and on many other factors. Furthermore, even if it were possible to provide some objective and unique definition of convenience, this would not necessarily be correlated with the truth-claims of scientific concepts and theories. True and useful theories are occasionally highly inconvenient. More objective characteristics of theories are involved in considerations of truth, and it is these we shall be principally concerned with here.

Among the types of objective simplicity, let us first distinguish and dismiss what may be called *notational simplicity*. Consider pairs of descriptive

[1] See for instance W. v. O. Quine, *From a Logical Point of View* (Cambridge, Mass., 1953), 17, 45; and *Word and Object* (Cambridge, Mass., 1960), 19f.

[2] One of the best investigations is to be found in the symposium on simplicity in *Phil. Sci.*, **28** (1961), 109, containing papers by R. S. Rudner, M. Bunge, N. Goodman, R. Ackermann and S. F. Barker. I am indebted to Elliott Sober for illuminating discussions of simplicity.

sentences, or of conjunctions of sentences, which are logical equivalents of each other in virtue of a set of definitions taken to be analytic. For example, the equation for a circle $r = a$ is logically equivalent to $x^2 + y^2 = a^2$, where the transformation equation $r^2 = x^2 + y^2$ from polar to cartesian coordinates is understood. The logical equivalence ensures that no facts can in principle be expressed by one expression which are not the same facts expressed by the other. But there is a clear sense in which the first expression is simpler than the second, although it is a sense which in itself can make no difference to the confirmation relations of a theory in which it occurs, since logical equivalents have identical confirmation relations. The notationally simpler expression may of course make a subjective or heuristic difference—for example the polar expression may be pragmatically more convenient than the cartesian, and perhaps such convenience may be measured. But we shall not pursue the question of how this might be made more precise, since it has no relevance to the confirmation of the theory. Again, the intrinsic representation of a circle as that curve for which every point is equidistant from a centre may suggest geometric analogies more clearly than the cumbersome cartesian expression. Ability to 'see' such analogies will no doubt influence relative initial degrees of belief, and is therefore not unimportant in considering scientific confirmation, but for present purposes we shall assume that the theories we are concerned with are already laid out in their most perspicuous form, so that the differences of simplicity involved are only those between theories that are logically distinct.

It must be added at once, however, that the situation of logical equivalence just described is very rare in interesting scientific examples. I have deliberately described it in terms of *logical*, not *observable* content, for I am presupposing that no useful distinction can be made in these contexts between the observable and unobservable content of a theory. A consequence of this is that many pairs of theories which have often been claimed as equivalent are not logically equivalent in the sense intended here, because the claim to equivalence has been confined to logical entailment of what is 'in principle' observable. Thus, for example, it has been held that a Ptolemaic system with sufficient epicycles is equivalent to a Copernican theory fitting the same planetary positions; that Maxwell's field theory with some *ad hoc* additions is equivalent to continental action-at-a-distance theories with some other *ad hoc* additions; that Einstein's special relativity is equivalent to Lorentz's theory plus a clock retardation hypothesis; and that Schrödinger's wave-mechanical formulation of quantum mechanics is equivalent to Heisenberg's matrix mechanics. But in all these cases the theories are not only notationally but also *conceptually* different. The Copernican system includes the essential hypothesis of the heliocentric universe; Maxwell's theory includes the presence of energy in 'empty space'; special relativity involves a radical modification in the concept of space–time; Schrödinger's theory postulates underlying continuity of the wave function, about which Heisenberg asserts nothing. In these cases the claim to equivalence can be no more than the assertion that, if either theory were true, then no observation entailed by that theory can in principle decide between the two theories. But there is great difficulty in delimiting what it is that is entailed by any given theory, since a theory is never precisely formulated in a deductive system, and there are many plausible suggestions and consequences which in a loose sense accompany a theory without being strictly deducible from it. Moreover, the notion of what is 'in principle' observable is by no means clear, and it is not profitable to try to distinguish rigorously between what a theory asserts of the facts and what can be observed among those facts. Maxwell asserted the presence of energy in the field; any attempt to observe this 'directly' can always be reinterpreted in terms of an action-at-a-distance theory without introducing field energy, but the two theories are not logically or factually equivalent.[1] Similar considerations emerge in connection with the Einstein–Lorentz example, which will be considered in more detail later.

## II. Content

Having set aside for present purposes subjective and notational concepts of simplicity, we are left with objective simplicities of the logical and conceptual structure of theories which do have distinctive factual consequences. These are the concepts of simplicity we shall attempt to explicate. We have immediately to notice a distinction within objective simplicities between what I shall call concepts of *content* and of *economy*. These are irreducibly different notions of simplicity, both of which are generally taken to be desirable characteristics of theories, but which sometimes conflict in application.

First, a theory is often required to have high power or content: to be at once *general* over many facts, and *specific* in its description of each of them, and to make precise, detailed and comprehensive claims about the state of the world. In Popper's terminology, this is to say that a theory should be *highly falsifiable*. Against all probabilistic theories of induction, Popper maintains that this requirement has the consequence that good theories should be in general *im*probable, since the more claims a theory makes on the world, other things being equal, the less likely it is to be true. On the other hand, as would be insisted by inductivists, a good theory is one that *is* regarded as more likely than its rivals to be true, and in particular it is frequently assumed that simple theories are preferable because they require fewer concepts and fewer premises, and hence would appear to make *fewer* claims about the state of the world, and hence be more probable. Since Popper has annexed the notion of

[1] I have discussed the field versus action-at-a-distance case in *Forces and Fields* (London, 1962), 216f.

simplicity to the first of these sets of conflicting requirements, it has been difficult to do proper justice to the second.[1]

The concepts of content and economy ought, however, to be considered separately. Take content first. In his account of simplicity, Popper has argued that the desirable characteristics of 'simple' theories consist in their having high empirical content, which can be explicated in terms of their greater generality over many states of affairs, and their greater specificity in describing such states. He proposes to measure both aspects of content by the notion of 'more potential falsifiers'. Taking the example of theories about the circular or elliptic orbits of heavenly bodies, including the planets, he argues firstly that since a theory about all heavenly bodies is more general in applicability than a theory about all planets, it is, *ceteris paribus*, to be regarded as preferable, and it has, of course, more potential falsifiers, namely all the heavenly bodies. Secondly, a circle is a special case of an ellipse with both axes equal; hence to assert that a body moves in a circle with parameter unspecified is more specific than to assert that it moves in an ellipse with parameters unspecified, and hence preferable. The circle has more potential falsifiers than the ellipse in the following sense: any three points can be made to lie on a unique circle, so that the hypothesis that the orbit is some circle may be falsified if false by the fourth observation, but some ellipse will accommodate any four points, and may only be falsified if false by the fifth observation. The same relation between specificity and number of potential falsifiers holds also in the case of a theory making a prediction of a specific value of a given parameter, in comparison with a theory that only makes a prediction within a certain interval including that value. The more specific theory, again, is said to be preferable.

To avoid confusion in the use of the term 'content', it should be noticed that the term does not refer to the *specific* content of a theory, that is, to what in particular that theory says, but rather to the *measure* of content—*how much* the theory says. In Popper's view, the absolute unit of content might be taken to be the single potential falsifier, but since there are grave difficulties about such an atomic view of empirical statements, his examples usually involve only comparisons of content, and these only as arising from relations of entailment between rival theories. This is the case in the examples just mentioned. 'Being a planet' entails 'being a heavenly body', and 'moving in a circle' entails 'moving in an ellipse' (with its axes equal). In these examples corollary (PC4) entails Popper's claim that the theory he holds to be preferable in point of content is also less probable than its rivals, as long as the probabilities of the entailed hypotheses are not zero.

But it must be noticed at once that this is the only sort of case in which the requirement of high content is *necessarily* correlated with improbability. If

---

[1] *Logic of Scientific Discovery*, chap. 7 and appendix *viii.

we are comparing expressions which do not stand in the relations of entailment exhibited by the examples, their relative probability values are not determined without some further assumption. For example, to compare the law 'All students are anarchists' with 'All young women are anarchists', where there are no such entailment relations, we may reasonably suppose that their contents are to be measured by counting each student and each young woman as one potential falsifier. But we are by no means obliged to assume in such a case that each is of equal probabilistic weight, that is, that any young woman is as likely to be an anarchist as any student, and hence that the law that applies to more individuals is less probable. Again, algebraically formulated curves may be ordered in content inversely as the number of data points required before the curve is falsified if false, but if the curves are not related so that one is obtained from the other by specifying one or more of its parameters, we are not compelled to order them in this way. Indeed many alternative suggestions have been made for a simplicity ordering of curves in which intuitive economy of parameters and of algebraic powers is directly correlated with probability, but in which falsifiability in Popper's sense does not correlate with this type of simplicity.[1]

In particular, we are not always dealing with mathematical expressions with *unspecified* parameters, as in the case of the general equations of the circle and the ellipse. Suppose we had to compare the straight line $y = a + bx$ with the second-order curve $y = a + bx + 4x^2$, where the coefficient of $x^2$ is specified. Both curves are falsifiable, if false, by the third point to be observed, and hence have the same content by Popper's definition; but the first would generally be held preferable, other things being equal, because it is more economical of parameters and algebraic powers.

Where there are no entailment relations between theories, therefore, and particularly where they may be held to have the same content, simplicity in the sense of economy may be ordered by probability quite independently of content.

With regard to content itself, it remains to be enquired whether and under what conditions it is a desirable characteristic of theories. Popper is undoubtedly right to object to an ordering of the preferability of theories directly with their probability, since this necessarily devalues their content by leading to an entailed theory always being preferred to an entailing theory, and, unless special restrictions on choice are adopted, leads to the data themselves being preferred to any theory, since at least in the deductive account the data are entailed by all adequate theories. This looks like a *reductio ad absurdum* of the probability account of theory choice.

But here the argument has gone too fast, for we have to remind ourselves of the conditions under which theories are to be compared. We may be

---

[1] For example, the theory of Harold Jeffreys discussed above, chapter 8.

concerned in theoretical contexts with attaining the widest possible *true theory*, or we may be concerned in more practical contexts with making *specific predictions* about uncertain outcomes of test situations. In either case there is what I shall call a *relevant content* in terms of which the theory is to be judged. In the first case it will contain as much of the furniture of the world as has at that point entered the considerations of scientists. In the second case it will include empirical evidence already available, together with the alternative possible outcomes of all those tests whose actual outcomes are to be predicted. In neither case is it profitable to try to delimit precisely what relevant content includes; what is important is to consider that theories which are to be compared for probability are theories which *conflict* over some part of this content. Thus the directive of a probabilistic account of theory choice is not 'Choose the most probable theory', but 'Of two *conflicting* theories of *equal* relevant content, generally choose that which yields the more probable predictions'. Lest this should seem unduly restrictive, consider the various logically possible content-relationships of two theories $h$ and $h'$.

(i) $h$ *includes* $h'$ as a proper part. Then they do not conflict, and no question of choice of conflicting assertions arises. Preference for $h$ may occur when it is desired to have as wide a domain of applicability as possible, or when it is desired to make a prediction in a domain not covered by $h'$. In particular, the conclusion that the evidence itself will always be chosen is false. In the simple deductive model where the evidence is entailed by every adequate theory, the evidence conflicts with nothing, and therefore does not enter the arena of choice. In more sophisticated accounts in which the evidence itself may be regarded as only loosely fitting the best theory, the evidence may conflict with some theories, and then the most probable conjunction of modified evidence and theory having the desired relevant content will be chosen. This may also be the simplest conjunction. As Harold Jeffreys puts it in discussing the preference of a physicist for a simple curve lying among the data points rather than a more complex curve which exactly fits all of them, '. . . . his predilection for the simple law is so strong that he will retain it when it does not satisfy the observations exactly'.[1]

(ii) The contents of $h$ and $h'$ are either *disjoint*, or *intersect but do not conflict*. Here again no conflict or question of choice arises, except perhaps in the pragmatic sense of deciding where to put money and research into developing one or other theory further. But we are here concerned with confirmation and not directly with utility.

(iii) Relevant content *intersects and conflicts*. This is the case where confirmation theory is most interestingly applicable. Two theories $h$ and $h'$ can always be made to have notionally the same content by conjoining with $h$ that part of the relevant content of $h'$ with which it does not conflict, and similarly

---

[1] *Scientific Inference*, 60.

for $h'$. Then supposing the conflicting answers to questions posed in the domain where they do conflict to have the same content, comparison of the probabilities of the conflicting predictions may be made wholly on grounds of simplicity in the sense of economy. Since it is this sort of case that is of primary interest here, I shall make no further attempt to specify a measure of content. Content and economy can be regarded as two independent dimensions of simplicity whose relative weightings in theory choice will be a matter of judgment and circumstances, to be reflected in initial probabilities and utilities. But even to consider such relative weightings we need to have an independent assessment of the effect of economy on probability. Let us therefore consider various aspects of economy.

## III. Economy and clustering

I shall argue that all the recognized types of economy used as simplicity criteria of good theories can be interpreted as indicating *clustering* among the consequences of the theories. It follows that economy criteria apply in the first place to clustering of *particular* instances or systems, and only secondarily to economy of *theories*. Given the intransitivity of confirmation, this is as it should be, for even if high probability can be given to theories on the basis of their economy, this will not indicate that the theories are also good predictors unless positive relevance can also be shown among their consequences.

First, a disclaimer. The discussion of kinds of economy which follows is not supposed to supply a definitive ordering of different theories in virtue of their economy or initial probability. It is supposed only to indicate that the general postulate of clustering or homogeneity of the universe is related in diverse ways to intuitions of what it is for a theory to be economical, and therefore to indicate how personalist initial probabilities may be built up consistently with those intuitions. But once we pass beyond simple inductive and analogical arguments to complex theories, it is unlikely that there will be uniformity of judgment among scientists about how to weight various economy factors. There should correspondingly be no single ordering of initial probabilities, for where there is in fact dispute about criteria, a personalist confirmation theory would *fail* as an explication if it demanded an over-determination of scientists' intuitions. The examples that follow are therefore included mainly with the negative purpose of showing that they do not constitute counterexamples to a probabilistic confirmation theory, not with the positive purpose of providing definitive criteria for such a theory, for in this domain definitive criteria are inappropriate. Further illumination on the complexity of conditions involved and their relation to the general account of confirmation presented here can only be gained from detailed case histories.

## (a) Universal generalizations and analogy arguments

The relation between content, economy and clustering can be seen most easily in the types of inference already discussed, namely enumerative and analogical arguments, where the clustering postulate has been invoked as a presupposition. Consider first evidence consisting of individuals $a_1, a_2, \ldots a_n$, all of which have properties $P$ and $Q$. Now consider an individual $a_{n+1}$ with property $P$. Does $a_{n+1}$ have $Q$ or not? If nothing else is known, the clustering postulate will direct us to predict $Qa_{n+1}$ since, *ceteris paribus*, the universe is to be postulated to be as homogeneous as possible consistently with the data, and a world in which $PQa$ (where $a$ is any one of $a_2, a_2, \ldots a_n$) and $PQa_{n+1}$, is more homogeneous than one in which $PQa$ and $P\tilde{Q}a_{n+1}$. But this is also the prediction that would be made by taking the most economical general law which is both confirmed by the data and of sufficient content to make a prediction about the application of $Q$ to $a_{n+1}$. For $h \equiv$ 'All $P$ are $Q$' is certainly more economical than the 'gruified' conflicting hypothesis of equal content

$h' \equiv$ 'All $x$ up to $a_n$ that are $P$ are $Q$, and all other $x$ that are $P$ are $\tilde{Q}$'.

Notice that we are here considering economy *in a given predicate language,* for the expression for $h'$ is of course a candidate for 'gruification' in some other language, where it may be more economical than $h$. But we have already noticed that application of simplicity criteria presupposes a prior solution of the grue paradox.

Simplicity criteria apply in a different way to analogy arguments. Suppose $a_1, \ldots a_n$ have equally weighted properties $PQRS$, and $a_{n+1}$ is known to have properties $PQ\bar{R}$. Does $a_{n+1}$ have $S$ or not? Since $a_{n+1}$ is more similar to than different from $a_1, \ldots a_n$, clustering would direct us to predict that it does have $S$. Now consider the competing general hypotheses in this case. The evidence supports equally three such hypotheses relevant to a prediction about $a_{n+1}$ having $S$, namely

$$h \equiv \text{'All } P \text{ are } S\text{'}$$
$$h' \equiv \text{'All } Q \text{ are } S\text{'}$$
$$h'' \equiv \text{'All } S \text{ are } R\text{'} \equiv \text{'All } \bar{R} \text{ are } \bar{S}\text{'}$$

Since the predicates are assumed equally weighted, these hypotheses have the same initial probability and the same content. Of the three hypotheses, $h$ and $h'$ both predict $Sa_{n+1}$, and $h''$ predicts $\bar{S}a_{n+1}$. Hence $Sa_{n+1}$ is predicted by $h \vee h'$, which on these assumptions has less content and is more probable then $h''$.[1] Therefore in comparing the three hypotheses, choose the prediction $Sa_{n+1}$. Any other hypothesis having the same content as $h \vee h'$ and predicting $\bar{S}a_{n+1}$, which is consistent with the evidence, will necessarily mention and

---

[1] Compare the 'pluralism' of otherwise conflicting theories favoured by Maxwell (below, page 280). If two different theories both give the same prediction, that prediction is generally better confirmed than if only one theory gives it.

distinguish between $a_1$, ... $a_n$ on the one hand, and $a_{n+1}$ on the other, and will be less economical than some hypothesis of the same content predicting $Sa_{n+1}$, by an argument similar to that used above for universal generalizations.

It follows in the two cases considered, that if a rule is adopted to choose the prediction resulting from the most probable hypothesis on grounds of content or, in case of a tie in content, the most economical hypothesis of those of equal content, this rule will yield the same predictions as the clustering postulate for both enumerative and analogical arguments. In a probabilistic confirmation theory the clustering postulate will ensure that these predictions have the highest confirmation. It is, moreover, important to notice that in such a confirmation theory predictions are justified by the initial belief in clustering of individual instances rather than in the economy of laws as such, since the probability of a law is untransferable to its instances in virtue of non-transitivity.

## (b) Economy of parameters and properties

Many examples of economy do, however, seem to involve hypotheses directly. A straight line hypothesis is preferred to a higher order curve when both are consistent with the evidence, and in general, hypotheses introducing fewer extra-logical predicates and fewer arbitrary numerical parameters are preferred to those that introduce more. As we have seen, Popper and Jeffreys both give accounts of the simplicity of curves which depend essentially on counting the number of arbitrary parameters and assigning simplicity inversely as this number. Let us see how this general proposal can be applied in the particular case of a predictive inference from data points to another specific point satisfying the hypothesis of a straight line rather than a higher order curve.

Let $f$ be the assertion that two data points $(x_1, y_1)$, $(x_2, y_2)$ are obtained from an experiment. These points can be written equivalently $(x_1, a+bx_1)$, $(x_2, a+bx_2)$, where $a$, $b$ are calculated from $x_1, y_1, x_2, y_2$. The two points are consistent with the hypothesis $y = a+bx$, and also of course with an indefinite number of other hypotheses of the form $y = a_0+a_1x+a_2x^2...$, where the values of $a_0, a_1, ...$ are not determined by $x_1, y_1, x_2, y_2$. What is the most economical prediction of the $y$-value of a further point $g$, where the $x$-value of $g$ is $x_3$? Clearly it is the prediction which uses only the information already contained in $f$, that is, the calculable values of $a$, $b$, rather than a prediction which assigns arbitrary values to the parameters $a_0, a_1, ...$ of a higher order hypothesis. Hence the most economical prediction is about the point $g \equiv (x_3, a+bx_3)$, which is also the prediction given by the 'simplest' hypothesis on almost all accounts of the simplicity of curves. Translated into probabilistic language, this is to say that to conform to intuitions about economy we should assign higher initial probability to the assertion that points $(x_1, a+bx_1)$, $(x_2, a+bx_2)$, $(x_3, a+bx_3)$ are satisfied by the experiment, than to that in which the third point is inexpressible in terms of $a$ and $b$ alone.

In this formulation economy is a function of finite descriptive lists of points rather than general hypotheses, and the relevant initial probability is that of a universe containing these particular points rather than that of a universe in which the corresponding general law is true (which probability may in any case be zero). Description in terms of a minimum number of parameters may therefore be regarded as another aspect of homogeneity or clustering of the universe.

There are several classic historical cases of theory choice where preference appears to be given in this way to economy of parameters in mathematical expressions of the theory. Consider the comparison of the Ptolemaic and Copernican theories. We have already seen that these cannot in any circumstances be said to be logically equivalent, but it is often held that they are observationally indistinguishable with regard to planetary positions, since for any data which is well-fitted by the Copernican theory (and indeed for any data at all), there is an equally well-fitting Ptolemaic theory involving a sufficient number of parameters specifying epicycles. However, if we consider a specific form of Ptolemaic theory with all parameters specified and limited in number, and consider predictions from *this* theory, there is no guarantee that they will coincide with predictions from the Copernican theory. And in this case there is no doubt that predictions from the Copernican theory would generally be preferred on simplicity grounds as involving fewer parameters in their description, and thus indicating higher clustering among different states of the system.[1] The desirability of this notion of simplicity is supported by the fact that, in so far as the Ptolemaic and Copernican theories coincide for particular planetary data, the Ptolemaic theory seen as a general cosmology requires more extra explanation than the Copernican. For example, the parameter in each planet's motion that corresponds in Copernicus to the relative motion of the earth round the sun, and is therefore the same for each planet, appears as an unexplained coincidence in Ptolemy but as a natural consequence of heliocentrism in Copernicus.

---

[1] The historical situation was of course a great deal more complicated than this. But if the attempt is made to abstract the qualitative geometrical features of the two theories from a multiplication of other influential factors, including metaphysics, theology, natural conservatism and empirical data involving what were later recognized as perturbations, then the suggested economy criterion remains as *one* necessary feature of the debate, though by no means the decisive one. T. S. Kuhn summarizes the mathematical situation thus: 'The seven-circle system . . . is a wonderfully economical system, but it does not work. It will not predict the position of planets with an accuracy comparable to that supplied by Ptolemy's system. . . . Copernicus can give a more economical *qualitative* account of the planetary motion than Ptolemy. But to gain a reasonably good *quantitative* account . . . Copernicus, too, was forced to use minor epicycles and eccentrics.' (*The Copernican Revolution* (Cambridge, Mass., 1957), 169.)

That in spite of this ill-fit, and many other scientific and non-scientific features counting against it, the qualitatively simpler theory eventually won out is an example of *post-hoc* interpretation of the data (as subject to 'perturbations') in order to fit an initially preferred simple theory.

Another classic case, which will be dealt with in more detail in a subsequent section, is that of the simplicity comparison of Einstein's and Lorentz's theories of electrodynamics. This introduces the notion of mathematical *invariance*[1] between different coordinate systems as an economy of parameters, and hence as a clustering characteristic of theories. In the period immediately following Einstein's 1905 paper, Lorentz's theory was consistent with all existing experimental results, including the Michelson–Morley experiment, and even had greater content than Einstein's theory in that it contained a micro-theory of electrons which Einstein did not attempt. But Einstein postulated the invariance of Maxwell's equations with respect to transformations between inertial coordinate systems, and this permitted him to abandon the ontological postulate of the aether, and eliminated velocity relative to the aether as a parameter in physical laws. Thus electrodynamic laws were shown to be homogeneous, or clustered, rather than heterogeneous, in their indifference to uniform relative motion, and in the absence of any privileged framework defining absolute rest.

In the development of Maxwell's theory itself, however, this property of invariance was not at first obviously related to economy of parameters or concepts.[2] The mathematical symmetry of Maxwell's equations, which yields the type of invariance later exploited in special relativity theory, was bought by Maxwell at the price of an extra term in the equations, and an extra postulate of the displacement current in aether—an absence of economy of both mathematical form and of postulated real properties. However, I shall try to show in the next chapter how in this case Maxwell attempted to derive the displacement current by *analogical* argument from other electric and magnetic systems. He was not wholly successful in this, but if he had been successful, the added economy of interpreting material dielectrics as essentially conductors-plus-aether, and thus unifying apparently diverse 'contact' and 'distance' transmissions of electromagnetic effects, might have been said to outweigh loss of superficial mathematical simplicity. Interpreted in this way, the displacement current was not an extra 'real property' but a generalization of the concepts of charge and current at a fundamental level, affecting the interpretations of these concepts in all their applications.

### (c) *Economy of mathematical form*
As well as economy of parameters, most accounts of mathematical simplicity have required low values of algebraic powers, and of orders of differential equations, tensors, etc. Apart from pragmatic convenience, at least two sorts of reasons can be given for this preference. First, additional powers or orders

---

[1] The best discussion on simplicity as invariance and 'optimal coding' is to be found in H. R. Post, 'Simplicity in scientific theories', *B.J.P.S.*, **11** (1960), 32; and 'A criticism of Popper's theory of simplicity', *ibid.*, **12** (1961), 328.

[2] Similar examples are pointed out by M. Bunge, *The Myth of Simplicity* (Englewood Cliffs, N.J., 1963), chaps. 5, 7.

sometimes involve additional parameters as coefficients, as in the general algebraic expression of the $n$th order. But this does not account for the preference for, say, $y = a^2x^2$ over $y = a^4x^4$. A second type of economy relevant here is limitation of the number of solutions of a given equation, for, in general, higher power or higher order equations will have more solutions, between which decisions have to be made which may be arbitrary. Thus $y = a^2x^2$ has two solutions, namely, $y = \pm ax$, while $y = a^4x^4$ has four, namely $y = \pm ax$, and $y = \pm iax$. In the absence of further information about the admissibility of imaginary solutions, the first equation indicates greater homogeneity among different systems to which it is applicable than the second.

Considerations of this kind seem to have weighed with Einstein in his choice of second—rather than higher—rank tensors to express general relativity theory, for, as he says, the requirement of covariance together with this type of simplicity 'determines the [gravitational] equations ... almost completely'.[1] However, Einstein regretted the consequent introduction into this theory of an undetermined parameter $\lambda$ representing the curvature of space, and welcomed theories, such as Friedman's continually expanding universe, which eliminated this arbitrary constant. In a letter to Ilse Rosenthal-Schneider, Einstein wrote

> In a reasonable theory, there are no (dimensionless) numbers whose values are only empirically determinable . . . this world is not such as to make an 'ugly' construction necessary for its theoretical comprehension.[2]

In other words, this is a case where individual judgments of weighting as between mathematical economy and economy of parameters, have to be made.

### (d) Ontological economy
The example of Einstein's rejection of the aether raises questions about economy in postulation of *entities* and *kinds of entities* that reveal further aspects of the notion of clustering. To multiply kinds of entities is to multiply predicates of the language; to multiply entities themselves is to multiply individual or mass referring terms. I have already argued that the purposes of confirmation cannot be served by introducing theoretical predicates having no paraphrasable relation with existing predicates in the language. Certainly theories introducing such predicates gratuitously must be regarded as in violation of economy and of the homogeneity of the universe. On the other hand, it is clear that many theoretical advances have occurred precisely by introducing new entities and new kinds of entities: Jupiter's moons, atoms, mesons, quasars. However, these are often introduced with a view to overall

---

[1] A. Einstein, 'Autobiographical notes', *Albert Einstein: Philosopher–Scientist*, ed. P. A. Schilpp (La Salle, Ill., 1949), 89.
[2] Quoted by Ilse Rosenthal-Schneider, *ibid.*, 144–5.

*reduction* of kinds of entities or of properties, as when a multiplicity of different chemical compounds are described in terms of a limited number of elements and their properties and relations. Given that two theories have the same content, that which introduces few *primitive* entities will generally be preferred.

It has been objected to this assumption that ontological economy is not always sufficient to ensure simplicity. Koslow, for example, has questioned whether removal by Einstein of a *singular* referring term (the aether) from electrodynamics represents a simplification, since such removal may only be traded for greater complexity of *predicates*, or what Quine calls 'ideological complexity'.[1] Certainly there will often be difficult questions of weighting in such cases, as there are between other aspects of simplicity. For example, rejection of substantial electric fluids was bought at the cost of introducing a new irreducible predicate of matter, namely electric charge, and there are many other such cases in the history of science. But Koslow seems to be mistaken in thinking that these issues might be settled by finding a definite 'proxy function', such as is defined by Quine, since comparison of the ontologies involved in different scientific theories is not necessarily a matter of 'ontological reduction' in Quine's sense. As examples of such reduction, Quine discusses Carnap's replacement of 'impure numbers', such as a temperature of $x$ degrees Centigrade, by the 'pure number' $x$ predicated by the term 'Centigrade-temperature', and also the Fregean reduction of numbers to classes. Scientific examples would presumably be reduction of chemical compounds to their molecular and atomic descriptions, and reduction of light to electromagnetic waves. Such examples may involve economy, but it is not an economy of existents, for the reduced entities do not cease to *exist* as a result of reduction, they still exist as transformed by Quine's proxy functions, that is, as what they are said to be in their reduced descriptions. But rejection of an entity like the aether, or a substance like phlogiston, is not a reduction in the same sense. Aether and phlogiston, just *do not exist* either as themselves, or by being described reductively in the theories which supersede them. In Einstein's theory, or in Lavoisier's, aether or phlogiston is not described at all; nothing in those theories corresponds with their existence, although various things may of course correspond with what were thought wrongly to be their properties and effects.

Koslow's suggestion regarding comparison of theories with different ontologies appears to be the following: in cases of Quine's ontological reduction, take the proxy function as indicating how the sentences of one theory are

---

[1] A. Koslow, 'Comment', *Minnesota Studies*, vol. 5, ed. R. Stuewer (Minneapolis, 1970), 356, in reply to K. F. Schaffner, 'Outlines of a logic of comparative theory evaluation with special attention to pre- and post-relativistic electrodynamics', *ibid.*, 311. See also Quine, 'Ontological reduction and the world of numbers', *Ways of Paradox* (New York, 1966), 199, and *Ontological Relativity* (New York, 1969), 55f.

mapped on to those of another, and how, as Quine puts it 'the objects of the one system must be assigned severally to objects of the other'. The theories are then guaranteed to be in some sense commensurable, at least in content, and their respective numbers of entities and primitive properties can be counted and compared. But where ontological economy is not a matter of ontological reduction, but a matter of removing entities or properties from the list of existents, we cannot define any proxy function, because there is no 'assignment' of, for example, phlogiston to the objects of post-Lavoisierian chemistry. The suggestion that there is might arise in an account of theories which both regarded theoretical terms as in some sense definable by observation terms, and also regarded the observation language as stable through theory change. But we have seen reason to reject both these assumptions. Where ontological economy is a matter of removing terms from the scientific language, theories with different ontologies cannot be as easily compared as in cases of ontological reduction. But, as with other species of economy, this kind of ontological economy will have to be weighed against others in ways that will doubtless always defy formalization.

The notion of economy of parameters and properties has been invoked in the foregoing sections as a measure of clustering among particular systems that is adequate to account for the high probability given to analogy arguments, to mathematically simple expressions, to mathematical invariances and to ontological economy. There are, however, serious objections to taking economy in these senses always to indicate high probability. In his discussions of the grue paradox Goodman has made us familiar with the fact that economy depends on the mode of description of data and hypotheses, and has claimed that it can give no decision between conflicting hypotheses, since it is generally possible to redescribe the hypotheses in a language with a different predicate or parameter base, so as to reverse any economy ordering. But I have already argued that this problem is soluble without abandoning the concept of confirmation of the rival hypotheses. Moreover both the statement and the solution of the problem depend on an antecedent theory of economy relative to a *given* predicate base.[1] Therefore the account of simplicity just given is prior to and independent of the grue problem, although the application of this account to rational theory choice will undoubtedly depend on a satisfactory resolution of that problem such as I have attempted to provide in an earlier chapter.

---

[1] Goodman has shown how to define a measure of the simplicity of the predicate-base itself of a given language, in terms of the minimum number of predicates the language needs to retain all logical equivalences. He does not, however, consider the relative simplicities of different sentences expressed *within* that language. Both measures are needed for a full account of ontological economy. *Cf.* a series of papers in *J.S.L.*, **8** (1943), 107; **14** (1949), 32 and 228; **17** (1952), 189; *J. Phil.*, **52** (1955), 709; also 'The test of simplicity', *Science*, **128** (1958), 1064; and 'Recent developments in the theory of simplicity', *Phil. and Phen. Res.*, **19** (1959), 429.

(e) *Economy of theoretical premises*

Another frequent source of claims to simplicity is economy in the number and character of the explanatory or theoretical premises of a theory. This type of economy seems to involve irreducibly the probability of theories themselves, rather than clustering among their particular consequences, and there is no doubt that arguments about theory choice in science often proceed as if it were the theories rather than their specific consequences that are being compared for economy and acceptability. If indeed general theories can be compared for non-zero probability, whether by restricting them to finite domains or by some other assignment of probability, then we may assume, consistently with the clustering postulate, that higher probability is assigned to theories exhibiting greater homogeneity of the universe, and hence we may be tempted to conclude that, if content is equal, theories that are more probable in this sense are also more acceptable as a basis for prediction. But if mere economy of theoretical premises is sufficient to ensure high probability, there seems to be an implicit contradiction between probability as a criterion of acceptability of theories, and the considerations regarding transitivity we have been discussing. For surely theories chosen for high probability, and of comparable content, should also be better predictors? At least this seems to be assumed in many arguments for theory choice.

However, I believe that the appearance of contradiction here is illusory, for if we examine the ways in which the condition of economy of theoretical premises is actually applied, we shall find that it can be accommodated to the various species of economy by the clustering postulate already introduced.

First of all it is clear that a proposal merely to count the number of postulates required by a theory does not provide an adequate measure of economy. Any postulate set can be reduced to one member by conjunction, and even if this move is forbidden by some *ad hoc* device, the same system can often be formulated by means of many different but equivalent postulate sets. Notational and logical equivalences, however, are not relevant here; we need as before to consider the empirically distinguishable content of postulates.

The matter is best approached by examining certain *avoidance prescriptions* which seem to constitute between them the intuitive criteria of economy of theoretical premises. We are enjoined to avoid large numbers of logically independent premises, redundant premises and *ad hoc* premises which have little support from the data. All these can be seen to be related to the desirability of clustering among consequences of the theory. Premises which are logically independent are possibly also probabilistically independent (indeed it is often avoidance of probabilistic independence that is enjoined rather than mere absence of entailment relations among the postulates). And probabilistic independence among postulates may[1] indicate probabilistic independence among their consequences, and hence gives no guarantee of positive relevance. On the other hand, avoidance of redundant premises can be seen as a prescription for high confirmation of the theory itself, since if conjoined premises are unnecessary for the deduction or confirmation of any available or possible evidence, they themselves can attain no confirmation by converse entailment or the relevance criteria, and they cannot affect the confirmation of any predictions of the theory.

The injunction to avoid *ad hoc* hypotheses needs a little further discussion, since it has been one of the main criteria of theory choice discussed by writers in the Popperian tradition.[2] Popper himself has indeed characterized all probability theories of confirmation as necessarily implying 'Choose the most *ad hoc* hypothesis'. It therefore behoves us to show that when hypotheses are *ad hoc* in an undesirable sense, they offend against one or other of the types of criteria adopted here, that is, they either reduce theory content gratuitously or they violate one or other of the species of economy.

What makes a hypothesis *ad hoc* is by no means precisely clear, and there are several distinct senses of the term in the literature. The original sense seems to apply to adoption of a hypothesis that is specially tailor-made to avoid refutation of a more general theory, some of whose consequences have been empirically falsified. But several cases should be distinguished here.

(i) Sometimes this move involves gratuitously restricting the content of the general theory, as when 'All gases satisfy Boyle's law' is reduced to 'All near-ideal gases within certain limits of temperature and pressure satisfy Boyle's law'. It seems to be this sort of case that Popper has in mind in his strictures against choosing an *ad hoc* hypothesis with lower content and higher probability, when the proper response might be to look for a more general replacement for Boyle's law. And in this case Popper is undoubtedly right to reject the hypothesis of lower content which ignores rather than attempting to explain refuting instances. The case is, however, not in itself a case of choice between *conflicting* theories such as we are considering, since a more adequate general law would not necessarily conflict with the *ad hoc* hypothesis, but would include it.

(ii) Hypotheses are also called *ad hoc* when they not so much reduce content as fail to increase it. This occurs when entities or properties are invoked as explanations, but are not testable by any observations other than

---

[1] 'May', not 'must', since probabilistically independent postulates may have probabilistically dependent consequences, and probabilistically dependent postulates may have probabilistically independent consequences.

[2] See especially Popper, *Conjectures and Refutations*, 61, 244, 287, and 'The aim of science', *Ratio*, I (1957), 24; and I. Lakatos, 'Changes in the problem of inductive logic', *The Problem of Inductive Logic*, ed. I. Lakatos (Amsterdam, 1968), 315, and 'Falsification and the methodology of scientific research programmes', *Criticism and the Growth of Knowledge*, ed. I. Lakatos and A. Musgrave (Cambridge, 1970), 124f.

those they were invoked to explain. For example, a special 'vital force' has sometimes been introduced to 'explain' organic processes, but without any indication how its presence could be otherwise detected. Here what I have called the relevant content of the theory is not increased, and the hypothesis also offends against ontological economy; hence the present confirmation theory would exclude such a hypothesis.

(iii) Sometimes, however, a move to save a general theory from refutation does involve *increasing* its content by specifying further conditions that explain why the expected consequences of the theory are not observed. Such conditions may themselves be independently testable, and may then acquire independent confirmation. Thus, apparent violation of energy conservation in certain nuclear disintegrations was explained by postulating a new particle, the neutrino, and this particle was subsequently independently detected. Another such case was the hypothesis of an as yet unobserved planet, first suggested to explain the anomalous orbit of Uranus, which led to the discovery, that is, the visual observation, of Neptune. In such cases the original '*ad hoc*ness' of the hypothesis cannot be objected to, and a probabilistic confirmation theory need not reject it.

## IV. The principle of relativity and classical electrodynamics

We shall now take up in more detail the example of Einstein's relativity theory versus Lorentz's modification of classical electrodynamics. In the years immediately following Einstein's first paper on special relativity in 1905, the problem of choosing between special relativity and Lorentz's electrodynamics makes an exceptionally good case history for simplicity criteria, since it involves practically all the types of criteria distinguished in this chapter, and also illustrates how different judgments were possible regarding the weighting of these criteria.[1]

Einstein states the principle of special relativity in the 1905 paper as follows.

> The laws by which the states of physical systems undergo change are not affected, whether these changes of state be referred to the one or the other of two systems of coordinates in uniform translatory motion.[2]

This principle can itself be interpreted as a principle of economy in the sense that, if it is violated, physical laws themselves determine a distinction between

[1] For this historical case see G. Holton, 'On the origins of the special theory of relativity', *Amer. J. Phys.*, **28** (1960), 627; 'Mach, Einstein, and the search for reality', *Daedalus*, **97** (1968), 636; and 'Einstein, Michelson, and the "crucial" experiment', *Isis*, **60** (1969), 133. For a careful analysis of senses of '*ad hoc*' in this case, see A. Grünbaum, 'The bearing of philosophy on the history of science', *Science*, **143** (1964), 1406; and for a methodological comparison of the theories of Einstein and Lorentz, see K. Schaffner, *Minnesota Studies*, vol. 5, 311.

[2] 'On the electrodynamics of moving bodies', reprinted in *The Principle of Relativity* (Eng. trans., London, 1923, and Dover Publications), 41.

different coordinate systems in relative uniform motion, and the expression of any law must therefore mention some parameter identifying the co-ordinate system relative to which it is expressed. In Lorentz's theory the identifying parameter is the velocity of the system relative to the aether, which is assumed to determine the coordinate system that is absolutely at rest. However, the economy permitted by the principle of relativity is bought at a price, for the question whether any particular physical law satisfies the principle cannot be answered until a transformation function is specified to carry the values of the physical variables measured in one inertial coordinate system over into their values in the other systems, and this transformation may itself be more or less economical. For example, Newtonian mechanics satisfies the principle of relativity with respect to the so-called Galilean transformation, which takes a measured distance $x$ from the origin $O$ of coordinates at time $t$ in one system, into a measured distance $x' = x - vt$ from the origin $O'$ of another system which coincided with $O$ at time $t = 0$, and is moving with velocity $v$ in the $x$-direction relative to $O$. The Galilean transformation equations for the rest of the space–time coordinates in this case are the identities $y' = y$, $z' = z$, $t' = t$. Under this transformation the Newtonian laws of motion have the same form when expressed in the coordinates of either system, and are said to be *invariant* with respect to Galilean transformation.

Maxwell's equations, however, are not invariant with respect to Galilean transformation, and hence cannot satisfy the principle of relativity in the same sense that Newton's laws do. This fact constitutes the main problem of Einstein's 1905 paper, which he introduces by posing a thought-experiment within Maxwell's theory. The theory seems to require a distinction to be ·drawn between the effects of motion of a conductor in a magnetic field on the one hand, and a stationary conductor in a changing magnetic field on the other, where the only difference lies in which system is regarded as being absolutely at rest and which in motion. This absence of symmetry of relative motion constituted for Einstein an offence against the internal perfection of classical electrodynamics, since 'motion relative to the aether' was picked out as being motion referred to an absolute standard of rest. There is what Einstein calls a 'disturbing dualism' in Maxwell's theory between the kinetic energy due to motion, and the potential energy of the field, although kinetic energy is formally transformable into field energy by merely changing the mode of coordinate description. Moreover, there appear to be no observable phenomena corresponding to this absolute standard.

> ... the unsuccessful attempts to discover any motion of the earth relatively to the 'light medium', suggest that the phenomena of electrodynamics as well as of mechanics possess no properties corresponding to the idea of absolute rest.[1]

---

[1] *Ibid.*, 37. There has been much debate about whether Einstein explicitly considered the null result of the Michelson–Morley experiment of 1885 as being among the inductive grounds for special relativity in 1905, or whether indeed he even knew of this experiment

Einstein requires that the principle of relativity should be satisfied by Maxwell's equations. It is satisfied only if Galilean transformations are replaced by the so-called Lorentz transformations, which yield in the case of the uniformly moving coordinate systems described above the mathematically more complex set of transformation equations

$$x' = \beta(x - vt), \; y' = y, \; z' = z, \; t' = \beta(t - xv/c^2)$$

where

$$\beta = \frac{1}{\sqrt{(1 - v^2/c^2)}}$$

The principle of relativity for Maxwell's equations has various powerful consequences. Firstly, the 'systems of coordinates in uniform translatory motion' referred to in the principle must be interpreted as coordinate systems related to each other by Lorentz transformations, and this changes the physical characteristics of space and time, most notably by requiring that a time-like measure $t$ in one system is partially space-like as measured in another system. Secondly, Newton's laws of motion are not now invariant with respect to uniformly moving systems, and therefore if mechanics is to satisfy the principle of relativity with respect to Lorentz transformations, the laws of mechanics must be modified. The relativistic equations that replace them yield empirically distinct consequences, such as the non-additivity of relative velocities, and the increase of the measured value of the mass of a body with its velocity relative to the measuring instrument. Thirdly, since the numerical constance $c$, equivalent to the velocity of light, appears *explicitly* in Maxwell's equations, the velocity of light must itself have the same value in any uniformly moving system in which it is measured. This consequence is experimentally supported by the negative results of both first and second-order experiments designed to measure the velocity of the earth relative to light transmission, which culminated in the Michelson–Morley experiment of 1885.

The principle of relativity effects various sorts of economy in the theory of electrodynamics. First, there is economy of description with regard to specification of the relative velocity of electrodynamic systems, since this velocity no longer appears in the laws themselves, but only in the transformation equations from one set of coordinates, or one relatively moving observer, to another. This in itself might not be judged as an overall economy, were it not for two further immediate ontological economies. First, kinetic energy and

---

at that date. Historical evidence, including Einstein's own later recollections, is conflicting and inconclusive on this point. However, the passage quoted here clearly indicates that he appealed to *some* inductive support, if only to the null results of earlier first-order experiments. See references in p. 239, note 1, and p. 246, note 1.

field energy become essentially the same physical concept, distinguished only by the form of description in different coordinate systems, and second, the concept of a material aether providing the standard of absolute rest becomes redundant. This same search for ontological economy is seen in Einstein's belief, expressed in another paper of 1905, that since inertial mass is shown in the Maxwell–Lorentz theory to be partly of electromagnetic origin, all inertia should be regarded as a form of electromagnetic energy: 'The mass of a body is a measure of its energy content'.[1] Again, in his development of general relativity, Einstein interprets the empirically discovered correlation of the values of inertial and gravitational mass as indicating ontological identity. In general, he describes the theory of relativity as 'an astoundingly simple combination and generalization of the hypotheses, formerly independent of each other, on which electrodynamics was built'.[2]

Before attempting further assessment of the simplicity criteria which are illustrated by special relativity theory, something must be said about the generalization of the principle of special relativity to a general *principle of covariance*. The principle of special relativity requires invariance only with respect to uniform translations of coordinate systems, but in passing to general relativity theory Einstein extends this requirement to *all* coordinate systems: the laws of physics can be expressed in a form which is independent of the coordinate system. Attempts to justify this covariance principle have since been made in various ways. First of all it may be held that coordinate systems are purely artificial aids to the expressions of measurements by an observer, whereas the essentials of a measurement involve only coincidences of separate events: the pointer with a mark on a scale, a flash of light with a clock reading, etc. If it is further assumed that physical laws are no more than codifications of such coincidences, it follows that physical laws should make no intrinsic reference to coordinate systems. An objection to this argument is that it gives an unduly operational interpretation to physical laws, for why should there not be some *theoretical* reason why one coordinate system should be preferred over another, even though measurements make no necessary reference to such a system? Bridgman, for example, has suggested that the system in which the experimenter himself is at rest should properly have privileged status.[3]

Non-relativistic arguments for a privileged status of one particular co-ordinate system, however, have not usually been based either on the idea of a

---

[1] 'Does the inertia of a body depend on its energy content?', *The Principle of Relativity*, 71. Note the ontological prescription: 'The combination of the idea of a continuous field with that of material points discontinuous in space appears inconsistent . . . the material particle has no place as a fundamental concept in a field theory'. ('On the generalized theory of gravitation' (1950), in *Ideas and Opinions* (London, 1954), 345.)

[2] *Relativity, the Special and the General Theory* (Eng. trans., London, 1920; first published 1916), 41.

[3] P. W. Bridgman, *The Nature of Physical Theory* (Princeton, 1936), 82–3.

particular *abstract* system, nor on the anthropocentric claims of a system accidentally coinciding with a particular observer, but rather on an assumed *physical* property of one particular system, for example that which is defined by the all-pervasive material aether and by the transmission through it of electromagnetic waves. The appeal to coincidences involved in measurement cannot rule out the possibility that there is such a physical medium, which might itself be an ingredient in these physical coincidences, and hence define not an abstract but a physically privileged coordinate system.

This possibility raises the question of how far the principle of covariance itself has any empirical consequences. Some confusion has arisen over this question, since it seems on the one hand that the principle of relativity, which is a restricted covariance principle, is empirically *supported* by the null results of attempts to detect physical effects on the aether and would be empirically refuted by positive results of such attempts, while on the other hand it has been proved that *any* physical law can be put mathematically into covariant form with sufficient exertion of mathematical ingenuity. Thus, for example, in a well-known textbook, Tolman first seeks to justify the principle of covariance by remarking that, if it were false, 'differences in form for different coordinate systems could be taken as an evidence for differences in the absolute motion of the spatial frameworks used in setting up the space–time coordinate system', which implies that the principle is falsifiable (for we might indeed observe such differences), while on the other hand he argues that the principle 'imposes no necessary restriction of the nature of [physical] laws', which implies that the principle in itself has no empirical content.[1]

This apparent paradox is resolved by noticing a logical distinction between the use made by Einstein of the principle of relativity in the special theory on the one hand, and his later general statement of covariance on the other. In the first place, it was Maxwell's equations themselves, as discovered to hold in local coordinate systems at rest with respect to the observer, which Einstein wanted to retain as invariant. He therefore adopted the Lorentz transformations under which they are invariant for all inertial frames. It is true as a matter of mathematics that any conflicting set of electrodynamic equations, or indeed Newton's laws of motion themselves, *could* have been rendered covariant, but in order to do so further functions of the physical quantities involved would have had to be introduced, thus sacrificing economy of mathematical form of the covariant equations for the property of co-variance itself. As Einstein puts it

> Someone should just try to formulate the Newtonian mechanics of gravitation in the form of absolutely covariant equations . . . , and he will certainly be

[1] R. C. Tolman, *Relativity, Thermodynamics, and Cosmology* (Oxford, 1934), 166–7.

convinced, that the principle (of covariance) excludes this theory practically, even if not theoretically![1]

The same choice of mathematical simplicity *among* covariant equations is seen in Einstein's adoption of second rather than higher-rank tensors in his development of general relativity, for tensor equations are necessarily all covariant.

In contrast to the general principle of covariance, then, the principle of special relativity does have empirical consequences. It can be seen as a principle of economy or clustering in the sense that no extra parameter or entity is required in the specification of electrodynamic laws in passing from one inertial coordinate system to any other. But in adopting the principle in 1905, Einstein sacrificed some confirmations and some simplicities for others. The Galilean transformation in mechanics is replaced by the Lorentz transformation, involving dependencies on the relative motion of observers, and on the numerical constant $c$, which were previously not found in mechanics. The Lorentz transformation involves also an intuitively complex notion of the four-dimensional space–time manifold. The high inductive confirmation accumulated over two centuries for Newtonian mechanics is in a sense sacrificed, although no experiments which had provided this confirmation were at the time sufficiently accurate to resolve the differences between the predictions of Newtonian mechanics and special relativity. Long before Feyerabend and others pointed out the ambiguity of all interpretations of experimental results, Einstein had remarked that 'essentially different principles' may 'correspond with experience to a large extent',[2] and this applies even where the interpretation of the space–time framework of the experience itself is radically transformed.

In exchange for these sacrifices of simplicity, special relativity indicated greater simplicity of electrodynamics, to the point where the result of the Michelson–Morley experiment, that scandal to classical physics, became a natural logical consequence of the new unified foundations of mechanics and

---

[1] Quoted in Ilse Rosenthal-Schneider, 'Presuppositions and anticipations in Einstein's physics', *Albert Einstein: Philosopher–Scientist*, 139n. Compare Einstein's implied contrast between the merely *heuristic* value of the general covariance principle, and the *heuristic and refutable* character of the principle of relativity: 'If a general law of nature were to be found which did not satisfy this condition, then at least one of the two fundamental assumptions of the theory would have been disproved' (*Relativity*, 43); and 'The general theory of relativity, accordingly, proceeds from the following principle: Natural laws are to be expressed by equations which are covariant under the group of continuous coordinate transformations. . . . In the first instance one may even contest [the idea] that the demand by itself contains a real restriction for the physical laws; for it will always be possible thus to reformulate a law, postulated at first only for certain coordinate systems, such that the new formulation becomes formally universally covariant. . . . The eminent heuristic significance of the general principles of relativity lies in the fact that it leads us to the search for those systems of equations which are *in their general covariant* formalation the *simplest ones possible*' ('Autobiographical notes', *Albert Einstein*, 69; italics Einstein's).

[2] 'On the method of theoretical physics' (1933), *Ideas and Opinions*, 273–4.

electrodynamics. If one asks why Einstein preferred simplicity of electro-dynamics to simplicity of mechanics, it must be replied that he saw electro-dynamics as the more fundamental and potentially comprehensive theory. In the course of many historical disputes about the extent of the experimental evidence which Einstein had in mind in formulating special relativity, it has not been sufficiently emphasized that for Einstein the empirical bedrock of the whole enterprise was acceptance of Maxwell's equations (in the form given by Hertz) as *successful* descriptions of the empirical facts. Not only did Maxwell's equations occur in the first line of the 1905 paper as posing by their acceptance the fundamental problem of the paper, but in the 'Autobiographical notes' of 1949 Maxwell's theory is explicitly likened to thermodynamics in the empirical security of its basis, differing only in that Maxwell needed one hypothetical concept: the displacement current. In his writings around 1920, Einstein is concerned to emphasize the *continuity* between the Maxwell–Lorentz theory and his own, and appeals to the successes of this theory to support his principle of constant light velocity: 'the confidence which physi-cists place in this principle springs from the successes achieved by the electro-dynamics of Maxwell and Lorentz'. The empirical status of the Maxwell–Lorentz theory is undoubted, and serves equally to support the theory of relativity:

> The experimental arguments... limit the theoretical possibilities to such an extent, that no other theory than that of Maxwell and Lorentz has been able to hold its own when tested by experience.[1]

Maxwell's electrodynamics, together with the negative results of experi-ments to determine physical effects of relative motion, constitute the two major inductive grounds of special relativity which can be said to have been strongly supported in 1905. Of these two grounds, Maxwell's equations were the common basis of all the current rival theories, while the absence of measurements of motion in the aether was regarded as an anomaly to be dealt with rather than as a decisive refutation of current fundamental concepts. It is impossible to escape the conclusion, to which all recent commentators accede with varying emphasis, that Einstein's own grounds for preferring relativity theory to its rivals in 1905 must be found in what he called 'internal' or simplicity criteria in their various forms.

These criteria were not, however, entirely based on the 'clustering' characteristics to which we have been led to give high probability and hence strong predictive power. Several students of the development of Einstein's theory have been so impressed by its counter-intuitive and even counter-inductive character that they have used it as a paradigm case of what they take to be the pre-eminent importance of unpredictable creativity and intuitive genius in science, as against entirely secondary and sometimes misleading

[1] 'What is the theory of relativity?' (1919), *Ideas and Opinions*, 229, and *Relativity*, 50.

attempts to follow 'rules of method'.[1] This view, however, neglects the duty of philosophers of science (which is shared by more practical judges of scientific work, including editors of journals and committees of scientific academies) to make some distinctions between fruitful creativity and plausible eccentricity, reason and accident, and to do this *before* and not after the success or failure of the project is evident.[2] The opposing, inductive, view of special relativity has been cogently argued by Adolf Grünbaum, who has patiently uncovered the necessary empirical presuppositions of the theory, and discussed the experimental support that was available for them in 1905. However, Grünbaum is driven by the logic of his own argument to locate the empirical justification of special relativity *primarily* in experiments done *after* 1905, for example the Kennedy–Thorndike experiment, published in 1932, and the detection of increase of mass of sub-atomic particles with relative velocity.[3] Thus the question remains open: what were the grounds of relativity theory in 1905?

Grünbaum's conclusions are based on a close examination of the *necessary and sufficient* empirical conditions for each element in the theory. This is a fundamentally operational approach, which is in harmony with Grünbaum's own rejection elsewhere of the Duhemian thesis of theoretical networks.[4] But if we take a less rigorous view of the concept of empirical support, we can appeal not only to direct inductive grounds, but also to clustering characteristics of the theory. These, however, as we have seen, give ambiguous results in this case, for it is by no means clear that Einstein's 'aesthetic intuition' that the Lorentz invariance of Maxwell's equations requires a radical change in our views of space and time can be interpreted as giving overall economy to his theory as compared with classical electrodynamics and

[1] An 'intuitive' view of Einstein's achievement is sketched in M. Polanyi, *Personal Knowledge* (London, 1958), 10ff, and a somewhat similar subjective view is supported with a wealth of historical detail by G. Holton in 'On the duality and growth of physical science', *Amer. Scientist*, **41** (1953), 89, and in the references given in p. 239, note 1. For a critical account of the Grünbaum–Holton debate, see G. Gutting, 'Einstein's discovery of special relativity', *Phil. Sci.*, **39** (1972), 51.

The tendency of such historical studies has been to decisively refute Reichenbach's claim that 'The physicist who wanted to understand the Michelson experiment had to commit himself to a philosophy for which the meaning of a statement is reducible to its verifiability. . . . It is this positivist, or let me rather say, empiricist commitment which determines the philosophical position of Einstein' ('The philosophical significance of the theory of relativity', *Albert Einstein: Philosopher–Scientist*, 291). On the other hand, S. Goldberg considerably overstates the contrast between Einstein and Poincaré in concluding 'Einstein's approach was that of deduction from first principles. Poincaré's approach was that of induction to first principles' ('Henri Poincaré and Einstein's theory of relativity', *Amer. J. Phys.*, **35** (1967), 944.

[2] Neglect of this temporal requirement greatly diminishes the value of the elaborate classification of research programmes into 'progressive' and 'degenerate' attempted in I. Lakatos, 'Falsification and the methodology of scientific research programmes', *Criticism and the Growth of Knowledge*, 91.

[3] *Philosophical Problems of Space and Time* (New York, 1963), 386ff.

[4] See for instance 'The falsifiability of a component of a theoretical system', *Mind, Matter, and Method*, ed. P. K. Feyerabend and G. Maxwell (Minneapolis, 1966), 273.

Newtonian mechanics. Economy comparisons can be made only relatively to the detailed development of classical theory by Lorentz, and Lorentz's theory, unlike Einstein's, was explicitly motivated by the need to accommodate the null result of the Michelson–Morley experiment.

Lorentz was faced with two apparently incompatible demands. He assumed, as did most classical physicists, that Galilean inertial frames and the Newtonian equations of motion must be retained. These, however, had to be made consistent with the constancy of the measured velocity of light apparently demanded by the Michelson–Morley and other experiments, whereas in Newtonian theory the velocity of light as measured in different inertial systems should of course depend on the velocity of the system. Lorentz resolved the apparent contradiction by postulating physical length contractions, and in later versions of his theory also time dilatations, in Galilean space–time as real physical effects of transfer from one coordinate system to another. The constancy of light velocity is therefore explained in terms of two independent and coincidentally cooperating effects of relative velocity which cancel each other out. These effects would be empirically undetectable unless it could be shown that they had some independently establishable physical basis. The contraction hypothesis was not necessarily *ad hoc*, however, for it is rare for such substantial assumptions to yield *no* further testable fallout, and we must include in the notion of empirical test not only the logical consequences of the hypothesis but also what may be called its analogical overtones. Lorentz appealed in support of the hypothesis to analogies between the 'molecular forces' he required to explain contraction, and electromagnetic forces whose physical characteristics were known. His assumptions were *ad hoc* in the sense of being postulated to solve a specific problem, but not *ad hoc* in the sense of being empirically untestable. They did indeed *increase* the empirical content of his theory compared with Einstein's, for they constituted an attempt at a unified electrodynamics, mechanics and theory of matter, which Einstein's theory at this stage did not attempt. But with regard to the particular questions with which Einstein was concerned, namely the effects of relative motion on electrodynamic systems, Lorentz required a number of logically independent premises, entailing a relatively low economy for his theory.

It was this sort of comparison, rather than the accusation of being *ad hoc*, that led to the abandonment of all the versions of Lorentz's contraction hypothesis during the years following 1905. In his discussion of Lorentz's theory,[1] Holton lists eleven hypotheses he describes as *ad hoc*, of which eight turn out to involve increased content and testability of the theory, and hence not to be necessarily undesirable except where the theory is competing with another of comparable content which does not need these special hypotheses.

[1] *Amer. J. Phys.*, **28** (1960), 630.

The remaining three hypotheses are: introduction of an upper limit of velocity of the earth to which the theory is applicable; introduction of a stationary material aether which convicts Lorentz of lack of economy compared to Einstein; and the introduction of the Lorentz transformation equations as independent and unexplained premises just for the purpose of explaining the absence of measured velocity relative to the aether. Only these last three hypotheses need be rejected as undesirably *ad hoc* (at least in comparison with another more economical theory), and all three would indeed be rejected on the criteria adopted in this chapter, either as having reduced content or as relatively uneconomical.

In order to point the moral of the Einstein–Lorentz case for an understanding of simplicity criteria, let us replace the question 'What were Einstein's grounds for special relativity in 1905?' by a sharper question suggested by our search for confirmation criteria for predictions. This is the question 'Supposing Einstein not to have been aware of the Michelson–Morley experiment in 1905, would he have had good grounds for predicting its result?' If 'good grounds' are here interpreted in terms of his own intuitive feeling for mathematical order, that is in terms of his own *subjective* beliefs,[1] the answer is undoubtedly Yes. But if 'good grounds' are interpreted as suggested in this book in terms of potentially intersubjective scientific beliefs based on the clustering postulate, the answer must be No. On these grounds, acceptance of Einstein's theory must be seen in terms of comparison with its only rival of comparable content, that is, Lorentz's developed theory, and for this comparison it is essential that the result of the Michelson–Morley experiment be accepted as part of any viable theory. Thus, whatever the status of the experiment was for Einstein in 1905, it is an essential ingredient in the empirical justification of Einstein's theory in the years immediately following 1905.

This conclusion may seem to detract from the rational value of Einstein's contribution. Grünbaum has suggested that if a theory such as Einstein's is epistemologically indistinguishable at its inception from a 'pathetically abortive speculation' such as, for example, the contemporary non-relativistic theory of Ritz, then scientific discovery cannot be distinguished from the psychology of imagination.[2] But this conclusion does not follow, for the justification of a theory may properly include relevant features of the contem-

---

[1] Einstein himself makes an explicit probability judgment in 1916 which is certainly based on abstract mathematical rather than general inductive grounds. After arguing that the principle of special relativity holds with 'great accuracy' in mechanics, although mechanics is not a sufficiently broad basis for a theory of all physical phenomena, he goes on 'But that a principle of such broad generality should hold with such exactness in one domain of phenomena, and yet should be invalid for another, is *a priori* not very probable' (*Relativity*, 14). But as we have seen, homogeneity with respect to the principle of relativity over different physical systems is in itself counterbalanced by the rejection of Newtonian mechanics and the concepts of classical space and time.

[2] *Philosophical Problems of Space and Time*, 378–9.

porary state of science as well as the subjective beliefs of an individual, however great a scientist he may be. To admit that, as it left Einstein's pen, special relativity was primarily an elegant speculation lacking inductive confirmation, is not to question his stature as a scientist. But that stature does not depend only on mathematical luck, but also on his assimilation of subsequent empirical evidence and his subsequent development of the theory. Moreover, as we shall see in the next section, Einstein himself was by no means neglectful of the need for inductive support of theories.

## V. Einstein's logic of theory structure

Einstein is the outstanding case of a physicist who not only explicitly appealed to simplicity criteria at crucial points of the development of his theories, but also reflected upon the status of these criteria in judging the acceptability of theories. The place of simplicity in Einstein's methodology is, however, frequently misunderstood, for he is usually taken as the paradigm case of a 'hypothetico-deductive' theorist, who introduced simplicity criteria merely to provide grounds for choosing among conflicting hypotheses arrived at by guesswork. But closer examination of his philosophical writings reveals a more complex understanding of the hypothetical and inductive components of theorizing and a more intimate connection between simplicity criteria and the experimental content of theories.

The basis for the common view that Einstein's methodology of physics is narrowly hypothetico-deductive is to be found mainly in his later writings, where he frequently refers to physical theories as 'free creations of the mind' to which there is no logical path from experience, and describes the basic principles of physics as 'purely fictitious' and 'logically arbitrary'. In his 'Autobiographical notes' of 1949 he also explicitly rejects the positivist anti-speculative epistemology of Ernst Mach which had greatly influenced him as a young man. There is, indeed, reason to think that Einstein's views on the nature of science did undergo some change towards deductivism in the period after 1905.[1] But throughout his writings there are also to be found brief statements of a methodology which is clearly different from deductivism, sometimes expressed even on the same page as assertions identifying theories with 'fictions' and 'free creations' of the mind.

[1] 'Autobiographical notes', 27, 29. *Cf.* Holton's papers cited above. The shift from experientially based theories to a view of theoretical concepts as independent of experience is well captured by a change of metaphor in writings dating from 1922 and 1936 respectively: 'For even if it should appear that the universe of ideas cannot be deduced from experience by logical means, but is, in a sense, a creation of the human mind . . . nevertheless this universe of ideas is just as little independent of the nature of our experiences as clothes are of the form of the human body' (*The Meaning of Relativity*, sixth edition (London, 1956; first published 1922), 2). '. . . I do not consider it justifiable to veil the logical independence of the concept from the sense experiences. The relation is not analogous to that of soup to beef but rather of check number to overcoat' (*Ideas and Opinions* (London, 1954), 294).

I

There are three elements in Einstein's discussions of method which go beyond deductivism. They are: firstly, a distinction between 'constructive theories' and 'theories of principle'; secondly, a hint that although general theoretical principles are not logically 'abstracted' from data, they are *suggested by* or *emergent from* data; and, thirdly and most importantly, a claim that it is possible to arrive at unique true general laws by using criteria of simplicity.

### (a) Constructive theories and theories of principle

In a short popular article written for *The Times* in 1919, Einstein distinguishes *constructive theories* from *theories of principle*. It is worth quoting the passage in full.

> We can distinguish various kinds of theories in physics. Most of them are constructive. They attempt to build up a picture of the more complex phenomena out of the materials of a relatively simple formal scheme from which they start out. Thus the kinetic theory of gases seeks to reduce mechanical, thermal and diffusional processes to movements of molecules—i.e., to build them up out of the hypothesis of molecular motion. When we say that we have succeeded in understanding a group of natural processes, we invariably mean that a constructive theory has been found which covers the processes in question.
>
> Along with this most important class of theories there exists a second, which I will call 'principle-theories'. These employ the analytic, not the synthetic, method. The elements which form their basis and starting-point are not hypothetically constructed but empirically discovered ones, general characteristics of natural processes, principles that give rise to mathematically formulated criteria which the separate processes or the theoretical representations of them have to satisfy. Thus the science of thermodynamics seeks by analytical means to deduce necessary conditions, which separate events have to satisfy, from the universally experienced fact that perpetual motion is impossible.
>
> The advantages of the constructive theory are completeness, adaptability and clearness, those of the principle theory are logical perfection and security of the foundations.
>
> The theory of relativity belongs to the latter class.[1]

In 1936 and 1949 Einstein reaffirms the distinction of the two types of theory, contrasting both with a third type, the 'phenomenological' or 'positivist' physics approved by Mach and Ostwald. Superficial reading might lead to the impression that Einstein is here making merely a distinction between hypothetical and inductive theories, with approval being given only to the former.[2] Consider the following characterization of 'phenomenological' physics.

---

[1] *Ideas and Opinions*, 228.

[2] Holton gives an interpretation of this type in *Isis*, **60** (1969), 166, where he emphasizes the role of 'scientific taste in deciding which theory or hypothesis to accept and which to reject'. In *Daedalus*, **97** (1968), 646, 656, Holton contrasts Einstein's post-1920 'speculative–constructive' view of science with Mach's phenomenalism, but, like Mach himself, he fails to see the full significance of the distinction between empirically based principles like the invariance of light velocity on the one hand, and hypothetical constructions of the kind Einstein still had doubts about as late as 'Autobiographical notes' on the other. The distinction is indeed mentioned in *Daedalus*, **97**, 667, and in *Isis*, **60**, 185, but not exploited.

It is characteristic of this kind of physics that it makes as much use as possible of concepts which are close to experience but, for this reason, has to give up, to a large extent, unity in the foundations. Heat, electricity and light are described by separate variables of state and material constants other than the mechanical quantities; and to determine all of these variables in their mutual and temporal dependence was a task which, in the main, could only be solved empirically. Many contemporaries of Maxwell saw in such a manner of presentation the ultimate aim of physics, which they thought could be obtained purely inductively from experience on account of the relative closeness of the concepts used to experience.[1]

Einstein goes on to contrast kinetic theory, 'which by far surpassed pheno-menological physics as regards the logical unity of its foundations', at the cost of the 'constructively speculative character' of the unperceived atomic entities. But it would be a mistake to conclude that all theories having unity of their foundations are thus speculative, for four pages later in the same essay Einstein asks how we can expect to choose hypothetical concepts and axioms so that 'we might hope for a confirmation of the consequences derived from them?', and replies

The most satisfactory situation is evidently to be found in cases where the new fundamental hypotheses are suggested by the world of experience itself. The hypothesis of the non-existence of perpetual motion as a basis for thermodynamics affords such an example of a fundamental hypothesis suggested by experience; the same holds for Galileo's principle of inertia. In the same category, moreover, we find the fundamental hypothesis of the theory of relativity.[2]

The same identification of thermodynamics and the theory of relativity as neither constructive nor phenomenological is made in the 'Autobiographical notes'. Describing how the situation after 1900 in mechanics and electro-dynamics led him to the principle of relativity, Einstein says

I despaired of the possibility of discovering the true laws by means of con-structive efforts based on known facts. The longer and the more despairingly I tried, the more I came to the conviction that only the discovery of a universal formal principle could lead us to assured results. The example I saw before me was thermodynamics. The general principle was there given in the theorem: the laws of nature are such that it is impossible to construct a *perpetuum mobile* (of the first and second kind).[3]

It has been remarked that several of the summary formulations of experi-ence used by Einstein as examples of theories of principle are in the form of what Whittaker has called 'postulates of impotence': 'it is impossible to detect absolute motion', 'it is impossible to specify absolute simultaneity', 'it is impossible to find causal processes travelling faster than light', and so on (compare Einstein's own paradigm for a theory of principle, 'it is impossible

---

[1] 'Physics and reality' (1936), *Ideas and Opinions*, 302.
[2] *Ibid.*, 307.
[3] 'Autobiographical notes', *Albert Einstein*, 53.

to construct a perpetual motion machine').[1] Whittaker considers that postulates of impotence must be distinguished from empirical laws, since they are not the direct result of experiment and do not mention any measurement, but are 'the assertion of a conviction, that all attempts to do a certain thing, however made, are bound to fail', and are indeed essentially *negative* knowledge. This is, however, a misleading way of putting the matter, for two reasons. First, it is unduly subjective, for any principle of impotence corresponds to a negative existential assertion about the world rather than about the scientist's convictions, and hence is logically equivalent to a universal generalization. Second, it neglects the 'experimental' character of the corresponding universal generalizations such as 'All motions are relative to arbitrary rest-frames', 'All causal processes travel at not more than the speed of light', and so on. These are testable generalizations which have confirmatory instances, and hence may surely be regarded as having experimental status rather than the subjective status of 'assertions of conviction'.

In suggesting that they represent merely negative knowledge, however, Whittaker may have in mind that the *absence* among our observations of a certain state of affairs—a standard of rest, a perpetual motion machine, etc.—does not constitute good confirmatory evidence for assertion of its absence as a matter of law. This would indeed be the case if no experimental efforts were made to observe its presence in conditions where one might initially expect to find it, for then its observed absence might be an accident of what happens to have been observed. But in most of the examples of 'principles of impotence' adduced, experimental efforts *have* been made to refute them; and these efforts count as 'severe' (because antecedently improbable) positive tests of the logically equivalent universal generalizations. On the other hand, the principle that no causal process travels faster than light, if regarded as an empirical generalization with no directly observed counter-instances, does not seem to be supported by severe tests, because it is not clear under what circumstances one could attempt to produce such instances.[2] This particular impotence principle seems to acquire support indirectly from the theoretical definition of simultaneity in relativity theory, and from the adoption of Lorentz transformations. The same can be said of the principle that there is no absolute simultaneity, for, as Grünbaum points out,[3] this rests on the empirical premise that there is no moving clock that retains absolute synchronization with a stationary clock under transport. For this negative existential assumption, however, it must be concluded that there is no *direct* experimental evidence, and hence it must be justified by the success of the theory as a whole.

[1] E. T. Whittaker, *From Euclid to Eddington* (Cambridge, 1949), 59, 104.

[2] Compare Grünbaum's criticism of Einstein's own thought experiment requiring travel *at* the speed of light (*Philosophical Problems of Space and Time*, 373). Whether this is possible, and what one would find it if were, can only be discussed *within* some theory or other.

[3] *Ibid.*, 346.

## (b) The relation of theory and data

The second element in Einstein's methodology consists in his remarks about the relation of theories of principle with experience. His insistence that theories cannot be derived by 'abstraction' (by which he means by some *strictly logical* method) does not prevent him asserting that the relation between general principles and the facts is more intimate than that between hypothetical theories and the facts. We have just seen that the fundamental principles may be 'suggested by the world of experience itself', in contrast to 'constructive speculative' theories. In an earlier paper he puts it like this: the scientist 'has to worm these general principles out of nature by perceiving in comprehensive complexes of empirical facts certain general features', and 'the principle of the constant velocity of light . . . *emerges* from the optics of bodies in motion'.[1]

It is true that Einstein's later emphasis is on the 'distance' of theories from the facts and on the complex chains of deductive reasoning required to connect them,[2] and also that he describes the connection as ultimately 'intuitive' rather than logical. It must be noticed, however, that the context in which he uses the term 'intuition' is that of the relation of sense experience with both every-day physical objects and with physics: 'The connection of the elementary concepts of everyday thinking with complexes of sense experiences can only be comprehended intuitively and it is unadaptable to scientifically logical fixation'.[3] The contrast here is not between intuition and the empirical, but rather between intuition and logical connection, for 'intuition' is just Einstein's label for the elusive connection between theoretical systems and the empirical world, whether the direction of inference is inductively to theories or deductively to tests of theories:

> Although the conceptual systems are *logically* entirely arbitrary, they are bound by the aim to permit the most nearly possible certain (*intuitive*) and complete coordination with the totality of sense experiences.[4]

[1] 'Principles of theoretical physics' (1914), *Ideas and Opinions*, 221, 222 (my italics). *Cf.* Einstein's letter to Besso, quoted by Holton, *Daedalus*, **97**, 646: '. . . a theory which wishes to deserve trust must be built upon generalizable facts. . . . Never has a truly useful and deep-going theory really been found purely speculatively.'

[2] 'Autobiographical notes', 27, and 'The problem of space, aether and the field in physics' (1934), *Ideas and Opinions*, 282.

[3] *Ideas and Opinions*, 292.

[4] 'Autobiographical notes', 13 (my italics). *Cf.* Einstein's 'Reply to criticisms', *Albert Einstein*, 673: 'A basic conceptual distinction, which is a necessary prerequisite of scientific and pre-scientific thinking, is the distinction between "sense-impressions" (and the recollection of such) on the one hand and mere ideas on the other. There is no such thing as a conceptual definition of this distinction (aside from circular definitions, i.e., of such as make a hidden use of the object to be defined). Nor can it be maintained that at the base of this distinction there is a type of evidence, such as underlies, for example, the distinction between red and blue. Yet, one needs this distinction in order to be able to overcome solipsism. . . . We represent the sense-impressions as conditioned by an "objective" and by a "subjective" factor. For this conceptual distinction there also is no logical–philosophical justification. But if we reject it, we cannot escape solipsism. It is also the presupposition of every kind of physical thinking.' The epistemology implied here is very similar to what I have presented in chapters 1 and 2 above.

This view gives no more aid and comfort to deductivism than to inductivism in respect of their relation to experience, for although Einstein believes there is no strictly logical connection between experience and theory in the inductive direction, he also believes there is none from theory to experience in the deductive direction either. Inductive arguments based on propositions representing the evidence are not forbidden any more than are deductive arguments issuing in such propositions as conclusions. In both cases the relation of intuition is between these propositions and the sense experiences.

### (c) *Simplicity*

In the 'Autobiographical notes', the burden of the distinction between constructive theories and theories of principle has passed to the third element in Einstein's methodology, the criteria of simplicity or, as he calls them, 'internal' criteria. Thermodynamics is 'the only physical theory of universal content concerning which I am convinced that . . . it will never be overthrown', not because, as in 1919, its elements are empirically discovered, but because 'A theory is the more impressive the greater the simplicity of its premises is, the more different kinds of things it relates, and the more extended its area of applicability'.[1]

Criteria of simplicity appear at all stages of Einstein's work, often in answer to the question 'How can we find a *uniquely* valid theory among the free constructions of the mind?' Here is an example from an essay of 1918.

> The supreme task of the physicist is to arrive at those universal elementary laws from which the cosmos can be built up by pure deduction. There is no logical path to these laws; only intuition, resting on sympathetic understanding of experience, can reach them. In this methodological uncertainty, one might suppose that there were any number of possible systems of theoretical physics all equally well justified; and this opinion is no doubt correct, theoretically. But the development of physics has shown that at any given moment, out of all conceivable constructions, a single one has always proved itself decidedly superior to all the rest. Nobody who has really gone deeply into the matter will deny that in practice the world of phenomena uniquely determines the theoretical system, in spite of the fact that there is no logical bridge between phenomena and their theoretical principles; this is what Leibnitz described so happily as a 'pre-established harmony'.[2]

The unique theory is characterized by completeness and economy:

> The aim of science is, on the one hand, a comprehension, as *complete* as possible, of the connection between the sense experiences in their totality, and, on the other hand, the accomplishment of this aim *by the use of a minimum of primary concepts and relations*. (Seeking, as far as possible, logical unity in the world picture, i.e., paucity in logical elements.)[3]

---

[1] 'Autobiographical notes', 33.
[2] 'Principles of research', *Ideas and Opinions*, 226.
[3] 'Physics and reality', *ibid.*, 293.

However, Einstein is not confident that it will be possible to explicate simplicity precisely:

> The problem here is not simply one of a kind of enumeration of the logically independent premises (if anything like this were at all unequivocally possible), but that of a kind of reciprocal weighing of incommensurable qualities. Furthermore, among theories of equally 'simple' foundation that one is to be taken as superior which most sharply delimits the qualities of systems in the abstract (i.e., contains the most definite claims) . . . We prize a theory more highly if, from the logical standpoint, it is not the result of an arbitrary choice among theories which, among themselves, are of equal value and analogously constructed.[1]

We may summarize Einstein's most fully developed criteria for 'theories of principle' in the following points.

(1) Theories of principle do not emerge logically from data, but are 'intuited' from, or suggested by, data. At any rate, these theories do not involve 'theoretical constructs'.

(2) Their elements are mathematically related, and these relations are discovered in experience.

These two points ensure secure inductive foundations of the theory.

(3) Theories should have precision, wide applicability, and be comprehensive over different kinds of things, that is in the terms adopted in earlier chapters, they should have high content and high consilience.

(4) Theories should be economical in premises, parameters and primary concepts.

These two points are intended to ensure that a unique theory is arrived at in a given physical domain. Their absence is aptly characterized, as Einstein characterizes Lorentz's theory, by an absence of 'depth'.[2] Depth, logical unity, clustering, are the key concepts for Einstein's internal criteria of theories.

## VI. Summary

The chief positive conclusion of this chapter has been that various species of economy can be seen as applications of the clustering postulate. In order to show that economy can be measured by probabilities, as assumed by the postulate, it was first necessary to refute claims that probabilistic confirmation cannot represent our preferences for simple theories. This refutation depends on distinguishing, on the one hand, cases where entailment relations between theories dictate an inverse relation between their content and their probability, from cases on the other hand where theories conflict in specific content, and

[1] 'Autobiographical notes', 23.
[2] Compare the report by M. Wertheimer, *Productive Thinking* (London, 1961; first published 1945), 230.

where therefore no initial probability relations are dictated for those conflicting contents by the probability axioms. In the latter cases various species of economy can be described which are independent of content. The economies generally accepted as desirable are

(a) clustering of properties among individuals, or the overall homogeneity or analogy of the universe;

(b) economy of properties and parameters used in specifying systems, including mathematical invariances;

(c) economy of mathematical form;

(d) ontological economy of entities and of the predicate-base of a language;

(e) economy of theoretical premises.

All these criteria have been interpreted as indicating clustering among different particular states of the world, in the sense that they all succeed in picking out the theory that asserts a minimum set of causal factors or relationships which are sufficient to account for the relevant content of the systems in question. The clustering postulate for economy is therefore a generalization of the postulate already invoked to explicate ordinary inductive and analogical arguments, and so succeeds in establishing some unity and systematization among our inductive intuitions.

With regard to the relative weighting of the different species of economy, no definite criteria can be found. It is only possible to show that where definite intuitions about the preferability of simple theories reveal themselves, these can in most cases be interpreted without contradiction as probability judgments. The probabilistic account of simplicity, like Bayesian methods in general, is *local* in application. That is to say, relative initial probabilities may be assigned in cases where simplicity comparisons of the same kind are possible, but in other cases uncertainty of relative initial belief will reflect uncertainty in practice about theory and prediction choice, and the moral will be to seek more data and more comprehensive theories.

If the question is now asked 'Why prefer economical theories?', the proposed explication of economy will take us no further than the modest programme for induction already discussed. For, apart from pragmatic considerations, the reasons for believing that economical theories are more likely to be true must be inductive reasons, and no more is claimed for confirmation theory in general than that it clarifies and systematizes our *de facto* inductive methods and reveals the ontological commitments presupposed by them. Certain of our intuitions regarding economy have also been modified by the present explication, since it has been argued that it is not the economy of theories as such which should be sought, but economy implied by certain definite kinds of clustering of instances.

Beyond the modest programme of explication, all that can be said is that if the world is clustered in various ways, then the present inductive methods of science are appropriate and will be expected to be generally successful. But this cannot itself be construed as any kind of inductive or hypothetico-deductive argument without begging the question. For, firstly, such an argument would require absence of counter-evidence, and in the case of simplicity there are clear counterexamples where the simplest theory has *not* proved successful: the planets do not move in circles; Newton's law of composition of velocities is false; protons and electrons are not the two irreducible elements which constitute all matter; and so on, and so on. Moreover, it may be that the sample of simple laws we do believe to be successful is not unbiased, for it may represent just that aspect of nature which has been found sufficiently simple to be manageable to the human mind, and no conclusion can be drawn from it to the rest of nature. Finally, any analogy with hypothetico-deductive argument breaks down because it is circular in the following sense. The justification for choosing one out of a number of hypotheses which yield the data deductively must be in terms of some comparison among the hypotheses which itself depends on various kinds of simplicity. Hence this deductive type of justification depends on the justification for assuming that the world is more likely to be simple. The justification for simplicity itself cannot therefore be given in terms of a hypothetico-deductive argument. There is no ontological evasion of the problem of induction by appeal to simplicity, and as always, however schooled we may be by natural selection, we have no guarantee that we have not just now reached the limits of inductive success in fundamental science.

# CHAPTER ELEVEN

# Maxwell's Logic of Analogy

## I. Hypothetical, mathematical and analogical methods

Maxwell is one of the few leading physicists to have given explicit and quite subtle analyses of scientific method, particularly in connection with his theory of electrodynamics.[1] The theory itself has often been cited in philosophical literature in support of hypothetical, mathematical and analogical modes of scientific inference, but it is not often noticed that Maxwell himself claimed that his method is the authentic Newtonian one of 'deduction from experiments' without the aid of unproven hypotheses. Even his displacement current, which looks on the face of it like a paradigm case of the 'theoretical concept' of later philosophical analysis, was intended by him to involve no hypothesis or physical 'model' as ordinarily understood, but rather to be justified by a generalized method of induction and analogy from experiments.

In several investigations Maxwell explicitly adopts the classic hypothetical and eliminative method. In his early work on Saturn's rings, for example, he lists three possible mechanical hypotheses as to their form: they are solid and uniform, or not solid, or not uniform. A solid uniform ring would be unstable, the form of solid irregular ring which would be stable is inconsistent with observation; hence the hypothesis of a fluid ring remains to be investigated. And in the first paper on the dynamical theory of gases, the hypothesis of small, hard, perfectly elastic spheres is to be explored:

> If the properties of such a system of bodies are found to correspond to those of gases, an important physical analogy will be established, which may lead to more

---

[1] These remarks occur scattered throughout his writings. Some of the most perceptive are found in an early paper written for the Apostles' Club at Cambridge in 1856: 'Analogies: are there real analogies in nature?', *Life of James Clerk Maxwell*, L. Campbell and W. Garnett (London, 1884), 347; and in the first pages of 'On Faraday's lines of force' (1856), *Scientific Papers*, ed. W. D. Niven (Cambridge, 1890), vol. I, 155. (References to the *Scientific Papers* will be to the volume and page number, thus: I, 155.) More extensive methodological discussions, especially of the significance of the generalized Lagrangean formulation of mechanics, are to be found in a series of papers from 1870 to 1879, for example, 'Address to the mathematical and physical sections of the British Association' (II, 215), 'On the mathematical classification of physical quantities' (II, 257), 'On the proof of the equations of motion of a connected system' (II, 308), 'Action at a distance' (II, 311), 'On the dynamical evidence of the molecular constitution of bodies' (II, 418), 'Thomson and Tait's natural philosophy' (II, 776).

accurate knowledge of the properties of matter. If experiments on gases are inconsistent with the hypothesis of these propositions, then our theory . . . is proved to be incapable of explaining the phenomena of gases. In either case it is necessary to follow out the consequences of the hypothesis. (I, 378)

In both these examples it is noticeable that the hypotheses are entirely mechanical in character, that is to say, no physical quantities are involved except those concerning masses in motion and the forces of weight, impact, pressure, friction. When the subject matter is less obviously mechanical, however, Maxwell is more critical of the hypothetical method. In electric and magnetic science, he says, if 'we adopt a physical hypothesis, we see the phenomena only through a medium, and are liable to that blindness to facts and rashness in assumption which a partial explanation encourages' (I, 155). Theoretical entities such as molecules and the aether should not be postulated without evidence (II, 253, 315), and should not be endowed with *ad hoc* 'attractive and repulsive forces whenever a new phenomenon has to be explained' (II, 339; *cf.* 223). And, in the case of chemistry, where the material systems are too small to be directly observed, the hypothetical method is only amenable to verification 'so long . . . as someone else does not invent another hypothesis which agrees still better with the phenomena' (II, 419). Not only was the hypothetical method regarded by Maxwell as undesirably pervasive in nineteenth-century physics, but his objections to it rest on *arguments*, not on timidity or prejudice.

On the other hand Maxwell is not content with theories composed of purely mathematical formulae in which 'we entirely lose sight of the phenomena to be explained; and though we may trace out the consequences of given laws, we can never obtain more extended views of the connexions of the subject' (I, 155). In a review of Thomson's *Papers on Electrostatics and Magnetism* he complains that no one has developed Thomson's theory of vortex molecules: 'Has the multiplication of symbols put a stop to the development of ideas?' (II, 307); and in a paper of the same period he writes

> We must retranslate [symbols] into the language of dynamics. In this way our words will call up the mental image, not of certain operations of the calculus, but of certain characteristics of the motions of bodies. (II, 308)

Explicit criticism of the purely mathematical method does not go very deep in Maxwell's writings, but it has been pointed out interestingly by G. E. Davie that dislike of pure analysis is ingrained in the Scottish tradition in which Maxwell had his first philosophical education.[1] Maclaurin's method of geometrizing the calculus was not dead in Edinburgh, and Maxwell's own interest in physical interpretations of Euclidean geometry is exhibited in

---

[1] G. E. Davie, *The Democratic Intellect* (Edinburgh, 1961), 192ff. I owe this reference to the unpublished PhD thesis of P. M. Heimann, 'James Clerk Maxwell, his sources and influence' (Leeds, 1970).

several of his early mathematical papers. We shall see later, however, that logical as well as genetic reasons for objecting to purely formalist methods in physics emerge implicitly in Maxwell's treatment of physical theory.

For Maxwell the middle way between the 'rash assumptions' of physical hypotheses and the 'analytical subtleties' of mathematical formulae consisted sometimes in a method of 'physical analogy', and sometimes in a Newtonian method of deduction of forces from phenomenal motions. The first of these methods appears in the introduction to his first paper on electricity and magnetism, 'On Faraday's lines of force'.[1] Faced with the dilemma between hypothesis and pure mathematics

> We must therefore discover some method of investigation which allows the mind at every step to lay hold of a clear physical conception, without being committed to any theory founded on the physical science from which that conception is borrowed . . .
>   In order to obtain physical ideas without adopting a physical theory we must make ourselves familiar with the existence of physical analogies. By a physical analogy I mean that partial similarity between the laws of one science and those of another which makes each of them illustrate the other. (I, 156)

Maxwell here gives four examples, which recur throughout his subsequent discussions of physical analogy: the laws of numbers on which all mathematical sciences are founded, the resemblance of form between both corpuscular theory and wave theory and the phenomena of light, William Thomson's analogy between electric and magnetic attraction and the equations of heat conduction and fluid flow, and the analogy which Maxwell develops in *FL*, between theories of electric and magnetic action at a distance, fluid flow, and Faraday's representation in terms of lines of force.

The precise nature and function of this kind of 'physical analogy' is not altogether easy to gather from Maxwell's explicit remarks about it, and I shall consider it in more detail below. As a preliminary statement it may be said that on the one hand Maxwell is concerned to insist that the existence of a 'formal' analogy of equations does not imply identity of physical process or substance, as when fluid flow is compared with heat flow, with current, and with electric induction, without any implication that these last processes in fact involve fluids in motion. The function of formal analogies is rather to aid the imagination in understanding formal relationships, and to enable transfer of mathematical results from one system to another irrespective of subject matter. But

---

[1] In what follows I shall refer to Maxwell's major writings in electromagnetism as follows:
  *FL*: 'On Faraday's lines of force' (1856), I, 155.
  *PL*: 'On physical lines of force' (1861/2), I, 451.
  *DT*: 'A dynamical theory of the electromagnetic field' (1864), I, 526.
  *Note*: 'Note on the electromagnetic theory of light' (1868), II, 137.
  *Treatise*: *A Treatise on Electricity and Magnetism*, first edition (London, 1873). (References will be to paragraph numbers)
  *Elem*: *An Elementary Treatise on Electricity*, ed. W. Garnett (Oxford, 1881).

on the other hand, systems of ideas which are 'really analogous in form' must be distinguished from those that are merely mathematical, and when such analogies are found, they lead 'to a knowledge of both [systems], more profound than could be obtained by studying each system separately' (II, 219), and 'It becomes an important philosophical question to determine in what degree the applicability of the old ideas to the new subject may be taken as evidence that the new phenomena are physically similar to the old' (II. 227).[1] Again, although *causes* of invisible processes cannot be identified with their formal analogues in observable processes, the relations between cause and effect are similar, and what we require are methods of representation so general that they express the real similarity of relations without introducing unwarranted hypothetical ideas into the expression of the cause.[2]

These general methods of representation are to be 'deduced' from experiments by the method used by Newton in deriving the law of gravitational force. Echoing Newton's own statement of method, Maxwell contrasts 'the true method of physical reasoning [which] is to begin with the phenomena and to deduce the forces from them by direct application of the equations of motion' with the 'too frequent practice' of inventing a particular dynamical hypothesis and deducing results, the agreement of which with phenomena 'has been supposed to furnish a certain amount of evidence in favour of the hypothesis' (II, 309).[3] In electrical science, however, the motions involved are not all observable, and the forces involve quantities which are not mechanical.

---

[1] Maxwell's own discussions of physical analogy have been considered by Joseph Turner in two pioneering papers: 'Maxwell on the method of physical analogy', *Brit. J. Phil. Sci.*, **6** (1955), 226, and 'Maxwell on the logic of dynamical explanation', *Phil. Sci.*, **23** (1956), 36. Turner does not recognize, however, the function of analogies in providing *inductive arguments* in Maxwell or elsewhere. Indeed, he explicitly denies (but without giving reasons) that the method of physical analogy permits inference from one physical system to another which is similar to it in some respects ('Maxwell on the method', 238). In this he seems to differ from Maxwell himself, as the remark just quoted from Maxwell in the text indicates.

In another analysis of Maxwell's method of analogy, 'Model and analogy in Victorian science: Maxwell's critique of the French physicists', *J. Hist. Ideas*, **30** (1969), 423, Robert Kargon has, almost alone among recent commentators, noticed that the analogical method is regarded by Maxwell as *inductively* preferable to the hypothetical and the mathematical. He has, however, interpreted analogy as a second-best to the method of deduction from experiments, when the latter is applicable. In this he differs from the interpretation given later in this paper, where an intimate relation is discerned between the analogical and the deductive methods.

[2] The most explicit discussion of the last point occurs in *Elem.*, section 64, and especially section 113: 'The effects of the current of conduction on the electrical state of *A* and *B* are of precisely the same kind as those of the current of convection. . . . In the case of the convection of the charge on the pith ball we may observe the actual motion of the ball. . . . But in the case of the current of conduction through a wire we have no reason to suppose that the mode of transference of the charge resembles one of [the] methods [of convection] rather than another. All that we know is that a charge of so much electricity is conveyed from *A* to *B* in a certain time.'

[3] The claim to deduction from experiments occurs not only in Maxwell's more discursive methodological writings, but also throughout his mature electromagnetic theory; for example: 'the laws of the distribution of electricity on the surface of conductors have been analytically

When we pass from astronomical to electrical science, we can still observe the configuration and motion of electrified bodies, and thence, following the strict Newtonian path, deduce the forces with which they act on each other; but these forces are found to depend on . . . what we call electricity. To form what Gauss called a 'construirbar Vorstellung' of the invisible process of electric action is a great desideratum in this part of science. (II, 419)

Gauss's 'consistent representation' seems to be introduced here as a means of dealing with what later came to be called the problem of 'theoretical concepts' —concepts used to describe unobservable entities and properties. But it is not easy to see how Maxwell can combine in one paragraph the demand for 'deduction' from observation with the suggestion that what is wanted in electrical science is a hypothetical model of invisible processes. Maxwell's dilemma in fact seems to be not only between the hypothetical and the mathematical methods, which might be resolved by a purely phenomenal type of science, but also between a science based on phenomenal concepts (which in the contemporary state of physics meant mechanical concepts) and the need to develop new types of concepts for unobservables without involving unwarranted hypotheses.

We have seen that applications of the Newtonian method of 'deduction from experiments' always involve some form of general premise regarded as more firmly based than mere hypotheses. For Maxwell in his later writings, the premise was supplied by the Lagrangean formulation of the laws of mechanics, and particularly the conservation of energy, where these principles themselves are regarded as immediate generalizations from experience and are powerful enough to enable specific laws to be deduced for particular systems. Maxwell developed the method explicitly in the *Treatise*, where he proposes

> . . . to examine the consequences of the assumption that the phenomena of the electric current are those of a moving system, the motion being communicated from one part of the system to another by forces, the nature and laws of which we do not yet even attempt to define, because we can eliminate these forces from the equations of motion by the method given by Lagrange for any connected system. . . . I propose to deduce the main structure of the theory of electricity from a dynamical hypothesis of this kind. (Section 552)

This preliminary account of Maxwell's analogical and deductive methods leaves their distinctive character obscure. For it may be argued that the method of physical analogy is nothing but the mathematical method in disguise, since many of Maxwell's remarks about it may be interpreted in the light of later

---

deduced from experiment' (*FL*, I, 155), 'the conclusions arrived at in the present paper are independent of this hypothesis [of motions and strains in the aether], being deduced from experimental facts' (*DT*, I, 564), 'I propose . . . to state [the electromagnetic theory] in what I think the simplest form, deducing it from admitted facts' (*Note*, II, 138). *Cf.* also Maxwell's correspondence at the time of the publication of *PL* and *DT*, in *Life*, 246, 255.

formalist analyses of science as implying no more than similarity of abstract mathematical relations, in which no real or physical similarity between the analogous systems need exist. On the other hand it may be held that the method of deduction from experiments is essentially the hypothetical method, since it admittedly depends on general premises which are logically indistinguishable from hypotheses. I think both these arguments are mistaken, and for the same reason. For in both cases it is assumed that elementary inductive generalizations do not constitute a distinctive form of scientific inference which is distinguishable from hypothetical inference. I shall try to substantiate this claim by showing that what Maxwell means by 'real physical analogy' is nearer to the relation recognized in the present account as a 'clustering' of particulars, and that when such analogies are established, hypothetical theories and concepts may be avoided.

## II. Experimental identifications

Some light is thrown on what Maxwell means by 'real physical analogy' by what he calls the 'mathematical classification' of physical quantities (II, 257). Classification into scalars, vectors and quaternions is an important example, as is the distinction among vectors between forces, represented by magnitude along a line, and fluxes, represented by the magnitude of an area, directed normally to that area. This classification of forces and fluxes conveys the mathematical distinction between many pairs of physical quantities: length and area, temperature and heat, electric potential and current. There is also a distinction among vectors between those referring to translation and those to rotation, as in the grad (Maxwell's 'slope'), div (Maxwell's 'convergence') and curl functions. These functions represent linear and vortex fluid motion respectively, and also, by analogy, electric and magnetic actions. It follows that the mathematical entities involved: scalars, vectors, forces, fluxes, translations, rotations, are not uninterpreted symbols, but have *at least* a spatio-temporal interpretation which is *identical* in all physical systems to which they apply. The 'flux' of current may not be flow of a substance, but it has direction in space and quantity related to cross-sectional area, and in these respects is exactly the same as flow of water along a tube. Translation and rotation have equally obvious spatial interpretation in all Maxwell's analogous systems—indeed in spite of his many disclaimers that the relata of an analogy need not themselves be at all similar, he assigns 'rotation' to magnetic force precisely because it causes spatial rotation of the plane of polarized light, detected by physically *turning* polarization gratings in space. But *what* it is that causes the rotation is (at least in his later papers) not specified in the case of magnetic force as it is in the case of material fluid vortex motion.

Transformation rules between formally analogous systems do not, however, always depend on identities of spatio-temporal properties. Sometimes they take the form of identifications of dissimilar causes which nevertheless have identical effects. The most obvious example is the identification of all forms of force with mechanical force, whether produced by mechanical or electrical or magnetic or any other physical system. All forms of force satisfy the dynamical laws of motion, and hence have the same effects. They are therefore generically identical properties of systems. A less obvious example in Maxwell's electrical theory is the identification of the potential function of electrostatic force with the 'electric tension' (potential difference) causing currents in conductors. This identification deserves scrutiny, because although physically elementary, it reveals a fundamental logical structure which has been little noticed in descriptions of scientific inference.

In *FL* Maxwell introduces the theory of current electricity by reference to the researches of Ohm and Kirchhoff, which had shown that in maintaining a current, 'pressure' must be different at different points of the circuit.

> This pressure, which is commonly called electrical tension, is found to be physically identical with the *potential* in statical electricity, and thus we have the means of connecting the two sets of phenomena. If we knew what amount of electricity, measured statically, passes along that current which we assume as our unit of current, then the connexion of electricity of tension with current electricity would be completed. . . . Thus the analogy between statical electricity and fluid motion turns out more perfect than we might have supposed, for there the induction goes on by conduction just as in current electricity, but the quantity conducted is insensible owing to the great resistance of the dielectrics. (I, 180)

The argument is this: electric tension is the analogue of static potential and also of fluid pressure, in the sense that the experimental laws satisfied by these three systems are of the same form. Moreover, electric tension and static potential are *physically identical*. By this identity Maxwell seems to mean that, given a difference of potential between two points, its physical effect depends solely on the type of material existing between the points: if it is an insulator the phenomena are those of electric induction; if a conductor, those of electric current.[1] In this passage in *FL* Maxwell is inclined to carry the analogy further and suggest that it indicates also identity of the unobservable processes involved, namely that both current and electric induction consist of flow of a substance. We have here the first hint that electric induction should be treated as physically similar to current flow, a step which leads eventually to postulation of the displacement current. But it is already clear that this extension of the analogy is a shaky one, for between the whole process of induction and of current there are obvious negative analogies: conductors and insulators have very different electric properties; in particular, current generates heat, and induction so far as we know does not do so.

---

[1] This argument nearly becomes explicit in *Elem.*, section 5: 'Definition. *Whatever produces or tends to produce a transfer of Electrification is called an Electromotive Force*'.

However, the identification of tension with potential does not depend on this further analogical extension, which is in fact abandoned in *DT*:

> In the case of electric currents, the force in action is not ordinary mechanical force, at least we are not as yet able to measure it as common force, but we call it electromotive force, and the body moved is not merely the electricity in the conductor, but something outside the conductor, and capable of being affected by other conductors in the neighbourhood carrying currents. (I, 539)

And, as we shall see later, even the model of current as 'something flowing' in the conductor is not inisisted upon in later expressions of the theory.

We now have a method of recognizing 'real analogy' as contrasted with mere mathematical similarity of equations. It is simply that if two apparently distinct properties of different systems turn out to be interchangeable in appropriately different physical contexts, they are to be considered the same property. I shall call this type of inference to identity *experimental identification*. There may be *generic* identification, when it is the *properties* of different systems that are identified, or *substantial* identification, when the same *entity* is found to be involved in apparently different systems. An example of substantial identification is Maxwell's conclusion in *PL*, on the basis of the identity of numerical value of the velocities of transmission of transverse electromagnetic waves and of light, that the aetherial media of electromagnetism and light are one medium (I, 492).

Another important example of generic identification is the assumption that all forms of energy are to be identified. This assumption is indeed essential for the method of 'deduction' of generalized laws of motion from experiments. It does not presuppose that all energy is reducible to the kinetic and potential energy of matter in motion, any more than the language of forces and fluxes implies that heat or electric current are so reducible. It does imply that some general terms can be substituted in Lagrange's equations for the 'kinetic' and 'potential' energy of mechanical systems, in such a way that the equations are satisfied by diverse physical systems, and that the real analogues of kinetic and potential energy in these other systems are observably identifiable. The identifications are somewhat more complex than the equivalence of effects which led to the identification of electric induction and potential, for they depend on *transformations* of energy according to quantitative conservation rules. The generic identity of energy justifies Maxwell's claim in *DT* that, whereas all other phrases inherited from the vortex model of the earlier *PL* 'in the present paper are to be considered as illustrative, not as explanatory', still 'In speaking of the Energy of the field, however, I wish to be understood literally' (I, 564). Moreover, by considering the observed character of magnetic and electric interactions, Maxwell concludes that it is possible 'without hypothesis' to identify the analogue of the kinetic energy term describing the medium with magnetic polarization, and that of the potential energy term with electric polarization (I, 533, 564). Analogical inference leading to experimental identifications goes very deep into the structure of science. Such inferences are not infallible—the identification of phlogiston in different chemical reactions, for example, turned out to be a mistake—but they are pervasive in all scientific theory, and may be supported by stronger or weaker inductive evidence. They are indeed involved even where no specific arguments for identification are adduced. Consider the identification of mechanical force due to magnets, and to electrified bodies, with mechanical force in general. No argument seems to be required to establish this, but it is not a logically necessary identification. It rests on the interchangeability of mechanical effects of different kinds of physical system, and the fact that 'force' in all cases satisfies the laws of motion. The possibility of such identifications rests on an assumption of relative independence among physical properties, that is to say, although the logical possibilities of variety in different physical systems are unlimited, it is not supposed that interdependence within physical systems is in fact so tight that all the properties of every system have to be considered as different from all of those of every other system.

Similar considerations regarding identity throw some light on the nature of the 'experiments' which form the starting point of Maxwell's (and Newton's) 'deductions'. In *DT* Maxwell claims to deduce his theory from Ampère's and Faraday's laws. These laws rest on such experimental identifications as that of all forces with mechanical force, magnetic force produced by permanent magnets with that produced by iron and by currents, and currents produced by batteries with current produced by friction or electric induction. The experimental laws are not only empirically vulnerable in the usual sense of being generalizations from instances, they are also vulnerable in being analogical generalizations over instances of different kinds, leading to experimental identifications of properties without which there could be no general descriptive language with which to carry on scientific inference. It may be noted parenthetically that the occurrence of these identifications makes the task of any logic of induction doubly difficult, for not only is there the usual problem of explicating evidence for inductive generalizations, but also the stock of basic distinguishable individuals and predicates in the language is continually changing as evidence for identifications accumulates or is found to be misleading.

We can now summarize the method which Maxwell claims is a preferable alternative to both the hypothetical and mathematical methods, and spell out exactly how it is distinguished from them. First, it should by now be clear that the methods of analogy and of deduction from experiments are not separate methods but aspects of a single method. The view that this is not so is due to a confusion between the method of analogy and the postulation of a hypothetical physical model, such as the vortex aether model in Maxwell's early electromagnetic theory. This model may exhibit some formal analogy with electromagnetic phenomena, but it is quite clear that even in *PL*

Maxwell thought that it failed to be a satisfactory theory: it is no more than a 'temporary instrument of research' which does not 'even in appearance, *account for* anything' (I, 207). There seem to be two kinds of reason for this failure, first that the analogy is not close enough to provide adequate experimental identifications, and second that it is not even clear that the model itself is a physically realizable system. With respect to the crucial identification of change of displacement of the 'idle particles' in the vortex aether with electric current ('A variation of [electric] displacement is equivalent to a current' (I, 496)), Maxwell says

> We have . . . now come to inquire into the physical connexion of these vortices with electric currents, while we are still in doubt as to the nature of electricity, whether it is one substance, two substances, or not a substance at all, or in what way it differs from matter, and how it is connected with it . . . why does a particular distribution of vortices indicate an electric current? (I, 468)

And with regard to the possible existence of the vortex mechanism

> I have found great difficulty in conceiving of the existence of vortices in a medium, side by side, revolving in the same direction about parallel axes . . . it is difficult to understand how the motion of one part of the medium can coexist with, and even produce, an opposite motion of a part in contact with it. (I, 468)

This problem is solved by the 'idle particles', on the analogy of epicyclic trains in which idle wheels are not rotating about fixed axes. But

> The conception of a particle having its motion connected with that of a vortex by perfect rolling contact may appear somewhat awkward. I do not bring it forward as a mode of connexion existing in nature. (I, 486)

The method of physical analogy, however, is different. It begins with two or more existing physical systems which are related in two ways.

(i) They satisfy formally analogous mathematical laws.

(ii) There are sufficient experimental identifications of their entities and/or properties to constitute a real physical analogy between the systems, and permit analogical inference from one system to the other.

Under these conditions it is possible, without specifying in detail the character of unobservable entities or causes, to use the general laws representing the formal analogy of the systems in deducing the particular form of the laws and their particular effects for particular systems. For example, in the case of Newton's theory, experimental identifications of the properties of various mechanical systems as constituting analogues related by the laws of motion, enable the particular form of the law of gravitation to be deduced from the observable laws of planetary motion. Both experimental identifications and the general laws of motion are necessary here, in order to permit laws derived directly from observable forces and motions to be transformed into laws concerning forces-at-a-distance, and bodies whose masses cannot be discovered by the usual operational means of scales and spring balances.

Enough has been said to indicate why the method of analogy and deduction from experiments is not a formal mathematical method. But it may be asked whether it is not after all just a particular form of the hypothetical method. It can indeed be represented in terms of the hypothetico-deductive schema, but there are the following crucial differences between it and the hypothetical speculations to which Maxwell objects.

(1) The general laws constituting the formal analogy between systems are themselves derived by direct inductive generalization from experiments; in other words they are 'experimental laws' rather than hypotheses.

(2) No hypothetical concepts or entities are postulated, since there is no specification of the detailed character of unobservable causes except in so far as this is justified by experimental identifications and analogical argument using the general laws.

(3) A logic of induction to laws and analogy between systems replaces the 'guess and test' method of the hypothetico-deductive account.

## III. The electrodynamic theory

In his third paper on electrodynamics Maxwell abandons the hypothetical vortex model, and claims that his conclusions are 'deduced from experimental facts'. It would be at least charitable to suppose that this claim is meant seriously, and that Maxwell now believes himself to have a valid and non-hypothetical inference from experiments to his electromagnetic equations, even though we might prefer to describe this inference as inductive or analogical rather than deductive.

In this paper Maxwell begins by arguing the substantial identity of the electromagnetic medium and the light aether, in contrast to *PL*, where this identity is held to *follow* from the postulates of the theory. The phenomena of light and heat, he says, give reason to believe 'that there is an aetherial medium filling space and permeating bodies, capable of being set in motion and of transmitting that motion from one part to another' (I, 528). This medium must carry energy, since heat and light radiation take time to traverse it, and the conservation of energy through time is assumed. It is also assumed that the energy is mechanical, in the sense that it is carried as the energy of motion of the material aether, and of its 'elastic resilience'.

> We may therefore receive, as a datum derived from a branch of science indepen-
> dent of that with which we have to deal, the existence of a pervading medium,
> of small but real density, capable of being set in motion. . . . (I, 528)

Later on in the paper, even this minimal specification of the material aether is withdrawn, and all that remains of the properties of the medium is whatever can be identified respectively with the kinetic and potential energy terms in Lagrange's generalized equations of motion.

The next step in *DT* is to consider the evidence from motion of conductors and dielectrics relative to a magnetic field. When a material body is moved across magnetic lines of force, whether by the motion of the body or change of the lines, its ends tend to become oppositely electrified, and it may even experience chemical decomposition. By the same inference that led to the identification of electric tension and potential, the electromotive force is defined as that single property which causes the observable polarization, whether in conductors, dielectrics or electrolytes, and since motion in a magnetic field is sufficient to cause any of these phenomena, the electromotive force must be conceived to be stored as energy in the medium along with the magnetic energy already postulated. Maxwell goes on to consider the action of the electromotive force in dielectrics:

> (11) . . . when electromotive force acts on a dielectric it produces a state of polarization of its parts similar in distribution to the polarity of the parts of a mass of iron under the influence of a magnet, and like the magnetic polarization, capable of being described as a state in which every particle has its opposite poles in opposite conditions. (I, 531)

A footnote here refers to the work of Faraday and Mossotti on polarization of material dielectrics.

On the face of it, Maxwell's argument in this section (11) is an appeal to the formal analogy of electric and magnetic polarization developed experimentally by Faraday, and mathematized by Poisson and Mossotti. After reading Faraday, Mossotti had realized that Poisson's theory of magnetic polarization could be directly applied to dielectrics by, as Maxwell put it, 'merely translating it from the magnetic language into the electric, and from French into Italian' (II, 258). But this formal analogy in itself is hardly what Maxwell would call a 'real physical analogy', sufficient to justify inference to the displacement current; indeed, the known relations between electricity and magnetism constitute rather a negative than a positive analogy. Poisson's own hypothesis had involved a 'magnetic fluid' in molecules of magnetic material similar to the electric fluid, but there is no evidence for the existence of this fluid, nor of 'magnetic conductors', either macroscopic or molecular, nor of 'magnetic displacement current' in the sense of magnetic poles migrating along magnetic particles however small—or at least we have no reason for making such a hypothesis. Moreover, as Maxwell has already argued (I, 503), when the interactions of magnetic and electric phenomena are taken into account, the two processes are found to be *complementary* rather than analogous, in the sense that one is found to be rotatory in character and the other linear. Ironically, the best analogy goes the other way—from electric currents to magnets, as in Ampère's hypothesis—for since all observable external effects of electric current circuits are the same as those of magnets, there is a good argument for the identification of unobservable magnetic particles with small electric circuits. It follows from these disanalogies between magnetic and electric polarization that their formal relationship provides no grounds for the displacement current.

In the paragraph following the one just quoted however, Maxwell sketches Mossotti's own representation of dielectrics, which involved more than the skeletal mathematical analogy with magnetic polarization. The physical content of Mossotti's hypothesis depended on another analogy, namely that between the behaviour of dielectrics and *conductors* in an electric field. Consider a large parallel plate condenser in which the plates are separated by free space. Its capacity is defined as the ratio of charge on one of the plates to potential difference between the plates, and is constant for constant geometric configuration of the plates. If macroscopic uncharged conductors are now introduced between the plates they are found to be polarized by the electric field, and the net effect on the potential distribution is found to be a decrease in the potential difference between the plates for given charge, or in other words, an increase in the capacity of the resulting condenser, the amount of the increase depending on the size and position of the small conductors. Now this increase in capacity is exactly what occurs when a material dielectric is inserted between the plates, the amount of increase being characteristic of the geometry and dielectric constant of the dielectric medium. Mossotti's hypothesis is that material dielectrics consist of microscopic conducting particles insulated from each other by space or aether, where the dielectric constant of each specific material depends on the unobservable geometric configuration of its microparticles. In this hypothesis there are properly speaking *no* material dielectrics; the only dielectric (insulator) that exists is aether. It therefore provides a simple, physically realizable macro-model for microscopic processes, and may be said to compare favourably in inductive status both with the formal analogy of magnetic and electric potential and with Maxwell's vortex model.

Maxwell does not mention the details of Mossotti's hypothesis in *DT*,[1] but he may have regarded it as implicitly strengthening his analogy between conducting currents and displacement currents *in material dielectrics*. It causes no difficulty so long as it is restricted to material dielectrics, but trouble arises as soon as, following Faraday, Maxwell begins to treat the aether itself as a dielectric. He is bound to take some such step, since his view is that energy processes are going on in the medium between conductors and magnets, whether this medium is free space or filled with material insulator. What more natural, then, than to identify the structure of the aether with that of material dielectrics, and to postulate a charge polarization of its parts under the influence of electromotive force by analogy with that of material dielectrics?

---

[1] He describes it in *Treatise*, section 62, where he says it 'may be actually true', provided there are no dielectrics with constant less than that of vacuum. He does not explicitly 'repudiate Mossotti's hypothesis,' as stated erroneously by Fitzgerald in 'M. Poincaré and Maxwell', *Scientific Writings of G. F. Fitzgerald*, ed. J. Larmor (Dublin and London, 1902), 284.

The cavalier manner in which Maxwell passes in *DT* from 'real' electric current and 'real' dielectric polarization to displacement current and polarization in aether, suggests that he has used this analogy quite uncritically, and regards it as needing no further support. After an interlude chiefly devoted to Ampère's and Faraday's investigations of the mechanical effects of *closed* currents, he moves straight on to the 'General equations of the electromagnetic field', and immediately asserts

> Electrical displacement consists in the opposite electrification of the sides of a molecule or particle of a body which may or may not be accompanied with transmission through the body. . . . The variations of the electrical displacement must be added to the currents $p$, $q$, $r$ to get the total motion of electricity. (I, 554)

Maxwell subsequently substitutes this 'total motion' for the ordinary conduction current $j$ in Ampère's experimental relation for closed circuits, so as to obtain (in our notation)

$$\operatorname{curl} H = 4\pi j + \frac{1}{c}\dot{D}$$

and he uses this equation, without further comment, to refer both to material dielectrics and to the aether (I, 557).

After so much insistence on deduction from experiments, the brevity of this most crucial part of the argument comes as something of a shock, and not surprisingly many critics found it unconvincing. It is worth noticing, however, that Maxwell might at this stage have supplemented the analogical argument by being more explicit about his interpretation of the nature of the displacement current. His unwillingness to specify the character of the cause here prevented him from examining the inductive force that might have been given to the argument, by analogies that lay close at hand, as I shall now show.

It was soon noticed by Maxwell's successors that the implicit interpretations of electric charge and current in his work are highly ambiguous and sometimes confused.[1] Broadly speaking, two distinct interpretations have been recognized, which I shall call simply the *first interpretation* and the *second interpretation*. The first interpretation regards electric charge as an incom-

---

[1] See for example P. Duhem, *Les théories électriques de J. C. Maxwell* (Paris, 1902); O. Heaviside, 'Electromagnetic induction and its propagation' (1885–7), *Electrical Papers* (1892), vol. I, 434, 477; H. Hertz, *Electric Waves* (1892), trans. D. E. Jones (London, 1893), introduction; H. Poincaré, *Electricité et Optique* (Paris, 1901), viii and *passim*; J. J. Thomson, 'Report on electrical theories', *Report of the 55th meeting of the British Association* (1886), 125ff. Among more recent commentators, see A. O'Rahilly, *Electromagnetic Theory* (New York, 1965; first published as *Electromagnetics*, London, 1938), 76ff; J. Bromberg, 'Maxwell's displacement current and his theory of light', *Arch. Hist. Ex. Sci.*, 4 (1967/8), 218, and 'Maxwell's electrostatics', *Amer. J. Phys.*, 36 (1968), 142; P. M. Heimann, 'Maxwell and the modes of consistent representation', *Arch. Hist. Ex. Sci.*, 6 (1969/70), 171, and 'Maxwell, Hertz and the nature of electricity', *Isis*, 62 (1970), 149.

pressible substance satisfying a fluid continuity equation, whose motion in conductors constitutes current; the second regards charge as an epiphenomenon of polarized lines of force in the aetherial medium, which becomes detectable when lines of force meet interfaces between insulators and conductors. The first interpretation is related to the traditional action at a distance view of electric force; the second is derived from Faraday's view of the primacy of the lines of force in the dielectric medium. There is no doubt, as several commentators remarked, that Maxwell's own use of the term 'displacement current' caused confusion at the outset, because it falsely suggests that he is consistently adopting some variety of the first interpretation, that is, that the postulated 'current in the aether' is an actual motion of substantial charge,

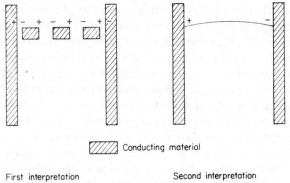

First interpretation
with material dielectric
particles indicated

Second interpretation
with no material dielectric
between plates

rather than a building up of lines of force in the medium. Another preliminary point to be noticed is that both these interpretations are *interpretations*; neither is 'directly read off' the phenomena, for no one has *observed* electric charges situated in conductors any more than he has observed tubes of force filling space. At best one observes such things as charged pith balls in relative motion, but this is an observation of the balls, not of the charges. To hold, as some critics have done, that the first interpretation is somehow a direct rendering of the 'facts', while the second is objectionably hypothetical, is epistemologically untenable.[1] There is here a genuine 'paradigm switch', in which observable pith balls, batteries, and so on, are read either in the first

---

[1] *Cf.* O'Rahilly: 'the ordinary working physicist remains convinced that a current really consists of something travelling along the wire. He remains frankly sceptical in face of the paradoxical hypothesis that it is everywhere except in the wire.' (*op. cit.*, 277.) J. H. Poynting, in 'On the transfer of energy in the electromagnetic field', *Phil. Trans.*, **175** (1884), 343, is more sensitively aware of the paradigm switch required to understand the second interpretation: 'It is very difficult to keep clearly in mind that this "displacement" is, as far as we are yet warranted in describing it, merely a something with direction which has some of the properties of an actual displacement in incompressible fluids or solids. . . . It seems to me then that our use of the term is somewhat unfortunate, as suggesting to our minds so much that is unverified or false, while it is so difficult to bear in mind how little it really means.' (360).

or in the second interpretation, and the way they are read is not, as we shall see presently, reducible without remainder to non-theoretical 'ordinary language' assertions about pith balls and so on, because use of one interpretation or the other presupposes commitments to a different set of experimental identifications and even a different descriptive language in each case, and consequently to different analogical relationships.

There is no doubt that the *prima facie* interpretation to be given to Maxwell's words in paragraph (11) is the first. Unfortunately, this makes the argument depend on an analogy between material dielectrics and aether which becomes increasingly implausible the more it is scrutinized. For in Mossotti's hypothesis, the aether is just that which intervenes between the conducting particles of the material dielectric, and if particles carrying moving charges are now postulated of aether itself, the question arises: what intervenes between the aether particles? At best the model reduces to one of action at short distances within the aether. There are admittedly places where Maxwell seems to allow this possibility, but it fits uneasily with his general insistence on the propagation of action through a medium.[1] A further physical difficulty of the first interpretation arises from the fact that charge is regarded as an incompressible fluid whose motion produces the total current of conduction and displacement, moving in closed circuits without sources or sinks.[2] In other words both current and electric induction are transmitted instantaneously, an assumption that is conceptually awkward in a theory of the electromagnetic aether in which wave disturbances are found to travel with the velocity of light.

In his discussion of various interpretations of Maxwell's equations in the introduction to *Electric Waves*, Hertz relates the first interpretation to Helmholtz's attempted formal derivation of Maxwell's equations, and in doing so uncovers another troublesome consequence of the first interpretation. Helmholtz had found that in order to eliminate longitudinal waves from the consequences of his theory, and to obtain the correct value for the velocity of transverse waves, a term representing the dielectric constant of aether has to be made indefinitely large. Hertz interpreted this in terms of an aether structured like a dielectric in which aether particles come to fill more and more space, while the distance forces between them become negligible. Helmholtz's result is then equivalent to the limiting case of this interpretation, which becomes formally identical with Maxwell's theory. But this interpretation is, to say the least, difficult to conceive physically, since the limiting process seems to cancel out exactly the physical distinction between conducting and insulating parts of the medium on which it is itself founded,

---

[1] '[I endeavour] to explain the action between distant bodies without assuming the existence of forces capable of acting directly at *sensible* distances.' (I, 527; my italics.) *Cf.* Maxwell's discussion of whether the aether is molecular or continuous in his encyclopedia article 'Ether' (II, 773f).

[2] *Cf. Treatise*, section 61. Since curl $B = 4\pi j + (1/c)\dot{D}$, and any div curl function is identically zero, div $(4\pi j + (1/c)\dot{D}$ is everywhere zero.

and it is not clear how an infinite dielectric constant of aether is to be understood.

The first interpretation, then, does not satisfactorily bridge the gap between material dielectrics and aether in the argument to the displacement current, both because it rests on no close analogy between dialectrics and aether, and also because it leads in itself to implausible physical consequences. When we consider the second interpretation, however, in which charge depends upon lines of force, many of the difficulties of the first interpretation disappear, but at the cost of reducing the plausibility of another part of the analogical argument.

As we have seen, Maxwell argues in many places that energy of electric polarization resides in aether in some fashion which reveals itself to observation as a tendency to motion along the direction of the electric lines of force. On the second interpretation, the charge on a conductor *is* just the interruption of a line of force by a conducting surface, not some *thing* whose presence *produces* the lines of force. We do not first 'observe' charges as particles, and then define lines of force, we observe motions of macroscopic bodies which have been treated in a certain way: by friction, by motion in the neighbourhood of magnets, by being connected to a battery, and so on. All these observable situations are reinterpretable directly into the second interpretation without intervention of the idea of charged particles in motion. Wherever a static or moving charge is mentioned in the first interpretation, it can be consistently reinterpreted as a state of the lines of force, at rest or in motion, in the second interpretation.

In particular, Mossotti's analogy from macrosystems of conductors may be given the second interpretation instead of the first. In the second interpretation, the material dielectric between the condenser plates consists of particles surrounded by aether in which lines of electric polarization appear when the condenser plates are charged (that is, when the battery is switched on). 'Charge' now resides at the ends of lines of force where they meet conductors, and 'displacement current' is just the process of establishing this polarization in the medium. Since its establishment in this interpretation is exactly the same physical process whether the conducting bodies are macroscopic conductors, for example condenser plates, or microscopic parts of a dielectric medium separated by aether, the analogical inference from the behaviour of material dielectrics to that of aether is immediate. But unfortunately the inference *to* Mossotti's hypothesis, which was supported by the first interpretation, has now become problematic. For in the first interpretation this step depended on considering dielectrics as conducting particles embedded in aether, in which the dielectric properties are derived from our knowledge of current carrying conductors and their observable models such as fluid-carrying pipes, whether open or closed. The step from closed currents to open currents *in conductors* is plausible, but the reinterpretation of open momentary currents in conducting particles embedded in dielectrics as momentary changing polarization of the aetherial medium of these conducting particles (which is the second interpretation's view of dielectrics) has no basis in observable analogues. Moreover, it seems to leave the *conductors* without currents, for if the charge on a conductor–aether interface is really only the interrupted end of a tube of force in aether, what is it that produces effects (such as heating) in the conductor?

This question was not answered, or even explicitly asked, by Maxwell himself, but before considering it, it is convenient to notice certain consequences of the second interpretation which are important in resolving difficulties in the first interpretation. In the first place no moving charges constituting currents are required in free space. A 'charge' is now identified with the end of a tube of force where it meets a conductor, and *no motion of a substance takes place along a conductor*, whether it be a macroscopic wire or a dielectric conducting particle. Secondly, there are *no* dielectric aether particles, and so no problem about what exists in the medium separating such particles, or about the physical meaning of the limiting process by which the particles come to fill all space. As in the first interpretation there is properly speaking only one dielectric constant, that of aether, which may now without contradiction be put equal to unity (as Maxwell explicitly does in *DT*, I, 572).

Again the notions of incompressibility and conduction of charge acquire new interpretations. There is no motion of particulate or fluid charge along a conductor, and nothing corresponding to pressure along it either; therefore there is no problem about instantaneous transmission. When a condenser is charged, the second interpretation identifies this process with the establishment of a system of tubes of force between the condenser plates, and between these and other conductors if there are any present in the space. As the charging process takes place, the tubes of force move into position at right angles to their length, so that they become established at the surfaces of all terminating conductors simultaneously. There is no instantaneous action from one condenser plate to the other; it is an action of the *battery* or other charging device, in which energy takes time to reach the conductors from the battery through the field. There is also implicit in the second interpretation a radical shift in the notion of a current-carrying conductor. This is no longer conceived on the model of material particles or fluids in pipes, but as a closed surface on the *outside* of which lines of force terminate, and along which they move, thus producing the effects of moving charge on the second interpretation. If *this* conception had had an observable model to replace fluids in pipes, the analogical argument to Mossotti's hypothesis, and hence to the displacement current, might have been said to be complete.

It cannot be pretended that Maxwell conceived anything like this detailed picture of the second interpretation in *DT*. Its full consequences were indeed not presented until the papers of Poynting in 1884 and 1885 and Heaviside in 1885–7, which showed that Maxwell's equations entail a continuous energy flow in the field where electric and magnetic actions are taking place.[1] In this they are only following out consistently Maxwell's original insight that energy is present in the field. In 'On the transfer of energy . . .' Poynting writes

> If we believe in the continuity of the motion of energy, that is, if we believe that when it disappears at one point and reappears at another it must have passed through the intervening space, we are forced to conclude that the surrounding medium contains at least a part of the energy, and that it is capable of transferring it from point to point. . . .
>
> According to Maxwell's theory, currents consist essentially in a certain distribution of energy in and around a conductor, accompanied by transformation and consequent movement of energy through the field. (p. 343)

> A conduction current then *may be said to consist* of this inward flow of energy with its accompanying magnetic and electromotive forces, and the transformation of the energy into heat within the conductor. (p. 351; my italics)

Poynting recognizes, however, that the inductive evidence for the theory is not increased by the new interpretation:

> We can hardly hope, then, for any further proof of the law [of energy transfer] beyond its agreement with the experiments already known until some method is discovered of testing what goes on in the dielectric independently of the secondary circuit. (p. 361)

And in the *Note* written four years after *DT* Maxwell himself reacts to the hiatus in the experimental argument for the displacement current by requiring the same direct verification:

> . . . the current produced in discharging a condenser is a complete circuit, and might be traced within the dielectric itself by a galvanometer properly constructed. I am not aware that this has been done, so that this part of the theory, though apparently a natural consequence of the former, has not been verified by direct experiment. The experiment would certainly be a very delicate and difficult one. (II, 139; *cf. Treatise*, Section 607)

The same assessment of the logical situation was later made by Helmholtz and Hertz. There is a well-known passage in the introduction to *Electric Waves*

---

[1] J. H. Poynting, *op. cit.*, and 'On the connexion between electric current and the electric and magnetic inductions in the surrounding field', *Phil. Trans.*, **176** (1885), 277; O. Heaviside, *op. cit.* Maxwell's first explicit statement of the second interpretation is in *Treatise*, section 111: 'The electrification therefore at the bounding surface of a conductor and the surrounding dielectric, which on the old theory was called the electrification of the conductor, must be called in the theory of induction the superficial electrification of the surrounding dielectric. According to this theory, all electrification is the residual effect of the polarization of the dielectric. . . . In the phenomenon called the electric current the constant passage of electricity through the medium tends to restore the state of polarization as fast as the conductivity of the medium allows it to decay. Thus the external agency which maintains the current is always doing work in restoring the polarization of the medium, which is continually becoming relaxed, and the potential energy of this polarization is continually becoming transformed into heat.'

in which Hertz seems to present the essence of a formal hypothetical view of Maxwell's theory:

> To the question, 'What is Maxwell's theory?' I know of no shorter or more definite answer than the following: Maxwell's theory is Maxwell's system of equations. Every theory which leads to the same system of equations, and therefore comprises the same possible phenomena, I would consider as being a form or special case of Maxwell's theory; every theory which leads to different equations, and therefore to different possible phenomena, is a different theory. (p. 21)

Closer inspection of the context of this passage reveals, however, that formalism was far from Hertz's intentions. What emerges in fact is his attempt to argue that Maxwell's equations, far from being hypothetical, are 'direct results of observation and experiment' (p. 28), in which the various elements are deducible from measured correlations. Indeed, as the first page of his introduction makes clear, Hertz's own experiments were prompted by the offer of a prize by the Berlin Academy for experimental investigation of the relationship between electromagnetic forces and dielectric polarization which would complete the deductive argument from experiment. Helmholtz had claimed to show in 1870 that Maxwell's equations follow from the universally accepted experimental laws of Faraday and Ampère, together with the following assumptions.

(1) Changes of dielectric polarization produce the same electro-magnetic forces as do the currents which are equivalent to them,

(2) Electromagnetic forces produce dielectric polarizations,

(3) Air and empty space behave in these respects like other dielectrics.

To complete the 'derivation from experiments' of Maxwell's equations, therefore, (1), (2) and (3) needed separate experimental verification.

In 1887 Hertz claimed to have shown (1) for some dielectrics, but despaired of verifying (1) and (2) directly for air. His eventual discovery of the finite propagation of electromagnetic waves in air is in fact derived from his earlier attempts at direct verification of (1) and (2) for air. The discovery was generally regarded as conclusive proof of the displacement equation, but since it was not the direct verification of the magnetic effects of displacement current that Maxwell, Poynting, Helmholtz and Hertz were all ultimately seeking, its logical force is a little puzzling. On the face of it, it seems to be no more than another successful test of a consequence of the displacement current hypothesis of a kind that Maxwell at least would not have accepted as 'proof'. Heaviside perhaps provides the best expression of the argument on which this proof depends as follows: taking the rest of the theory (effectively Ampère's and Faraday's experimental laws) as justified experimentally, only the displacement current term is in doubt. But now suppose no term corresponding to the displacement current is added. Then there is no propagation in aether of transverse electromagnetic waves, or rather, these waves would have infinite wavelength. But waves of finite length are observed; hence the extra term in Ampère's equation can after all be deduced, though by a different route from the one Maxwell first attempted.[1]

## IV. Meaning variance and experimental identifications

It must be concluded that in neither the first nor the second interpretation can Maxwell's analogical inference to the displacement current be made cogent.[2] A more fundamental question about his method, however, remains to be asked, namely, whether *any* such argument could be made to work in the light of objections stemming from the historicist or 'meaning variance' account of science. The conclusion 'There is a displacement current in aether', the historicist will claim, means something quite different in the two interpretations; how then can it be said to be inferred without theoretical assumptions from the experimental facts? Consistently with Maxwell's discussion of formal analogy and deduction of generalized equations from experiments, the reply would seem to be this: the conclusion 'There is a displacement current in aether', understood as asserting the occurrence of an extra term in Ampère's equation, does not depend for its experimental meaning or validity upon either interpretation of current. It is not an assertion of the *nature* of the unobservable cause of observable magnetic effects due to a changing electric field; it is a generalized expression of the laws relating these effects with other observables, namely the macroscopic set-up which constitutes the changing electric field. The displacement current has in this respect a status similar to that of the energy of aether, whose relations with observables are expressed by general laws of motion without specification of the *kind* of energy involved in mechanical or other microscopic terms.

This reply leaves two further points to be clarified. First, if the assertion of the displacement current is intended to be independent of either interpretation, what force could the attempted analogical arguments have, even if they were successful? For the arguments certainly do depend upon adopting one or other interpretation. It should be noticed, however, that if either or both of the analogical arguments *had* worked, the corresponding interpretation would not be objectionably hypothetical, because it would have rested on

[1] Heaviside, *op. cit.*, 477. For an analysis of the significance of Hertz's experiments in relation to Maxwell's theory, see S. d'Agostino, 'Hertz's discovery of electromagnetic waves', forthcoming.

[2] Joan Bromberg seems to be mistaken in her suggestion that in the second interpretation Maxwell has 'cut [his equations] off from the physical ideas [the first interpretation] upon which he had founded them' ('Maxwell's electrostatics', 151), if this implies that the first interpretation has some inductive cogency which the second lacks. At best it may be said that Maxwell's first understanding of the *meaning* of the displacement current was probably in terms of the first interpretation. But in the absence in most commentators of any distinction between the meaning of different models and their inductive force, it is difficult to know how to interpret judgments such as Bromberg's.

observable analogies of behaviour of just the same kind as those leading to identifications of the forms of energy, and of electric induction and potential. There would also have been no need to regard the interpretations as *alternatives* with respect to the argument for the displacement current term in Ampère's equation, for an argument from one set of observed analogies is surely strengthened by an argument from another set just in so far as the arguments lead to the same conclusion, even though in other respects they may lead to different conclusions.[1] An inference to the sine law of refraction, for example, is stronger in virtue of being an inference in both the corpuscular and the wave theories.

The second point is more fundamental. The historicist has doubtless been waiting throughout the last two paragraphs to object that the whole force of his analysis has been missed by talking about the 'observables' which are said to be related by the generalized 'neutral' displacement current equation. These observables themselves, he will maintain, are pervaded by theoretical interpretations. Here there is, however, an ambiguity. If it is being asserted that there are theoretical assumptions hidden in descriptions of experimental observations, such as those concerning electric potential for example, this has already been taken account of. These theoretical assumptions have been specified above in terms of empirically vulnerable experimental identifications which underlie not only the experimental laws assumed, but also the descriptive predicates in terms of which they are expressed. In the latter case the experimental identifications may not rest on any explicit argument—they are directly recognized as acceptable uses of whatever descriptive vocabulary is currently in use. No undesirable regress is thereby created, because *some* empirically vulnerable identifications of 'the same property again' must be presupposed in the use of any descriptive language. It does not follow, of course, that some such identification may not later have to be brought to consciousness, questioned and possibly abandoned in the light of further evidence, as was the case with the tacit identification of electric current with 'something flowing' (presupposed in the very metaphor of 'current').

In order to describe the concept of displacement current neutrally as expressing a relation between observables, it must be assumed that in both interpretations of current the same experimental identifications are assumed in expressing the observables. It follows that if the historicist objector wishes to claim that no distinction can be made between observables which are neutral relative to the two interpretations and the interpretations themselves,

---

[1] This logical point about a multiplicity of analogies gives a rational ground for Maxwell's *pluralism* of interpretations, an aspect of his physics which is remarked upon by d'Agostino, 'La pensée scientifique de Maxwell et le développement de la théorie sur champ électromagnetique dans le mémoire "On Faraday's lines of force" ', *Scientia*, **103** (1968), 7. It is also enough to put in their right perspective Duhem's many accusations of *contradiction* in Maxwell's work (especially *op. cit.*, 11, 101). A pluralism of models does not imply a self-contradictory theory.

he is in conflict with the present account. Without entering into this dispute in detail, it may be remarked that the historicist's mistake here lies in his assumption that the only alternative to 'radical meaning variance' between theories is a radical distinction between theories on the one hand and an absolutely neutral observation language on the other. In the present account an intermediate position is adopted: theory is not to be conceived, as in the hypothetico-deductive account, as an external hypothesis *imposed*, however intimately, upon independent observations; it is rather to be understood as *constituted by* the fundamental experimental identifications which control both the inference to general laws, and the basic descriptive language in which the laws are expressed. In this, the present account agrees with the historicist. But it does not follow in this account that all theories are self-contained and incommensurable, as some historicists have maintained, for some experimental identifications span more than one theory, and can be used as the relatively neutral ground of comparison between theories.

K

# CHAPTER TWELVE

# A Realist Interpretation of Science

## I. The aims of science

We began in the Introduction by noting how recent 'revolutionary' interpretations of science have undermined the entrenched belief that science is the paradigm case of well-grounded knowledge and cumulative discovery. If the extreme revolutionary thesis were correct, theories of natural science would be unconstrained by empirical criteria of verifiability or falsifiability, or by any logical criteria of theory choice, or by any requirement of continuity in meaning or truth value of theoretical statements. Apart from more general objections to the relativistic overtones of this thesis, it will not do as an adequate explication of natural science, because it is clear that scientists *do* accept some constraints on theories, whatever the proper construal of these constraints may turn out to be.

Traditionally, the constraints have been thought to be of two kinds, formulated at the outset of the seventeenth-century scientific revolution by Francis Bacon in terms of the attainment of 'light' and of 'fruit'. In other words, the true and hidden structure of the natural world is to be revealed in scientific theory, and in virtue of this discovery science is to exploit nature for the benefit of man. In recent philosophy of science these two aims have been discussed in terms of the *realist* and *instrumentalist* aspects of science, and the evaluative overtones of 'fruit' and 'benefit' have been excluded from philosophy of science, and examined, if at all, in moral philosophy or in sociology of science.

In this book I have not been concerned with value judgments upon the applications of science as such. In concluding, however, I shall examine the implications of the inductive network model developed here for the goals of science as discovery and instrumental control, and any assessment of that model as an adequate account of science will partly depend upon the relative value placed upon different aspects of these goals. In brief, we shall see that the network model tends to stress instrumental goals at the expense of realism, if realism is interpreted in terms of universalizable theoretical explanation.

First, let us recall how the network model has incorporated various sorts of constraints upon scientific systems. Science has been interpreted as a learning process, in which empirical data are processed by means of a certain descriptive language, and by certain coherence conditions which determine what theories are acceptable in the light of the data. A theory yields testable empirical predictions, and the results of these tests become data which are processed in their turn, and which either reinforce or undermine the theory which yielded the prediction. Science is subject to a self-regulating mechanism by which theories are formed, judged, and possibly changed. If the self-regulating mechanism is to be a good learner, however, this continuous corrective process is not enough; theories will also be expected to *succeed* in their predictions on more occasions than not, and this expectation will be relevant to what counts as an acceptable theory, and will therefore introduce constraints upon the system beyond those of empirical testing and correction.

These constraints have led us to examine the inductive assumptions that are usually held to give good grounds for expecting predictions to be successful more often than not, and without attempting to justify these assumptions as such, we have seen how a probabilistic confirmation theory can be developed to systematize and explicate them. The explication turns out to be not just systematically expository, but also corrective, in that it reveals the paradoxical consequences of demanding of any confirmation theory, whether probabilistic or not, *both* that theories should be construed in hypothetico-deductive fashion as universal in scope in infinite domains, and also that any prediction entailed by such a theory, if highly confirmed, should itself have high confirmation. I have suggested a way of resolving this transitivity paradox by construing theories as expressing *clustering* properties among data, and requiring that they should refer only to *finite* sets of individuals and properties. Various applications of the clustering postulate have been shown to be sufficient to explicate inference to and from lawlike generalizations, analogical argument from theoretical models, and simplicity criteria for theories.

The network model is therefore shown to satisfy various requirements of a logic of science that have been neglected in other post-deductivist analyses. In contrast to the non-inductive Popperian tradition, it defines the goals of science primarily in terms of expectations of successful prediction, rather than in terms of the search for more and more powerful testable theories. Other things being equal, the preference for more powerful theories is understood here in terms of preference for more comprehensive predictive content, rather than as a search for better approximations to an ideal universally true theory. Adoption of this goal is found to dictate certain criteria of theory choice and theory change, namely the search for finitist theories that exhibit the world as homogeneous or clustered in various respects that are consistent with presently accepted data. The traditional Popperian problem of *demarcation* of science from non-science and pseudo-science is solved in terms of the character of science as a successful learning device, incorporating mechanisms of prediction, test and self-correction.

All these features of the network model concern the *epistemological* aspects of science: what we understand by data and what by theory, in what senses data are accepted and theories inferred, how theories are tested and corrected, and the criteria for 'good' theories. Both the goal of science as successful learning and prediction, and the criteria for good theories that have been found to follow from it, seem at first sight to dictate an interpretation of science in purely instrumental terms. But we have not, except incidentally, considered the questions of *ontology*, namely, 'What, in the light of science, can we say about the real furniture of the world?' and 'What kinds of things is science competent to discover?'. I shall devote this closing chapter to an investigation of these questions, first briefly outlining their history since Bacon, and then examining how far the network model still permits us to give a realistic interpretation of science which is consistent with both its instrumental goals, and the relativity of its theories to the meanings of a currently accepted language.

## II. From naïve realism to pluralism

In seventeenth-century science two generally accepted assumptions provided such close connection between the realist and instrumentalist aspects of theories that these two goals could effectively be regarded as one and the same. First, it was assumed that true theories could be attained in practice. Even where it was recognized that theories are not true *a priori*, and that strictly speaking the verification of empirical consequences or effects could at best convey only a high probability to their antecedents or causes, nevertheless if a sufficient number of true consequences of a theory were observed, it was held to be so improbable that the theory should be false, that at least *moral certainty* could be ascribed to it.[1] It followed that this certainty of a theory flowed into all its predictions, and indeed that the justification of a general theory and its as-yet-untested consequences was one and the same thing. The second assumption might be called the assumption of naïve realism, namely that the hidden entities and processes of nature that are to be discovered by science are of the same kinds as observable entities and

---

[1] See for example the discussion of Leibniz's theory of science in L. Couturat, *La Logique de Leibniz* (Paris, 1901), chap. 6, paras. 37–41: when a hypothesis (1) is simple, (2) explains many phenomena, (3) permits new predictions, it has '. . . a "physical" or "moral" certainty, that is to say an extreme probability, like that of a postulated key which permits complete decipherment of a long cryptogram, yielding intelligible and coherent senses' (*ibid.*, 268). See also C. Huygens, *Treatise on Light* (1690; Eng. trans., Chicago, 1945), Preface, vi: 'It is always possible to attain . . . to a degree of probability which very often is scarcely less than complete proof. To wit, when things which have been demonstrated by the Principles that have been assumed correspond perfectly to the phenomena which experiment has brought under observation; especially when there are a great number of them, and further, principally, when one can imagine and foresee new phenomena which ought to follow from the hypotheses . . . and when one finds that therein the fact corresponds to our prevision.'

processes, and hence describable in the same descriptive vocabulary and satis-fying the same laws. Even Locke, who questioned the assumption that true theories could be attained in practice, never doubted that observable primary qualities are the qualities which universally belong to the fundamental particles of nature.[1]

Later developments in physical science, however, undermined both assumptions. During the eighteenth century, explanatory theories of physical and chemical phenomena and the structure of matter proliferated, and became more and more indirectly related with phenomena. As a consequence unobservable entities and processes began to be regarded rather as heuristic models than as discoveries of the real world. This development was encouraged by situations in which there were too many theories in the field as well as others in which there were too few. Sometimes there were a number of alternative theories more or less agreeing in their experimental consequences, between which it was very difficult or impracticable to devise crucial experi-ments. This was the case over long periods with one or two-fluid theories of electricity and magnetism, with fluid versus dynamic theories of heat, and with Newtonian force models in chemistry versus Daltonian atomic theories. On the other hand, in the nineteenth and early twentieth centuries, especially in electrodynamics and later in quantum theory, *no* theories could be found which satisfied the second naïve realist assumption that theoretical entities and processes are of the same kind as observables. These developments culminated in rejection of even the two most firmly held theories in classical physics: Euclidean geometry of physical space and Newtonian mechanics, thus finally discrediting the first realist assumption that science could, and had, actually attained some true general theories.

The result was a widespread retreat to various forms of positivism or instrumentalism. This did not at first entail abandonment of some of the features of realism, particularly the belief that science yields *some* knowledge that is firm, stable and cumulative. But such a belief had now to be seen in terms of laws relating observables, rather than as discoveries of unobservables, and even these laws (including Newtonian mechanics) had to be regarded as stable only in *approximation* in limited empirical situations, not as univers-ally applicable.[2] On the other hand, reduction of the cognitive aspects of science to lawlike correlations of observables in which theories merely

---

[1] 'If a great, nay, far the greatest part of the several ranks of bodies in the universe escape our notice by their remoteness, there are others that are no less concealed from us by their minuteness. These insensible corpuscles being the active parts of matter and the great instruments of nature, on which depend not only all their secondary qualities, but also most of their natural operations, our want of precise distinct ideas of their primary qualities keeps us in an incurable ignorance of what we desire to know about them.' (*Essay concerning Human Understanding*, book 4, chap. 3, 25).

[2] *Cf.* Duhem's rejection of theoretical models in science as unstable and pseudo-metaphy-sical, in favour of the network of theoretical representations of observables, i.e. experimental laws (*Aim and Structure of Physical Theory*, Part I).

functioned as the imaginative glue between observable data and predictions, or as the computer mechanism of a black box processing data into
predictions, seemed inadequate to account for the significance of theories,
and even of the attitudes which scientists themselves adopt towards theories.
A careful and influential survey of this stage of the argument between
realism and instrumentalism was given in 1960 by Nagel,[1] and his discussion
forms a convenient starting point for our examination of the present state of
the question.

Nagel makes two unargued assumptions about the nature of realism. The
first is that the realist position is identical with the view that statements of a
scientific theory are true or false. I shall accept that this is a necessary characteristic of realism, although we shall see later that there is more to be said
about it in the light of the relativity of correspondence truth to the current,
and changing, descriptive language. The second assumption made by Nagel
is that there are certain criteria of the physical reality of theoretical entities
and properties which concern both logically and empirically significant
features of the theories in which terms denoting them occur. Thus, to find
out whether scientists ascribe physical existence to given entities, we must
ask whether these entities function in certain specific ways in theories. For
example, 'Is the theory in which the term occurs well-supported and accepted
by the scientific community?', 'Does the term occur essentially in two or more
logically independent experimental laws, so that it is not merely introduced
*ad hoc*?', 'Does the term enter causal relations?', 'Does it satisfy conditions of
conservation or invariance?'. In all these cases, Nagel has no difficulty in
showing that positive answers to these questions can be given as well in an
instrumentalist as in a realist interpretation of theories, and he concludes

> It is therefore difficult to escape the conclusion that when the two apparently
> opposing views on the cognitive status of theories are each stated with some
> circumspection, each can assimilate into its formulations not only the facts
> concerning the primary subject matter explored by experimental inquiry but
> also all the relevant facts concerning the logic and procedure of science. In brief,
> the opposition between these views is a conflict over preferred modes of speech.[2]

Nagel's conclusion is, however, undermined by two major difficulties. The
first arises from the connection between a realist view of theories and the need
for theories to somehow mediate inductive inferences between data and
predictions. I have argued in chapter 9 that if there are to be inductive
grounds for making predictions from theories, these must be construed as
analogical arguments from data concerning sufficiently similar *observable*
systems. Theories must be regarded as having semantic interpretation in
terms of models, or, as I have put it in the light of the transitivity paradox,
theories must be construed as statements of the relevant analogies between

---

[1] *The Structure of Science*, chap. 6.
[2] *Ibid.*, 152.

observable systems, in virtue of which inference is possible from data to predictions. This construal presupposes that statements of a model, and hence of a theory, have truth value, and hence, in Nagel's terms, it presupposes a realist interpretation of theories. Nagel does not mention theoretical models among his criteria for realism, but this requirement does constitute a distinction between the two views, since instrumentalists have in general *denied* that the interpretation of theories into observable models is necessary for scientific explanation or inference. To modify instrumentalism in order to accommodate this requirement would indeed be to trivialize the traditional distinction.

This objection to Nagel's 'irenic instrumentalism'[1] is one aspect of a deeper objection to his strategy in examining the dispute. It can be argued that he has loaded the scales against realism from the outset by interpreting it as a view or set of views that must have some definite outcome in the 'logic and procedure of science'. If this is all the distinction between instrumentalism and realism amounts to, it is not surprising that the procedures of scientists can be alternatively interpreted instrumentally, for instrumentalism can be made into a sufficiently flexible theory to accommodate any logical or experimental activity engaged in by scientists. But, it is claimed in this objection, the dispute between realism and instrumentalism is not a procedural but a philosophical dispute about *what exists*, and it is not shown to be a pseudo-problem merely because instrumentalism can be made to account more or less adequately for the logic of science. The question remains, however, how to characterize the sense of 'realism' that Nagel is here accused of omitting from his discussion.[2]

One way of characterizing this sense derives from rejection of another of Nagel's major presuppositions, namely that of the stability of the observation language and of experimental laws relating observables. It is significant that Feyerabend's first paper in this debate was entitled 'An attempt towards a realistic interpretation of experience',[3] and that he there construed 'realism' as the view that a fundamental theory so permeates the perceptual and linguistic representation of observables that the world is wholly interpreted in terms of that theory. We have seen in chapters 1 and 2 how far this view

---

[1] The phrase is Sellars' ('Scientific realism and irenic instrumentalism', *Boston Studies in the Philosophy of Science*, vol. 2, ed. R. S. Cohen and M. Wartofsky (New York, 1965), 171), although he points to a possible misprint in *The Structure of Science*, 151, where 'irenic' should probably be read for 'ironic'. For Sellars' own realistic interpretations of scientific theory, see his 'The language of theories', *Current Issues in the Philosophy of Science*, ed. H. Feigl and G. Maxwell (New York, 1961), 57, and 'Theoretical explanation', *Philosophy of Science: the Delaware Seminar*, vol. 2, ed. B. Baumrin (New York, 1963), 61.

[2] Nagel deliberately excludes 'richer' connotations of realism: '[Physical reality] must not be understood as implying that a thing so characterized has a place in the scheme of things to be contrasted with certain other things having the invidious label of "mere appearance", or that in addition to satisfying the requirements specified by the corresponding criterion the thing is in some way more valuable or more fundamental than everything not so characterized.' (*The Structure of Science*, 150).

[3] *Proc. Aris. Soc.*, **58** (1957-8), 143.

can be adopted consistently with retaining some empirical constraints upon science. But the question remains as to how far the thesis of 'theory-laden' observation can be said to be consistent with realism in any of its traditional senses.

In the first place this thesis undermines the assumption of the stability of the observation language and of the truth of experimental laws, which is an aspect of realism that even Nagel tacitly retains. For on the thesis of theory-ladenness there is no perennial and accumulating corpus of observation statements and experimental laws which are independent of changing theories. It is true that this also counts against instrumentalism in its original form as much as against realism, for the instrumental aspects of science are no longer captured in a body of true and theory-independent observation statements, but must be interpreted at best as the set of instrumental manipulations whose linguistic descriptions will change from theory to theory. Feyerabend has likened this aspect of science to the operations of a robot for whom responses and manipulations are purely behavioural and devoid of meaning.[1] A robot might, presumably, come to have something analogous to green fingers or craft wisdom, or 'knowledge-how' in Ryle's terminology,[2] but since he has no propositions, no truth and no meaning, he cannot be said to have knowledge in the sense of 'knowledge-that', that is propositional knowledge. His truncated science cannot be said to consist of a corpus of propositional truths about the real world.

We will set aside here the difficult question of whether, if a robot manifested *all* the behaviour expected of a person, including linguistic behaviour and intelligent anticipation of the features of his world, we should still be justified in denying that he contemplated propositions and had knowledge. If a robot did become that human, the same problems about the nature of his theoretical knowledge would arise as for a person, and the conclusion that instrumental operations as such, whether of a robot or a person, cannot constitute propositional knowledge is unaffected.

The construal of truth, and of the existence or reality of theoretical entities, that follows naturally from the interpretation of all descriptive sentences in the categories of a particular theory, is more properly called an idealist or pluralist than a realist position. It has sometimes been labelled the 'super-realist' view,[3] because it entails that theoretical entities are somehow

---

[1] 'Explanation, reduction and empiricism', *Minnesota Studies*, vol. 3, 94.

[2] G. Ryle, *The Concept of Mind* (London, 1949), chap. 2.

[3] The phrase is due to D. H. Mellor ('Physics and furniture', *Studies in the Philosophy of Science*, ed. N. Rescher (Oxford, 1969), 171), and S. Morgenbesser ('The realist–instrumentalist controversy', *Philosophy, Science, and Method*, ed. S. Morgenbesser *et al.*, 201). Morgenbesser suggests, following Dewey, that the thesis that theoretical entities are real (that is, that they *alone* are real) should be distinguished from the thesis that they exist (that is, that they may be relational or functional with respect to real observable entities). I shall not use 'real' in this restricted sense, but as synonymous with 'exist'. J. Margolis ('Scientific realism, ontology, and the sensory modes', *Phil. Sci.*, **37** (1970), 114) speaks of

'more real' than observables, and that they should actually replace observables in interpretation and description of the world. There are no tables, it is held, there are only fundamental particles, fields and so on, in various configurations, which show some of our previous assumptions about tables to be false and tables to be non-existent. But, as Mellor has pointed out, even if this view were correct, it is not a sufficient condition for any traditional sense of 'realism' that current fundamental theories should imply the replacement of observables by theoretical entities in true descriptions, unless the current theories are *true*, and we can never know them to be true. Indeed by induction from the history of science they are very likely to be false (I have maintained here that, if taken universally, they are *certain* to be false). Moreover, one of the principal theses of Feyerabend and the other historical relativists is that theories both have changed in the past and ought to change in the future, radically and unendingly. The succession of spectacles through which we see the world is discontinuous and exhibits no convergence or accumulation.

The only possible construal of truth and existence in such a view is idealist, in the sense that judgments of truth and existence can be made only *within* each particular fundamental theory. Such a theory carries with it its own internal ontology of entities, its own meanings of terms, and its own assignment of truth values to sentences. But since idealism has usually been monistic, that is, it has claimed that only *one* global theory is ultimately self-consistent and therefore true, the view under discussion now is rather pluralist, since it holds that there is an indefinite plurality of possible ontologies and theories, any of which may be the vehicle of interpretation of the world. In any case it is misleading to speak of it as a 'realism', super or any other sort.

## III. Realism and relativity

Our discussion so far indicates three conditions that ought to be satisfied by an account of science that claims to be realistic in anything like the traditional sense. These are

(1) Theoretical statements have truth value.

(2) It is presupposed that the natural world does not change at the behest of our theories.

(3) The realistic character of scientific knowledge consists in some sense in the permanent and cumulative capture of true propositions corresponding to the world.

These three conditions will now be taken to define the sense of realism we

---

the strong version of realism as *scientific realism*, defined as 'a position which denies the existence of macro-physical objects and of persons'. That historical relativist positions, requiring complete replacement of descriptions in successive paradigms, should be regarded as 'idealist' is argued by G. Buchdahl, 'Is science cumulative?', *New Edinburgh Review*, no. 13 (1971), 4.

seek in science. Pluralism does not qualify, because although it may accept (1) in a purely coherence sense of 'truth', and may even accept (2) as a basis for the instrumental aspects of science, it rejects (3) absolutely as a consequence of its rejection of the correspondence analysis of truth.

Apart from radical pluralism there have been at least three other responses to the problem of realism and instrumentalism in recent philosophy of science. The first is the frank abandonment of any form of accumulation in science other than that of manipulative non-propositional skills, and hence the abandonment of realism as I have defined it. This view is of course consistent with a non-realist pluralism, and differences of emphasis between instrumentalists who hold it, and pluralists like Feyerabend, generally concern only the relative values put upon the manipulative and the theoretical aspects of science respectively.[1] An instrumentalist view of this kind entails abandonment not only of realism, but also of any commitment to forms of inference in science, and hence to any explicatory model of science such as I have attempted here.

A less radical response has features both of traditional realism and of traditional instrumentalism. This is the proposal to interpret observables and descriptions of observation in 'common-sense' language as constituting the 'real' world, and to decline to regard these as replaceable by the entities and laws of any theory, on the grounds that theories are demonstrably unstable and can be accepted only as heuristic aids to systematization and prediction of observables. This view is held not only by latter-day instrumentalists and operationalists, but also by all 'ordinary-language' analysts who resist the claim that scientific theory may *change* 'what it is correct to say' in ordinary language, and by all phenomenologists who hold that some phenomenological reduction of immediate human experience is more fundamental than the 'objectifications' of science. Ironically, this view also in its way implies a relativity of truth to theory, for as soon as it is admitted, as it must be in the light of the findings of history of science and history of ideas generally, that conceptually very different 'common-sense' languages may be viable, and that a given language may radically change, the language appealed to by the 'common-sense' school must be conceived to change *irrationally* with external circumstances, and not as a result of any discovery or rational consideration of empirical truth yielded by science. This view gives no account

---

[1] 'Instrumentalism' does not of course necessarily refer to a science whose interest is confined to technological application, but to a particular view of the cognitive status of theory. Where Feyerabend argues explicitly against cognitive instrumentalism, he does so primarily on the grounds that it discourages development of new theories in face of the sufficient instrumental success of the old. But this pragmatic prescription does not touch the cognitive question—continuous revolutions are inconsistent with traditional realism. See P. K. Feyerabend, 'Realism and instrumentalism', *The Critical Approach to Science and Philosophy*, ed. M. Bunge (London, 1964), 280, and W. Kneale, 'Scientific revolution for ever?', *B.J.P.S.*, **19** (1967), 27.

of an accumulation of objective descriptions of the world any more than does pluralism; it therefore cannot as it stands be called a realist view as defined here. But we shall be led to a modified version of this view below, in considering what form of realism might be consistent with the network model of science.

A third possibility is the realist position adopted by Popper.[1] Popper regards his view as realist, mainly because for him science is an attempt to make true statements about the world (in the sense of correspondence truth), and although he believes there are no inductive methods which would allow us to assert any theory as true or even as probable, theories may be asserted to be false, at any rate relative to their own test statements. Popper's criteria for progress in scientific theorizing are that successive theories have survived increasingly many attempts to falsify them, and that they become ever more powerful and comprehensive. He expresses these criteria in terms of what he calls the *verisimilitude* of a theory, which increases with its truth content (that is, with the set of true statements which are its entailments) and decreases with its falsity content. Since the set of entailments of any universal theory is infinite, we cannot in practice assign truth or falsity to all of them, and therefore we cannot know what the verisimilitude of a theory is, but only make estimates of comparative verisimilitude in particular cases, for example when one theory entails all the true consequences of another and more besides, or when one theory has been refuted and another has not.

In interpreting this measure of verisimilitude as an 'approach to truth' about the real world, Popper has not taken sufficient account of the changes in theoretical language that occur between successive theories. In his latest book he makes it explicit that his construals of truth and falsity are always *'relative to some "given" test statements'.*[2] He goes on: 'I do not raise the question, "How do we decide the truth or falsity of test statements?" '. His view therefore bypasses all the problems raised by the meaning-variance thesis, which would require a realist account of 'approach to truth' to include comparisons between successive theories which contain radically different concepts and assign 'truth' to radically different sets of test statements. In the light of the claim that the historical succession of scientific theories is always related by such meaning changes, Popper's account of 'increasing truth content' becomes inapplicable. It is inapplicable even to the more moderate meaning-variance thesis adopted here, for even if only some, and not all, test statements change in meaning from theory to theory, no simple logical comparison of truth or falsity content is possible.

Let us now examine the conditions for realism in more detail, and consider how far the network model of science is consistent with them.

---

[1] For his latest and in some ways most explicit statement of this view, see *Objective Knowledge* (Oxford, 1972), especially chaps. 2 and 8.

[2] *Ibid.*, 8; Popper's italics.

In the network model, all sentences of a theoretical system have truth value in a sense which has been defined as the correspondence with the world of statements expressed in a given descriptive language. The correspondence character of this concept of truth is not affected by the fact that only probability and not truth values can be definitely assigned to any descriptive statements, for the probability values are functions of our epistemological beliefs, whereas the truth values are functions of the ontological relation between language and the world in virtue of which our beliefs apply to the world. I have argued in chapter 9 that the requirement in the network model of high probability values for those predictions of a theory in which we have high confidence entails a realistic interpretation of models yielding these probability values in analogical inferences from particulars to particulars. It is this inductive construal of theoretical systems that has dictated realism, and realism is represented by the fact that all statements of a theoretical network have truth value and can therefore be assigned probability value as a measure of our belief. It is tempting to pass immediately to a definition of a 'realist' theory as one whose statements are either true or false, but unfortunately this would overlook the relativity of truth, and hence of realism taken in this sense, to the language in which the theory is expressed. And this relativity makes the claim to realism increasingly implausible the further removed the fundamental concepts of different languages are from each other.

Consider for example a language $L$ in which the majority of descriptive terms essentially involve the concepts of witchcraft and magic. I have already argued in chapter 2 that to ascribe truth relative to this language to a majority of its sentences is acceptable, since the meanings of all the terms are so different from *prima facie* similar terms in our language that our criteria of truth and falsity are irrelevant. But it appears less acceptable to hold that the real world for the *speakers of L* contains witches, spells and the like, while ours does not. In the first place a notion of realism deriving from a correspondence conception of truth should not make essential reference to a world 'for the speakers of $L$', and in the second place such an account would be indistinguishable from theoretical pluralism which we have already rejected.

Before considering how the network model differs from theoretical pluralism, let us look more closely at the relation between the truth of statements and their realistic interpretation. Quine[1] has made this relation more precise by construing the 'reality' of a theoretical entity or property as 'being a value in the domain over which some individual or predicate variable of a theory ranges (that is, over which the variable is quantified) in order to make the theory true'. This construal presupposes that the theory is formalized at least in terms of a first-order logic quantifying over individual variables and that individual constants and variables are specified and distinguished from

---

[1] 'On what there is', *From a Logical Point of View*, 1.

predicates. For example, Newtonian mechanics might be formalized by taking *particles* to be the individuals, and mass, space and time coordinates, and force, to be the monadic and relational predicates. The formalization of the laws of motion would involve quantification over 'all particles'; hence, according to Quine's account, particles are the real or existing entities in this theory. A further question, of interest to logicians, may arise about the 'reality' of the properties, mass, space, time and so on, which reduces in Quine's formulation to the question whether Newtonian mechanics requires quantification over predicates in a second-order logic. But whether it does or not, and whether in either case this is a satisfactory construal of the reality of properties, need not concern us here, because a more serious question already arises for the reality of individual entities in the light of the revolutionary character of theory change.

First of all, even one given theory may be formalizable in different ways with modification or interchange of what are taken to be individuals and what are taken to be properties. For example, at a certain period in the nineteenth century, electrodynamics was alternatively formalizable in terms of individual particles, with mass, charge and magnetization as monadic properties, and space, time and force as relations, on the one hand; or in terms of space and time points as individuals, and mass, charge, magnetic densities, and electric and magnetic force vectors as properties, on the other hand. According to Quine's account, therefore, what exists: particles, or space and time points? The answer, like the question, must be *relative to the formalization*. Now I have already discussed cases where a scientific theory is formalizable in different ways, and argued that even if it appears to yield identical sets of experimentally testable laws and predictions by entailment, yet a complete account of the theory will show it to be dependent also on the analogies suggested by the way it is formulated, and that different formalizations therefore constitute empirically different theories having different probability values. In the case of nineteenth-century electrodynamics, for example, the particle formulation is appropriate to a theory which takes its fundamental analogies from particle mechanics, whereas the field formulation takes its analogies from continuous distributions of force in a spatio-temporal continuum. It follows that the question of what exists is determined by which set of analogies and expectations is adopted *prior* to the formalization. Quine's analysis does not in itself decide this question; it merely gives the question precise sense when a particular formalization has been adopted for other than formal reasons.[1]

This clarification, however, does not help us to understand realism as committed to an accumulation of truths throughout successive theories which may be conceptually discontinuous and non-convergent. If the individual

---

[1] This point is made by D. H. Mellor, 'Physics and furniture', 181.

entities which are claimed in a certain theory to exist are radically different from those that are claimed to exist in succeeding theories, there seems to be no sense in which science yields an accumulating body of discoveries of the real world.

I believe this conclusion is mistaken. The mistake arises in part from a confusion imported from the irrelevant logical controversy about the existence of *properties*. If the notion of 'existence' is tied, as in Quine's account, to that domain over which the theory requires quantification, and if in addition the logic of the theory is first-order, whatever are taken to be individual entities in a given theory have a privileged status with regard to 'existence'. But there is no need to pass to a higher order logic in order to make sense of the notion of the physical existence of properties and relations. For example, a realistic interpretation of a theory about electrons requires only that assertions informally expressed as 'Electrons exist', or 'All electrons have mass $m$', should correspond to some true statement of the formalized theory. If the individuals of the theory are taken to be 'particles', the first of these expressions can be formulated in first-order logic as

$$(\exists x)(Ex) \tag{12.1}$$

where $E \equiv$ 'is an electron', and $x$ ranges over all particles, and the second expression can be formulated as

$$(x)(Ex \supset Mx) \tag{12.2}$$

where $M \equiv$ 'has mass $m$'. If (12.1) is true, electrons exist, and both (12.1) and (12.2) represent quantifications over all electrons without resorting to second-order logic. Such a construal of existence would not be sufficient for all the expressions in mathematical analysis that are required by physics, for some of these must be formalized in second or higher order logics; but this fact is not relevant to the present account of scientific theories, because in this account only statements referring to a finite domain of individuals need have finite confirmation values anyway. If every quantification in an infinite domain is taken to be false, no question arises about the physical existence of such domains. Mathematical expressions using such quantifications must be regarded with respect to the logic of science as conventional devices only.

## IV. The cumulative character of science

Having elucidated the notion of existence in relation to theoretical statements which have some probability of being true, we can now examine the relation of these statements to other theories which either appear to be in contradiction with these statements, or which are accepted as being false. Three sorts of cases can be considered. The first are cases where two theories presuppose different ontologies of individuals, but are so far as is known sufficiently consistent with observation, and so may both have some appreciable probability of being true. Secondly, two theories which presuppose different ontologies of individuals may be such that one of them, usually the earlier, is now accepted as being false. Thirdly, two theories may have the same ontology of individuals, but contain statements which *prima facie* contradict, and yet neither theory need on that account alone necessarily be taken to be false.

(1) Suppose present evidence permits theoretical formulations $T_1$, $T_2$ of so-far adequate theories in terms of individual particles and individual space–time points respectively. We seem to be committed by the language-relative construal of existence to saying that, according to $T_1$, only particles exist and only statements ascribing properties to particles can be true, and conversely that, according to $T_2$, only space–time points exist and only statements ascribing properties to these can be true. Are these two assertions incompatible? Does it follow that the two theories are inconsistent with regard to assertions of existence, and that at most one of them can be true? If these consequences do follow, and if it can be shown (as it certainly can with considerable generality) that *any* theory is capable of formalization in terms of alternative ontologies of individuals and sets of predicates, our assertions with regard to existence seem to be caught within the confines of particular languages, and to be empty of content as theories of what there is.

If we make use of the extended concept of existence described above, however, the two theories need not be construed as making conflicting assertions merely in virtue of their formalizations, for there will be some adequate translations which will take true statements of $T_1$ into true statements of $T_2$. For example, the assertion in $T_1$ to the effect that there are particles is represented by quantification of variables over the domain of particles. If $T_1$ and $T_2$ are on present evidence experimentally adequate, there will certainly be some statements of $T_2$ with some probability of being true, which correspond to the assertion 'There are particles', in the same sense as this is intended in $T_1$. If $T_1$ says, informally, 'All particles have mass, space and time positions, momentum, ... and some particles have charge, magnetic pole strength, ...' $(S_1)$, then a corresponding formulation in $T_2$ will be 'There are at least some space–time points that are predicated by mass-density, and this mass-density is continuously distributed between space–time points in such a way as to satisfy certain conservation principles of mass, momentum, ...' $(S_2)$. The formulation in $T_2$ is a little cumbersome, but there is no doubt that it can be carried out in such a way as to be an adequate translation of $S_1$, where the meanings of 'mass', 'space–time point' and so on are intensional references that are the same in both theories. 'Translation' here is not of course to be taken as logical equivalence, for as we have seen in chapter 2 there are no *logical* equivalences of meaning between one theoretical language and another, but there are intensions whose behavioural and local meanings are sufficient to constitute an adequate translation between some pairs of statements in the two theories. Both $S_1$ and $S_2$ may therefore be true in the correspondence sense without inconsistency, and since 'existence' is to be understood in terms of true statements, the entities and systems of entities referred to in both $S_1$ and $S_2$ can be taken to exist, and indeed to be the *same* systems of entities, though differently described. The difference is that in $S_1$ particles are construed as the 'carriers of properties', and in $S_2$ they are classes of space–time points. But *this* is a merely conventional or heuristic difference, not a significant difference of physical ontology.

This construal of existence has no implications with regard to the possibility of *reducing* particles to space–time points, or conversely. It may be compared with alternative formulations of the existence of a person: either 'existing' as an individual having properties and entering into relationships, or as a higher-order system of relationships of varying sets of molecules which reproduce a certain pattern as they float in and out of a certain space–time neighbourhood and thereby constitute the person as a continuously existing system. It may be thought that the alternative formulations, either of particles or of persons, imply the reductionist thesis that particles are 'nothing but' collections of space–time points, and that persons are 'nothing but' systems of molecules. But nothing that has been said about the alternative formulations excludes the possibility of 'emergent properties', of both particles and persons, that are irreducible merely to classes of space–time points or of molecules. There is a sense in which every *relational* property of individuals is emergent, since it requires at least two individuals to realize it, and it may be that there are many-adic relations which do not emerge as properties of systems of space–time points until there are, as we would say in $T_1$, many particles in interaction with one another, and similarly for persons as complex systems of many molecules. The predicate set may easily be made rich enough to represent, for example, a principle of uniqueness for each person if that is required, by building in a kind of exclusion principle to the effect that no two person–molecule complexes have identical sets of properties. Alternative formalizations of so far experimentally adequate theories do not beg any substantial reductive questions. This is not to deny, of course, that alternative formulations may be associated with different analogies and hence with different expectations—for example, a molecular representation of persons may undesirably suggest *restriction* of the relevant predicate set to those predicates required in physics, chemistry and sociology ('rights' are nothing but 'food, clothing, shelter, and culture sufficient for social integration')—but no formal language can *require* such poverty in its predicate basis.

(2) Suppose, secondly, that the theories involved in making existence claims are related as successive scientific theories or cosmologies often are, in that the later theory $T_2$ *denies* the existence of entities presupposed by the earlier theory $T_1$. For example, eighteenth-century chemistry asserted the existence of phlogiston gas, post-Lavoisierian chemistry denies it;

seventeenth-century popular belief asserted the existence of witches, modern science denies it.

In discussing the 'truth' of such assertions in chapter 2, I concluded that, with respect to a language sufficiently different from ours, we may regard most of its statements as true relative to that language, although *prima facie* false in ours, because all or most meanings must be held to have changed. Are we then committed to the conclusion that phlogiston existed 'for them' and not 'for us'? Put like this, the question seems to presuppose that 'existence' has a tense, and not just in the sense that something has in the meantime ceased to exist in the course of history, like pterodactyls. The tensed expression is misleading, however, because if 'existence' is to be taken relative to a language or society, then the emphasis is not on the *past*ness of beliefs in phlogiston or witches, but on the relativity to 'them' and to 'us'. 'For us' there is a sense in which we want to say witches neither exist now nor ever existed. The relativity to language need not be construed as mere subjectivity, however. To see why not, the cases of witches and of phlogiston ought to be distinguished in important respects.

First of all, 'witchcraft' is partially a social phenomenon, in the sense that 'royalty' is. That is to say, although it may have a physical substratum (the existence, in all societies, of hysterical persons, charismatic powers, and the like), it is in some societies an *institution*, defining individuals with *roles*.[1] Understood in this way not only may the seventeenth-century Inquisition truly say 'there are witches', but also *we* may truly say this, and mean by it 'There were people who played certain roles, satisfied certain judicial tests', and so on. Doubtless the Inquisition also meant 'There are persons who have had commerce with the Devil', and that this may have been regarded by them as part of the essential characteristics of a witch. But we have already seen that a word need not wholly lose its referential meaning by losing what were at one time its defining characteristics. Though we may now deny the truth of the Devil proposition, that is, we may deny that there ever was, for anybody, 'commerce with the Devil', we are not committed to denying many other propositions about witches in their social roles, for these propositions are true in our language as well as in theirs.

The question of 'social reality' is a complex one into which we need not enter further here. The case of phlogiston is more relevant and also more straightforward. Whatever may be the case with witchcraft and other esoteric concepts which are quite foreign to our society, in the case of phlogiston, caloric, aether and other substances in past science now believed not to exist, there is no *total* incommensurability of meaning between theories in which they appear and theories in which they do not. Many names of physical and

---

[1] For an elementary discussion, see P. Berger and T. Luckmann, *The Social Construction of Reality* (London, 1967). For the question 'Are there witches?', see the references in chapter 2, note 1 on p. 60.

chemical substances and processes, and most descriptive language for classifying macroscopic objects are either intensionally equivalent, or more or less
easily translatable. We may therefore distinguish various senses in which it is
possible to say that phlogiston does (and did) or does not (and never did)
exist.

First, there is, and always has been, a gaseous substance which is given off
when hydrochloric acid is poured on to zinc, and which is highly inflammable,
and there is, and always has been, a substance which eventually so saturates
the air in a confined space that organisms cannot breathe in it. Pre-
Lavoisierian chemists were correct in making these and many other assertions
about chemical substances, but they were mistaken (in their language as well
as ours) in believing that one and the same substance, namely phlogiston, is
referred to in both the processes just described and that this was also a
substance given off in the calcification of metallic zinc. Pre-Lavoisierian
chemists were also mistaken in believing that most of the substances they
identified with phlogiston are chemically primary or elementary. So were
their immediate Daltonian successors in believing that hydrogen and oxygen
atoms are primary and unanalysable entities. But since, as we have seen,
there is no need to restrict the notion of 'existence' to the individuals of a
given formulation of a theory, there is no reason to deny of pre-Lavoisierian
chemistry that substances called by them phlogiston did and do exist, or that
hydrogen and oxygen atoms did and do exist. What those substances or those
atoms essentially *are*, is something whose description changes from theory to
theory, and will never be finally settled as long as science continues to
develop.

Theories about essences are neither stable nor cumulative, and are therefore not part of the realistic aspects of science. But many of its other assertions
may be both approximately stable and cumulative, because they are translatable from theory to theory in virtue of recognizable identities of intensional
reference. Such assertions are of various kinds.[1] They may be approximate
forms of laws, true within certain experimental limits, like Newton's laws of
motion, Boyle's law, Hooke's law. They may be assertions of properties and
processes which exist but are ascribed to wrongly identified individuals, as in
the case of phlogiston. They may be assertions of analogy between different
entities or processes which remain significant from theory to theory, but
where the nature of the entities so related is described differently from
theory to theory. For example, that the planets, the earth, stones and seas are

---

[1] Forms of stability presupposing some translation from theory to theory are discussed
in W. Sellars 'The language of theories', *op. cit.*, p. 41, and in H. R. Post, 'Correspondence,
invariance and heuristics: in praise of conservative induction', *Studies in Hist. Phil. Sci.*, 2
(1971), 213. Sellars expresses the relation between a superseded theory and its successor as
follows: '[micro-] theories about observable things *do not explain empirical laws, they explain
why observable things obey, to the extent that they do, these empirical laws*' (*op. cit.*, 71; Sellars'
italics).

constituted by bodies that are significantly similar in mechanical properties to each other and satisfy the same laws of motion in each theory, was a discovery made in the seventeenth century that has been maintained through the revolution óf modern physics, although what 'massive bodies' essentially *are*, and the exact form of the laws they satisfy, has changed from theory to theory. The greater part of scientific theory and ordinary description of nature consists of assertions like these, and they form the accumulating core of scientific discovery.

(3) Finally, there are apparent contradictions between theories involving not different existence assertions with regard to entities, but different ascriptions of fundamental properties to the same set of entities. Any model of science deriving from Duhem's analysis must recognize that such ascriptions are not necessarily inconsistent. Consider for example the pair of assertions 'Physical space is Euclidean' ($R_1$), and 'Physical space has non-zero curvature' ($R_2$). These are formally inconsistent, and yet $R_1$ may be part of one theoretical formulation of geometry, mechanics and optics, which is experimentally equivalent to another formulation containing $R_2$. The two theories will, of course, contain correspondingly different assertions in mechanics or optics, but each theory faces experiment as a whole, not statement by statement.

The two theories can be regarded as consistent only if the meanings of the terms in $R_1$ and $R_2$ are taken to be different. These meanings will therefore not be among the stable intensional references that indicate that the two theories refer to the same entities, and in terms of which the truth values of the theories can be compared. Sufficient intensional meanings for such comparisons will, however, be provided by other relatively observational terms, and particularly in the case of physical geometry by 'operational definitions' of length by means of experimental equipment described equivalently in both theories. In general, theoretical descriptions such as $R_1$ and $R_2$ are unstable and non-cumulative, just as identifications of primary individuals and their essences have been seen to be, but accumulations of approximate forms of law are not thereby eliminated. In this example, indeed, any currently acceptable relativistic theory containing $R_2$ gives a value of the curvature so small that in local situations $R_1$ remains a sufficient approximation, and it is likely that this approximate stability will be maintained in any subsequent physical geometry.

If such accumulation of approximations is thought insufficient for 'realism', then this account of science may be called instrumentalist, but there are other respects in which it is nearer to realism. In summary we can distinguish several senses of 'realism' in terms of which to answer the question 'Is the network model of science realistic?'

(*a*) Theoretical statements have truth value. This is satisfied by the network model, in the correspondence sense of truth, and it has been shown to be a necessary condition if acceptable theories are to give high probability to their predictions.

(*b*) Theoretical entities are the only real entities, and theoretical statements the only true statements; all observation statements should therefore be *replaced* by theoretical statements, which are conceptually incommensurable with them ('super-realism'). The network model does not accept this account, and it has been argued that it is in any case not a realist but essentially a pluralist position.

(*c*) Confirmable theories are essentially universal in scope in potentially infinite domains. This is not satisfied by the network model, which ascribes finite probability of truth only to statements quantified over finite domains of individuals and containing finite sets of predicates. Universality in infinite domains is not regarded as an essential characteristic of realistic science.

(*d*) Theoretical entities are real entities, and may replace observable entities as primary individuals, but only in the sense that observable entities may be reinterpreted as classes or systems of theoretical entities. This requirement implies some translatability of meanings from theory to theory, and it is satisfied by the network model in virtue of intensional predicates which retain referential meaning from theory to theory. However, the notion of what are the 'primary' individuals is recognized as being theory-relative; the primary individuals of one theory may be superseded in another theory, and so on indefinitely.

(*e*) There is accumulation of truths, or rather of a body of statements with high probability, from theory to theory. This also implies some translatability between theories, and is satisfied by the network model for approximate forms of law and assertions of significant similarities between entities and systems of entities. But it may not be satisfied for identifications of primary individuals and their essential properties, nor for theoretical statements that are relatively remote from observation.

The network model of science is therefore non-realistic in so far as it has negative implications for the universal ontological and cosmological consequences that have sometimes been held to derive from natural science. There has been a constant tendency for the prestige of instrumental success to flow back into temporary ontologies and analogies, and to infect social and metaphysical thought about the nature and destiny of man and the universe. This tendency is a natural accompaniment of an epistemology which takes natural science to be the paradigm of knowledge, but it should be resisted even by the epistemology of science itself, for as we have seen a realistic appraisal of science with respect to its accumulating and approximate discoveries does not entail the realism of its primary entities and their properties as described in any given theory. There have on the other hand always been points of view which have looked elsewhere for organizing principles of knowledge. Two contrasting examples are provided in other traditions of philosophy of science by Duhem and by Habermas. On the one hand Duhem wishes to replace the pseudo-metaphysics of scientific theory by true metaphysics derived from Aristotelian and Thomist theology.[1] Habermas, on the other hand, wishes to replace it by a critical understanding of the *social* reasons why in particular situations particular types of scientific metaphysics or positivism flourish: for example, seventeenth-century scientific realism depended on a social communication situation in which

> ... individuation has progressed to the point where the identity of the individual ego as a stable entity can only be developed through identification with abstract laws of cosmic order. Consciousness, emancipated from archaic powers, now anchors itself in the unity of a stable cosmos and the identity of immutable Being.
>
> Thus it was only by means of ontological distinctions that theory originally could take cognizance of a self-subsistent world purged of demons. At the same time, the illusion of pure theory served as a protection against regression to an earlier stage that had been surpassed.[2]

Whatever the extra-natural-scientific principles subsequently adopted, the realistic view of science put forward here is consistent with the positions of both Duhem and Habermas in rejecting the cognitive value of such metaphysical interpretations.

That natural science has no realistic implication in this metaphysical sense does not, however, entail that the present account is a devaluation of the element of rational 'discovery' in science as against purely manipulative control. The element of control required is in any case not a rule-of-thumb technology, but the systematic self-correction of unified theories in an empirical learning process. This account recognizes that discoveries are made in interaction with the world using a particular language, and that this process has its limits, both with respect to theoretical truth, and with respect to the conditions under which learning takes place. Natural scientific inference has rational grounds, but these are essentially finite and local in application, and determined by empirical conditions of testability and self-correction. If we wish to go beyond this form of rationality, we must look to the studies of man, society and history, which in all European languages except English are still called 'sciences', and whose methods and aims are not exhausted by those of natural science. But that would be another story.

[1] *The Aim and Structure of Physical Theory*, appendix.

[2] *Knowledge and Human Interests*, 307. Habermas's discussion raises the question of the essential *irreducibility* of social and historical science to natural science. He adopts a model for natural science in terms of feedback and instrumental control which is not unlike that presented here, but his thesis of irreducibility needs supplementing in the more explicit climate of Anglo-Saxon philosophy. This supplementation can easily be provided if the 'machine analogy' for science is developed a little (*cf.* p. 51 above). No learning machine can operate in an environment which is either very unstable, or with which it is in very strong interaction. Natural science may be defined as that domain of knowledge where these conditions of instability do not obtain; the environment of the human sciences, on the other hand, is unstable in both respects. What alternative methodology is appropriate and justifiable for the human sciences remains an open and under-analysed question. *Cf.* my 'In defence of objectivity', *Proc. British Academy*, **58** (1972).

# Index of Names

# Index of Subjects